Signal Processing Using Non-invasive Physiological Sensors

Signal Processing Using Non-invasive Physiological Sensors

Editors

Imran Khan Niazi
Noman Naseer
Hendrik Santosa

MDPI • Basel • Beijing • Wuhan • Barcelona • Belgrade • Manchester • Tokyo • Cluj • Tianjin

Editors

Imran Khan Niazi
Centre for Chiropractic Research, Centre for Chiropractic Research,
Auckland
New Zealand

Noman Naseer
Department of Mechatronics Engineering
Air University
Islamabad
Pakistan

Hendrik Santosa
Department of Radiology
University of Pittsburgh
Pittsburgh
United States

Editorial Office
MDPI
St. Alban-Anlage 66
4052 Basel, Switzerland

This is a reprint of articles from the Special Issue published online in the open access journal *Sensors* (ISSN 1424-8220) (available at: www.mdpi.com/journal/sensors/special_issues/noninvasive_physiological_sensors).

For citation purposes, cite each article independently as indicated on the article page online and as indicated below:

LastName, A.A.; LastName, B.B.; LastName, C.C. Article Title. *Journal Name* **Year**, *Volume Number*, Page Range.

ISBN 978-3-0365-3720-7 (Hbk)
ISBN 978-3-0365-3719-1 (PDF)

Contents

About the Editors

Imran Khan Niazi

Dr Imran Khan Niazi is currently working as Dean of Innovation and Technology and Director of Centre for Chiropractic Research at New Zealand College of Chiropractic, New Zealand. He leads a multidisciplinary team. He has been working as an active researcher in physical and neural rehabilitation, focusing on non-pharmacological/conservative modes of treatment to optimize the patient's rehabilitation journey and improve the quality of life. He has extensive experience working in multidisciplinary teams, including chiropractors, physiotherapists, medical doctors, engineers, and neuroscientists, with a solid patient-centred, evidence-informed approach. He holds an adjunct position at Auckland University of Technology, New Zealand and Aalborg University, Denmark. He has authored 91 peer-reviewed journal papers and >100 conference papers (proceedings and extended abstracts). According to Google Scholar, his work has been cited more than 2800 times and has an h-index of 27.

Noman Naseer

Dr Noman Naseer is a Tenured Associate Professor, Senior Member IEEE,Head of Neurorobotics Research Group,Department of Mechatronics Engineering, Air University, Islamabad and Vice President of IEEE Robotics and Automation Society, Pakistan. Dr Noman Naseer holds a bachelor, Masters and Ph.D. in Mechatronics Engineering.Dr Naseer has published over 100 peer-reviewed papers and has received more than 2500 citations from all around the globe. Four of his journal papers have received the highest citation awards from their respective journals.Dr Naseer serves as a reviewer of above 85 SCI Indexed journals and as Associate Editor of 5 SCI-indexed journals.Dr Naseer has the honour to be selected as a Marie-Curie Post-Doctorate fellow.He is a recipient of the prestigious Bill and Malinda Gates foundation award. His name has been included in the directory of productive scientists of Pakistan for three consecutive years. He is ranked among the top 1% reviewers in his research field by Publons, Clarivate Analytics and among the top 2% scientists of the world as per the list published by Stanford University and Elsevier. He is also the founder and CEO of two companies, Alpha Automation and Emaila.

Hendrik Santosa

Dr Hendrik Santosa is a Faculty in the Department of Radiology, University of Pittsburgh. He received Ph.D. in Cogno-Mechatronics Engineering from Pusan National University, Korea, in 2016. Currently, he is working at the University of Pittsburgh Medical Center on the development of experimental paradigm, data collection, and analysis methods in brain imaging including NIRS, EEG/MEG, MRI, and multimodal techniques. His research interest includes a statistical method, brain-computer interface, hyper scanning, advance brain signal processing (e.g., ICA/PCA, PLS)

Preface to "Signal Processing Using Non-invasive Physiological Sensors"

This Reprint is a collection of research articles published in *Sensors* (MDPI) Special Issue Processing Using Non-invasive Physiological Sensorsand concentrated on two aspects: non-invasive biomedical sensors for monitoring physiological parameters from the human body and the advanced signal processing of the data recorded.

Our Reprint is addressed to medical technology enthusiasts, from undergraduate engineering students to academia and industry professionals.

We are also thankful to Mr Eloise Zhou, who coordinated on bel fog MDPI publisher and the authors who contributed to editing this book and the incredible research participants who made all this research possible.

Imran Khan Niazi, Noman Naseer, and Hendrik Santosa
Editors

Article

Estimation of the Respiratory Rate from Localised ECG at Different Auscultation Sites

Xinqi Bao [1], Aimé Kingwengwe Abdala [2] and Ernest Nlandu Kamavuako [1,2,*]

[1] Department of Engineering, King's College London, London WC2R 2LS, UK; Xinqi.bao@kcl.ac.uk
[2] Faculté de Médecine, Université de Kindu, Kindu, Democratic Republic of the Congo; a.abdala@univ-kindu.ac.cd
* Correspondence: ernest.kamavuako@kcl.ac.uk; Tel.: +44-207-848-8666

Abstract: The respiratory rate (RR) is a vital physiological parameter in prediagnosis and daily monitoring. It can be obtained indirectly from Electrocardiogram (ECG) signals using ECG-derived respiration (EDR) techniques. As part of the study in designing an early cardiac screening system, this work aimed to study whether the accuracy of ECG derived RR depends on the auscultation sites. Experiments were conducted on 12 healthy subjects to obtain simultaneous ECG (at auscultation sites and Lead I as reference) and respiration signals from a microphone close to the nostril. Four EDR algorithms were tested on the data to estimate RR in both the time and frequency domain. Results reveal that: (1) The location of the ECG electrodes between auscultation sites does not impact the estimation of RR, (2) baseline wander and amplitude modulation algorithms outperformed the frequency modulation and band-pass filter algorithms, (3) using frequency domain features to estimate RR can provide more accurate RR except when using the band-pass filter algorithm. These results pave the way for ECG-based RR estimation in miniaturised integrated cardiac screening device.

Keywords: respiratory rate (RR); Electrocardiogram (ECG); ECG derived respiration (EDR); auscultation sites

Citation: Bao, X.; Abdala, A.K.; Kamavuako, E.N. Estimation of the Respiratory Rate from Localised ECG at Different Auscultation Sites. *Sensors* **2021**, *21*, 78. https://dx.doi.org/10.3390/s21010078

Received: 18 November 2020
Accepted: 22 December 2020
Published: 25 December 2020

1. Introduction

Respiratory rate (RR) is the physiological indicator of breaths per minute, which is commonly used as an early warning sign in disease detection. The normal RR of a healthy adult at rest is between 12–16 bpm [1]. Compared with adults, children's RR is higher. For an infant, it ranges from 30–60 bpm, and with growth, the RR will gradually reach the adult level [2]. The resting RR of older people may slightly increase. For the healthy independent seniors, it is 12–20 bpm, and those who need long-term care will reach 16–25 bpm [3]. Generally, a resting RR outside of these ranges may indicate a potential disease. An increased RR (tachypnea) may suggest fever, dehydration, asthma, chronic obstructive pulmonary disease, heart disease, etc. [4]. A low RR (bradypnea) may reveal the use of narcotics, alcohol intake, abnormal body metabolism, sleep apnoea, etc. In critical care (or intensive care, ICU), RR is also a vital parameter in the monitoring of respiratory failure. It could be measured by the gas exchange using a ventilator, capnography monitors, or spirometry devices, and chest electrical activities using electrical impedance tomography (EIT), inductance plethysmography, or impedance pneumography [5,6].

The current measurement of RR outside of the critical care still relies on manually counting the chest undulations in one minute by the medical staff [7]. Although this practice is easy to conduct without using extra medical devices, it has some drawbacks such as low accuracy. Subject's awareness, poor visibility of a breath, and other interruptions will greatly affect the measurement. Besides, in practice, the manual counting is not completed in one full minute by the medical staff due to the heavy workloads. They usually multiply the 30 s or 15 s measurement by 2 or 4 to assess the RR, which will lead to further inaccuracies [8]; (2) it is labour-consuming, as the medical staff can only conduct the measurement on one patient at one time; (3) the measurement is not continuous. As an early sign of physical deterioration, real-time and continuous monitoring can help alert the staff to emergencies, such as heart failure, shock, diabetic coma, etc. However, the intermittent measurement cannot provide such information timely, so the RR is always underutilized. External devices to automate the RR measurement can remedy the deficiencies associated with manual counting to a certain extent. Despite this, there are still respective limitations to each method. For the gas exchange-based techniques, they are accurate methods to reflect the respiratory condition, but have no portability, which requires the patients breathing in the external tube of the devices. So, these techniques are generally only available in critical care [9]. The bioimpedance-based techniques such as impedance pneumography can measure the electrical activities on the chest during inhalation and exhalation. However, it requires the patients to wear a tight chest strap, which may cause discomfort [10]. Additionally, patient movement, bad contact, and obstruction of breath will cause inaccurate measurements. Acoustic sensors are also used in the measurement of RR, however, their performance will be affected by the environmental noise and skin friction [11]. Therefore, wearable devices for automatic RR measurement are in great need to effectively monitor the breath in real-time and detect the first sign of physical deterioration promptly.

Extracting respiratory signals from the Electrocardiogram (ECG) signals is a potential surrogate measurement of RR. In recent years, ECG devices are becoming miniaturised, and sensors have been integrated with sport bands, smartwatches, and other portable monitors. This provides the feasibility and potentiality to design wearable ECG-based RR measurement devices. The first study on respiration-induced ECG variation was proposed by Einthoven et al. [12]. Flaherty and Riekkinen further analysed the respiration influence on children and cardiac patients by isopotential surface-mapping and vectorcardiography (VCG) [13,14]. Nowadays, it is well known that respiration-induced ECG variations are caused by (1) Respiratory Sinus Arrhythmia (RSA) that refers to the cyclic variation that the heart rate accelerates during inhalation and decelerates during exhalation [15]. It can be reflected in the ECG signals as the frequency modulation (FM) of the R-R interval between the R peaks as shown in Figure 1a. (2) Respiration-induced electrical axis rotation. During the inspiration, the filling of the lungs stretches the heart apex towards the abdomen, and in expiration, the emptying of the lungs compresses the heart towards the breast. Due to the displacement of the heart, the electric cardiac vector will change during respiration [16]. In the ECG signal, this process can be indicated as amplitude modulation (AM) of the R peaks as shown in Figure 1b. (3) Baseline Wander (BW) is the artefact caused by body movement, including breathing. The expansion and contraction of the thoracic cavity due to respiration will cause a slow and undulating baseline in the ECG signals as depicted in Figure 1c [17].

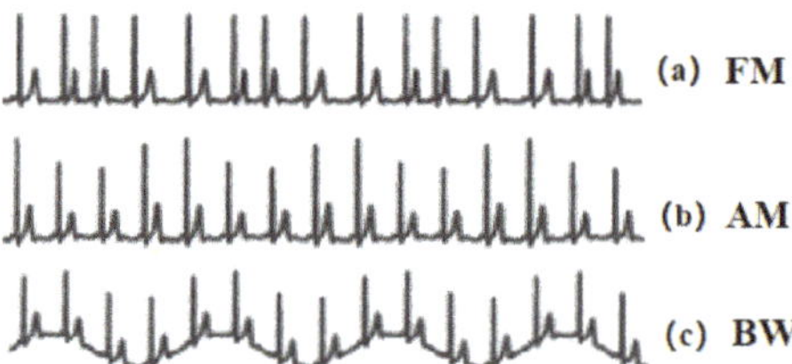

Figure 1. (**a**) Respiratory Sinus Arrhythmia (RSA) induced frequency modulation (FM). (**b**) Electrical axis rotation caused amplitude modulation (AM). (**c**) Baseline wander (BW) caused by chest movement.

Several techniques to extract respiratory signals from the ECG, the so-called ECG-derived respiration (EDR), have been proposed according to the respiration-induced ECG variation mentioned above. Some techniques are based on multi-leads ECG signals [16–20], while others attempt to extract respiratory information from one-lead ECG [21–26], as well as direct band-pass filtering (BP) of the ECG within the respiratory frequency band [17,26,27]. For the multi-leads EDR techniques, they mainly use the rotation angles of VCG from multiple ECG leads, while the one-lead EDR methods focus on the features related to the QRS complex, such as amplitude, interval, area, slopes, etc. There is no consensus on which is better in the performance; however, for a wearable device, one-lead ECG has the advantage in the system complexity and size. As part of our long-term project to design an integrated device for early cardiac screening, the final aim is to propose a small integrated device (around 8 cm^2) that can provide multiple physiological parameters including heart sound, ECG, and RR. The device will measure the ECG locally with heart sound rather than at different body parts. In our previous study, we did experiments to analyse the time property between ECG and heart sound when the ECG is captured at different auscultation sites [28]. Additionally, it indicates that the location of the ECG will cause the morphological variation of its signal, which may affect the obtainment of the EDR signal, therefore an important motivation of this study is to further analyse if these ECG variations will affect the performance of the EDR algorithms under this condition.

The aim of this study is threefold: (1) To investigate if the location of the electrodes at auscultation sites will affect the EDR algorithm accuracy; (2) to compare the performance of one-lead EDR algorithms based on the mentioned respiration-induced ECG variation; (3) to compare time-domain and frequency-domain features for RR estimation. All the findings will contribute to providing more accurate RRs for the integrated cardiac screening device.

2. Methodology

2.1. Subjects

The experiments were conducted on 12 healthy human subjects (8 male/4 female, age range 21–29 years, mean 25.9 years) with no history of heart diseases or respiratory issues. The procedures were approved by the King's College Research Ethics Committee (Approval No.: LRS-18/19-10673). Subjects gave written informed consent before the experimental procedures.

2.2. Experimental Setup

The standard Lead I ECG (as reference ECG), auscultation site ECG (captured at auscultation site A, P, T, M with 10 cm inter-electrode distance), and respiratory signals were recorded simultaneously during the experiment. A simple block diagram of the experimental setup is shown in Figure 2. The sensors for ECG signals were solid gel electrodes (Ambu WS, size: 36 $\times$ 40 mm, Medico Electrodes International LTD., Uttar Pradesh, India), and the respiratory signal was captured by a small microphone (developed at the Centre for Robotics Research (CORE) at Kings College London, UK) placed under the subject's nose. The recording used the commercial acquisition system (iWorx, model RA834, iWorx Systems Inc, Dover, NH, USA) and ECG devices (iWire-BIO4, iWorx Systems Inc, Dover, New Hampshire, US). The sampling frequency was 1 kHz and the analog filter for the ECG was 0.05–40 Hz [29].

During the experiment, subjects should keep supine and remain calm. Besides, subjects were required not to make sound from the larynx to ensure the sound captured was only respiration. The Lead I ECG and different auscultation site ECG signals were measured in pairs together with the respiratory signals. The duration of each recording group was three minutes, and two minutes break was given between different auscultation site trials.

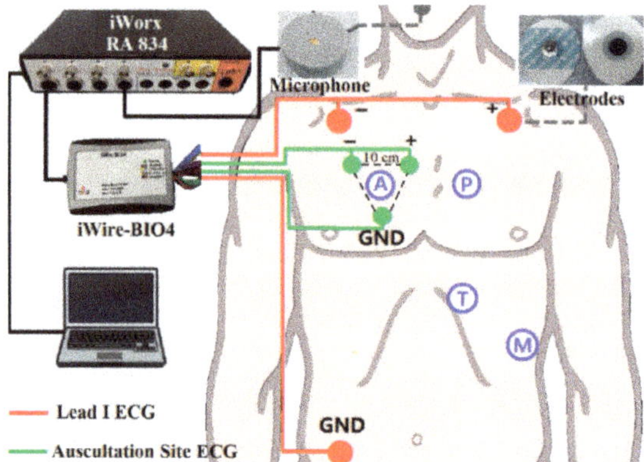

Figure 2. Block diagram of the recording setup: Red dots are Lead I Electrocardiogram (ECG) as a reference, green dots are auscultation site ECG. The grey dot is the microphone for respiration recording. iWire BIO4 is for ECG recording. All the data is transferred to the computer for processing through iWorx RA 834.

2.3. Signal Processing

In this study, EDR signals were obtained using BW, AM, FM, and BP algorithms from the reference (Lead I) and auscultation sites ECG signals, respectively. The RRs were estimated from the EDR signals using time and frequency domain features as detailed later. The performance of the algorithms and the effect of the locations were analysed by comparing it with the measured respiratory rate. The processing was conducted in the Matlab® R2018b environment, and the statistical analysis was performed using IBM® SPSS version 26.

2.3.1. Signal Filtering

The captured ECG signals and respiration sounds were filtered first to remove the unwanted artifacts and noise. For the ECG, a zero-phase 3rd-order Butterworth high-pass filter at 0.1 Hz was used to eliminate the large artifacts which were not related to respiration [30]. For the respiration sound, a 3rd-order Butterworth band-pass filtered (0.1–0.5 Hz) was used to smooth the waveform.

2.3.2. EDR Signals Extraction

In AM, BW, and FM algorithms, *R*-peak detection was a vital step, as all the features to be captured were related to R peaks. In this study, the Pan–Tompkins algorithm was used to detect *R*-peaks in the ECG signals [31].

1. AM algorithm: The amplitude changes due to the respiration in the ECG signals was obtained by connecting the captured *R*-peaks.
2. BW algorithm: Based on the *R*-peaks, Q points were found using the gradient descent method. Then, the baseline wander could be generated by connecting the middle points between *R*-peaks and Q points [32].
3. FM algorithm: The intervals between the R peaks were calculated. The resulting signal was the frequency modulation caused by respiratory sinus arrhythmia.

Afterward, all the signals generated by the algorithms above were interpolated to the same sample size of its raw ECG signals to increase the resolution.

4. BP algorithm: A band-pass filter (0.1–0.5 Hz) was used to capture the EDR signals. Although the normal RR for a healthy adult ranges between 0.2–0.35 Hz at rest, in our processing, we appropriately expanded the range to enable it to respond to special situations, such as the subjects' occasional deep or rapid breaths. Besides,

a wider band can help to further analyse the frequency components when there are no dominant peaks.

Representative derived respiration signals by the methods above are shown in Figure 3.

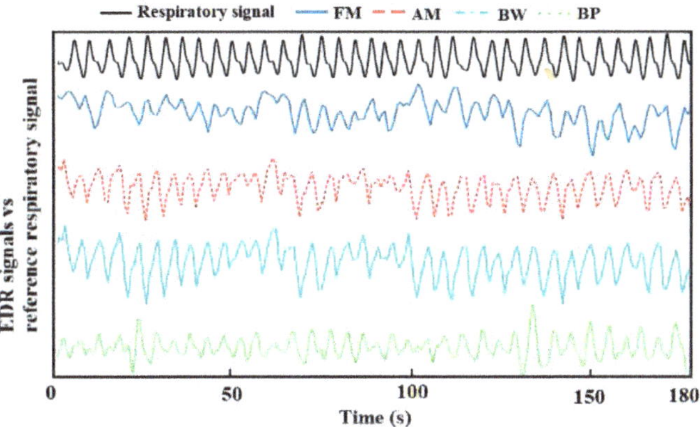

Figure 3. A representative derived respiration signals from auscultation site ECG and reference respiration signal.

2.3.3. Respiratory Rate Estimation

The reference RRs were obtained from the filtered respiration sound recorded using a nostril microphone. It was manually counted in the waveform to ensure accuracy. The estimated RRs from EDR signals were calculated by automatically counting in the time domain and using the median frequency (between 0.1–0.5 Hz), respectively. For the counting method, a moving average filter (window length: 50 ms) was used first to smooth the EDR signals and eliminate sub-peaks. Then, peak detection with the threshold of the signal mean value provided the estimated RR. The median frequency was chosen according to our previous study, which was proven to be the best feature in the frequency domain to estimate RR from EDR signals [27].

2.4. Statistical Analysis

The mean absolute errors (MAE) between the EDR-based estimated RR and reference RRs was used as the performance measure provided as mean ± standard error (SE). A three-way repeated-measures analysis of variance (ANOVA) was used to compare MAE. Factors were the features (counting and median frequency), EDR algorithms (AM, BW, FM, BP), and ECG locations (A, P, T, M, Lead I). A P-value of less than 0.05 was considered significant. Data were log-transformed to obey normality and variance homogeneity was satisfied.

3. Results

3.1. ECG Morphological Variation among the Auscultation Sites

Figure 4 shows a representative local ECG morphological variation compared with the lead I ECG from one subject. From (a) to (d), it can be seen that the amplitude of the R-peak, s-wave, and T-wave become larger from auscultation site A to M. Besides, it is also found that the R-peak of the site A ECG is normally on the left-hand side of it on Lead I ECG, which means the R-peak is advanced (approximately 10 ms by average). However, it will shift to the right-hand side when measured at site M, which means its onset is delayed (approximately 15 ms by average). Another phenomenon could also be observed that in the site A ECG, there is a J-point elevation shown as grey dots in (e). This happened on five subjects, and in 3 of them, the J point is even higher than the R-peak.

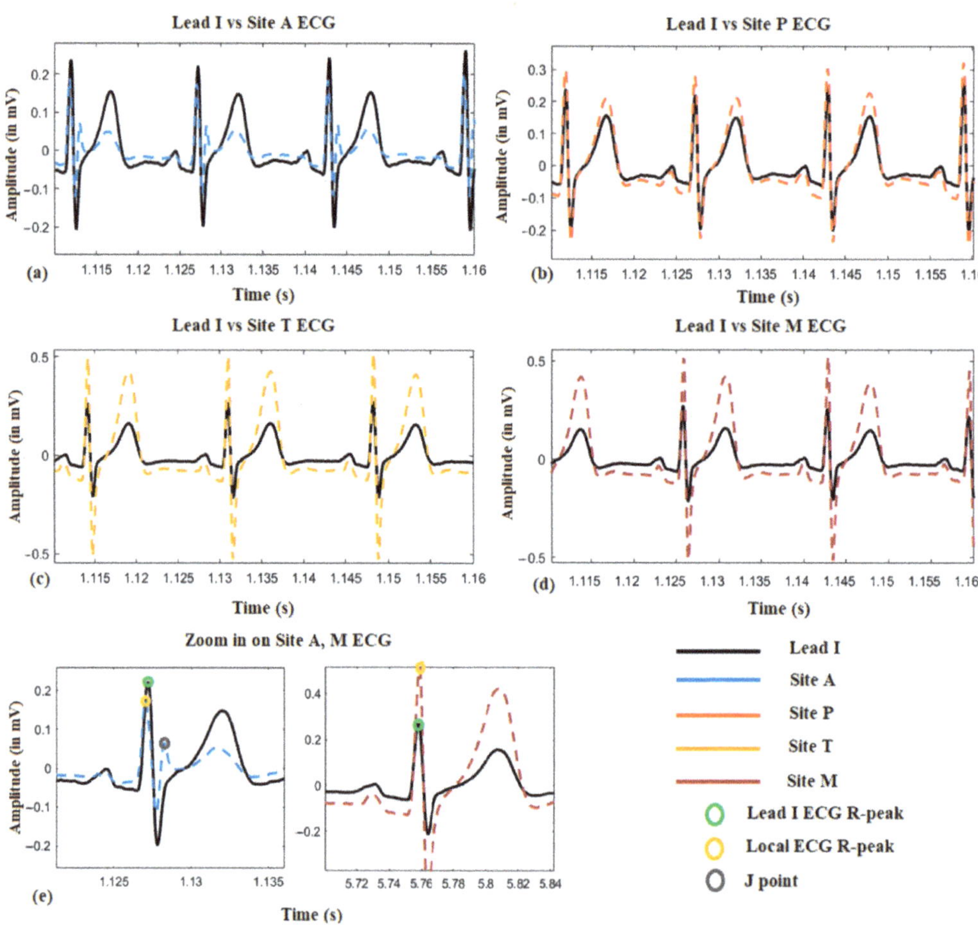

Figure 4. The local ECG morphological variation compared with reference Lead I ECG.

3.2. Location Effect on EDR among the Auscultation Sites

Table 1 summarized the EDR MAE of each subject averaged across estimation techniques and given per auscultation site. ANOVA results indicate that there is no statistical difference between the five sites (p = 0.746), and there was no interaction between EDR algorithms and sites (p = 0.516). All four EDR algorithms have quite close MAE between each auscultation sites, including average MAE at A: 1.656 ± 0.351, P: 2.297 ± 0.476, T: 1.733 ± 0.461, M: 1.467 ± 0.326, and reference ECG (Lead I): 1.834 ± 0.378 bpm. This indicates that RR can be harvested using ECG anywhere on the chest with negligible location effect. Figure 5 further visualized the location effect with different algorithms.

Table 1. The ECG-derived respiration (EDR) mean absolute errors (MAE) of each subject for each auscultation site and Lead 1 ECG signals, averaged across estimation techniques.

	A	**P**	**T**	**M**	**Lead I**
Subject1	0.16	0.06	0.14	0.12	0.08
Subject2	2.27	3.74	1.77	0.86	2.00
Subject3	1.88	1.37	3.17	2.15	2.60
Subject4	0.50	1.69	0.54	0.82	0.83
Subject5	1.27	1.10	0.28	0.14	0.47
Subject6	5.79	4.61	6.91	3.23	5.45
Subject7	1.60	2.60	0.38	2.60	2.24
Subject8	2.54	5.41	3.16	3.47	2.24
Subject9	0.73	0.20	0.67	0.53	1.16
Subject10	0.36	1.62	1.69	1.25	1.11
Subject11	0.85	2.10	1.03	1.46	1.45
Subject12	1.93	3.06	1.07	0.99	2.39
Mean	1.66	2.30	1.73	1.47	1.83

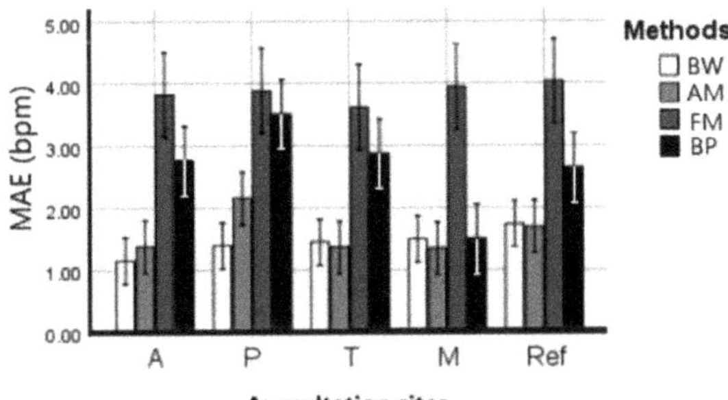

Figure 5. The performance (MAE ± SE bpm) of the EDR algorithms on different auscultation sites and Lead 1 ECG signals.

3.3. The Performance of the EDR Algorithms

After statistical analysis of the MAE on each subject with different EDR methods shown in Table 2, there was a significant difference between the four EDR algorithms ($p < 0.001$). The BW algorithm performed with MAE = 1.446 ± 0.181 bpm, closely followed by the AM algorithm with 1.589 ± 0.1966 bpm. Post hoc analysis revealed no statistical difference between BW and AM ($p = 0.31$), however, they were both significantly better ($p < 0.05$) than BP (MAE of 2.656 ± 0.258) and FM (MAE of 3.855 ± 0.329 bpm).

Table 2. The EDR MAE of each subject on different EDR methods, averaged across auscultation sites.

	BW	**AM**	**FM**	**BP**
Subject1	1.61	1.92	5.86	1.93
Subject2	1.65	2.88	3.26	2.45
Subject3	1.33	1.82	2.04	0.47
Subject4	2.19	1.99	4.06	2.19
Subject5	0.91	0.92	1.54	2.93
Subject6	0.49	0.41	3.55	3.67
Subject7	0.34	0.66	0.73	0.86
Subject8	0.76	0.75	3.23	1.19
Subject9	3.98	3.54	11.93	6.75
Subject10	0.45	0.38	2.08	0.38
Subject11	1.81	2.02	3.36	3.59
Subject12	1.84	1.78	4.61	5.89
Mean	1.45	1.59	3.85	2.69

3.4. Time vs. Frequency Domain

Deriving respiration rate using the median frequency (overall MAE 1.80 ± 0.223 bpm) outperformed the counting method (overall MAE 2.98 ± 0.312 bpm) in the time domain ($p < 0.001$) suggesting stability of the frequency domain, although a significant interaction ($p < 0.001$) with the applied method was observed. From Figure 6, it can be seen that the median frequency can provide a more accurate estimated RR on BW, AM, and FM algorithms. However, counting in the time domain is more accurate for the BP algorithm.

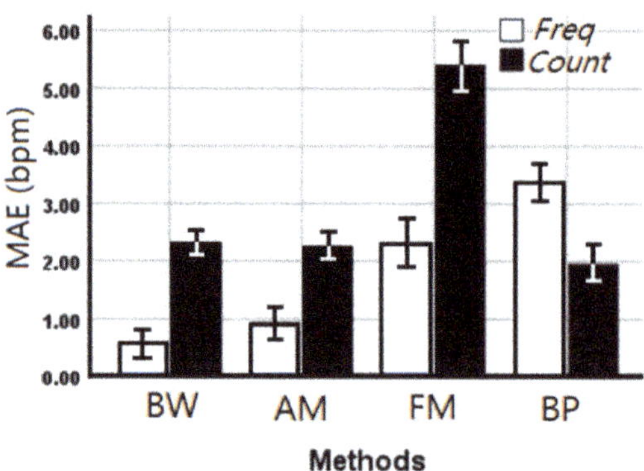

Figure 6. The mean absolute error (MAE ± SE bpm) of the EDR rates between EDR algorithms and estimation methods. Freq: Median frequency in the frequency domain, Count: Automatic counting in the time domain.

4. Discussion

This study aimed at analysing the performance of one-lead EDR algorithms in auscultation site ECG signals and EDR rate estimation in both time and frequency domain. The results show:

Firstly, it is found that the location effect on the obtainment of EDR between auscultation site and Lead I ECG signals is negligible in our experimental data. The result revealed that the ECG morphological variation between auscultation sites happened on the onset and amplitude of the ECG components including the *R*-peak delayed from site A to M, and the amplitude increase of *R*-peak, *s*-wave, and *T*-wave. These won't directly affect the EDR signals extraction, but it is worth noticing in cardiac researches. It is still unclear on the occurrence of J-point elevation or RSR' (An ECG finding in which there are two R waves) in five subjects' site A ECG signals. Normally they are pathological, but the subjects were confirmed healthy with no heart conditions, and this can be normal for the age group. The high J-point or double *R*-peaks may interfere with *R*-peak detection when the fake *R*-peak is higher than the true one. In our study, the performance of the four chosen EDR techniques was not affected, however, it may have an impact on the QRS area or slope based EDR methods. In the study of Sakai, it indicated that the location of the electrodes affected the quality of EDR signals and the more accurate RR estimation was obtained when the electrodes were attached near the heart [22]. The best placement was a negative electrode at the bucket-handle and a positive electrode at pump-handle movements of the ribs. However, in our experiment, electrodes were placed at auscultation sites on the upper chest, which were already close to the heart. Besides, as we want to design a miniaturised device, the inter-electrode distance was fixed and short (10 cm). Therefore, from the physiological mechanism, the locations in our study barely have an effect on the respiratory sinus arrhythmia, and the effect on the respiration-induced electrical axis rotation and chest undulation-induced baseline wander are minimal. This result verifies that the location effect on RR estimation can be ignored in designing an integrated cardiac screening device.

Secondly, the BW and AM algorithms outperformed FM and BP algorithms. Although BW has a slightly smaller MAE (1.446 ± 0.181 bpm) than AM (1.589 ± 0.1966 bpm), the difference ($p = 0.315$) is not statistically significant in our experiment data, which cannot confirm that the performance of BW is better than AM so far. This is in contrast with some previous work. In Charlton's study, it was shown that the BW performed better than AM without statistical analysis [7]. The performance of FM and BP methods are in line with previously reported MAE using the PhysioNet's MIMIC-II database, while the results of AM obtained in this study are similar to the MAE reported by Widjaja et al. [26] using their experimental data. It seems like the experimental setting for the database had a significant impact on the performance of EDR algorithms. Because our experiments were conducted under ideal conditions where the subjects were required to lie down calmly without any movement, the MAE was much lower than studies that have made use of the database.

For respiratory sinus arrhythmia induced FM, the magnitude of the oscillation varies from individual to individual, so that the obtained EDR signal is not that conspicuous sometimes [33]. For example, the FM waveform of 0–50 s is shown in Figure 3, the EDR signal in that period is messy, thus it will dramatically affect the peak detection in the time domain, causing inaccurate RR estimation. That should be the reason for FM's poor performance. For the BP algorithm, the choice of the frequency band is the current limitation. Though the frequency band (0.1–0.5 Hz) used in this study is appropriately extended, it is still not enough to capture RR from young children and stress tests. Besides, the use of a simple band-pass filter cannot remove unwanted interferences completely. The low-frequency component between 0.1–0.2 Hz, which is related to the baroreceptor reflex (blood pressure is regulated by the baroreceptors through the autonomic nervous system) and the high-frequency harmonic between 0.4–0.5 Hz will interfere with the RR estimation in the frequency domain [34]. Therefore, an adaptive frequency band is essential to improve the performance of using a band-pass filter.

Thirdly, the RR estimation in the frequency domain is found to be better than the time domain for BW, AM, and FM [35]. This result is the opposite of Charlton's result, which said Fourier analysis was inferior to breath detection in the time domain [7]. As discussed above, there are conditions where the EDR is not conspicuous enough, thus in the time domain, it is hard to detect the corresponding respiration related peaks, while still possible to capture it based on the power spectral density function. Besides, at the beginning and end of the EDR signals, there may be incomplete breathing, this will lead to the error for counting in the time domain. As there are not many breaths per minute, these errors are considerable for the RR estimation. Using frequency features will reduce this error moderately. However, it is also noticed that the performance of frequency estimation for the band-pass filter is worse than counting in the time domain as there are mentioned lower-frequency and higher-frequency components in the spectrum which weaken the domination of the respiratory band. Therefore, further analysis of frequency components is needed to improve accuracy when using the BP algorithm.

In this study, our research focused on the EDR of healthy adults at rest, and the experimental conditions were ideal that the subjects kept supine and breathed evenly without any movement. However, there are conditions of practical application that need to be considered including EDR performance on irregular respiration, such as deep breath or an increased respiration rate. Improvement still can be done to improve the RR estimation accuracy. The current validation study has compared four algorithms of the existing algorithms. More algorithms and fusion methods could be tested to improve the accuracy for clinical use. Future studies will include RR estimation throughout monitoring via a Holter-like monitor.

5. Conclusions

This study analysed the location effect on EDR algorithms' performance between auscultation sites and compared four EDR algorithms to estimate RRs in the time and frequency domain. The results showed that, firstly, the location of the ECG electrodes between auscultation sites barely affects the estimation of RR. Secondly, the BW and AM algorithms outperformed than FM and BP algorithms in generating the approximation of the respiratory signal. Thirdly, RR estimation in the frequency domain is more reliable except on BP algorithms. All the findings will contribute to building chest-based multiple physiological parameter monitors and providing more accurate RR estimation using EDR algorithms.

Author Contributions: Conceptualization: All; data collection: X.B.; Data analysis: X.B. and E.N.K.; critical review of the manuscript: All. All authors have read and agreed to the published version of the manuscript.

Funding: This research was funded by The King's China Scholarship Council.

Institutional Review Board Statement: This study was peer reviewed and approved by the King's College Research Ethics Committee (Approval No.: LRS-18/19-10673, 01/04/2019).

Informed Consent Statement: Informed consent was obtained from all subjects involved in the study.

Data Availability Statement: Data sharing not applicable.

Acknowledgments: The authors are grateful to the graduate students in the Prosthetics and Health Engineering Lab (PHE) and colleagues in Kings College London who gave their time to participate in our experiments.

Conflicts of Interest: The authors declare no conflict of interest.

References

1. Flenady, T.; Dwyer, T.; Applegarth, J. Accurate respiratory rates count: So should you! *Australas. Emerg. Nurs. J.* **2017**, *20*, 45–47. [CrossRef]
2. Fleming, S.; Thompson, M.; Stevens, R.; Heneghan, C.; Plüddemann, A.; Maconochie, I.; Tarassenko, L.; Mant, D. Normal ranges of heart rate and respiratory rate in children from birth to 18 years of age: A systematic review of observational studies. *Lancet* **2011**, *377*, 1011–1018. [CrossRef]
3. McFadden, J.P.; Price, R.C.; Eastwood, H.D.; Briggs, R.S. Raised respiratory rate in elderly patients: A valuable physical sign. *Br. Med. J. (Clin. Res. Ed.)* **1982**, *284*, 626–627. [CrossRef]
4. Park, S.B.; Khattar, D. Tachypnea. In *StatPearls [Internet]*; StatPearls Publishing: Treasure Island, FL, USA, 2019.
5. Brochard, L.; Martin, G.S.; Blanch, L.; Pelosi, P.; Belda, F.J.; Jubran, A.; Gattinoni, L.; Mancebo, J.; Ranieri, V.M.; Richard, J.C.M.; et al. Clinical review: Respiratory monitoring in the ICU-a consensus of 16. *Crit. Care* **2012**, *16*, 1–14. [CrossRef] [PubMed]
6. Wheatley, I. Respiratory rate 3: How to take an accurate measurement. *Nurs. Times* **2018**, *114*, 21–22.
7. Charlton, P.H.; Villarroel, M.; Salguiero, F. Waveform analysis to estimate respiratory rate. In *Secondary Analysis of Electronic Health Records*; Springer: Cham, Switzerland, 2016; pp. 377–390.
8. Ansermino, J.M.; Dumont, G.; Ginsburg, A.S. How uncertain is our reference standard for respiratory rate measurement? *Am. J. Respir. Crit. Care Med.* **2019**, *199*, 1036–1037. [CrossRef] [PubMed]
9. Chan, A.M.; Ferdosi, N.; Narasimhan, R. Ambulatory respiratory rate detection using ECG and a triaxial accelerometer. In Proceedings of the 2013 35th Annual International Conference of the IEEE Engineering in Medicine and Biology Society (EMBC), Osaka, Japan, 3–7 July 2013; IEEE: Piscataway, NJ, USA, 2013; pp. 4058–4061.
10. Yabuki, S.; Toyama, H.; Takei, Y.; Wagatsuma, T.; Yabuki, H.; Yamauchi, M. Influences of environmental noise level and respiration rate on the accuracy of acoustic respiration rate monitoring. *J. Clin. Monit. Comput.* **2018**, *32*, 127–132. [CrossRef] [PubMed]
11. Einthoven, W.; Fahr, G.; De Waart, A. On the direction and manifest size of the variations of potential in the human heart and on the influence of the position of the heart on the form of the electrocardiogram. *Am. Heart J.* **1950**, *40*, 163. [CrossRef]
12. Flaherty, J.T.; Blumenschein, S.D.; Alexander, A.W.; Gentzler, R.D.; Gallie, T.M.; Boineau, J.P.; Spach, M.S. Influence of respiration on recording cardiac potentials: Isopotential surface-mapping and vectorcardiographic studies. *Am. J. Cardiol.* **1967**, *20*, 21–28. [CrossRef]
13. Riekkinen, H.; Rautaharju, P. Body position, electrode level, and respiration effects on the Frank lead electrocardiogram. *Circulation* **1976**, *53*, 40–45. [CrossRef] [PubMed]
14. Malik, M.; Camm, A.J. Heart rate variability. *Clin. Cardiol.* **1990**, *13*, 570–576. [CrossRef] [PubMed]
15. Travaglini, A.; Lamberti, C.; DeBie, J.; Ferri, M. Respiratory signal derived from eight-lead ECG. In Proceedings of the Computers in Cardiology (Cat. No. 98CH36292), Cleveland, OH, USA, 13–16 September 1998; IEEE: Piscataway, NJ, USA, 1998; Volume 25, pp. 65–68.
16. Boyle, J.; Bidargaddi, N.; Sarela, A.; Karunanithi, M. Automatic detection of respiration rate from ambulatory single-lead ECG. *IEEE Trans. Inf. Technol. Biomed.* **2009**, *13*, 890–896. [CrossRef] [PubMed]
17. Pinciroli, F.; Rossi, R.; Vergani, L. Detection of electrical axis variation for the extraction of respiratory information. *Comput. Cardiol.* **1985**, *2*, 499–502.

18. Mazzanti, B.; Lamberti, C.; De Bie, J. Validation of an ECG-derived respiration monitoring method. In Proceedings of the Computers in Cardiology, Thessaloniki Chalkidiki, Greece, 21–24 September 2003; IEEE: Piscataway, NJ, USA, 2003; pp. 613–616.
19. Zhao, L.; Reisman, S.; Findley, T. *Derivation of Respiration from Electrocardiogram during Heart Rate Variability Studies*; IEEE: Piscataway, NJ, USA, 1994; pp. 53–56.
20. Behbehani, K.; Vijendra, S.; Burk, J.R.; Lucas, E.A. An investigation of the mean electrical axis angle and respiration during sleep. In Proceedings of the Second Joint 24th Annual Conference and the Annual Fall Meeting of the Biomedical Engineering Society, Engineering in Medicine and Biology, Houston, TX, USA, 23–26 October 2002; IEEE: Piscataway, NJ, USA, 2002; Volume 2, pp. 1550–1551.
21. Dobrev, D.; Daskalov, I. Two-electrode telemetric instrument for infant heart rate and apnea monitoring. *Med. Eng. Phys.* **1999**, *20*, 729–734. [CrossRef]
22. Sakai, M.; Sekine, R.; Zhu, X. Single-channel ECG suitable for ECG-derived respiration. *Biomed. Phys. Eng. Express* **2019**, *5*, 055002. [CrossRef]
23. Mason, C.L.; Tarassenko, L. Quantitative assessment of respiratory derivation algorithms. In Proceedings of the 23rd Annual International Conference of the IEEE Engineering in Medicine and Biology Society, Istanbul, Turkey, 25–28 October 2001; IEEE: Piscataway, NJ, USA, 2001; Volume 2, pp. 1998–2001.
24. De Chazal, P.; Heneghan, C.; Sheridan, E.; Reilly, R.; Nolan, P.; O'Malley, M. Automated processing of the single-lead electrocardiogram for the detection of obstructive sleep apnoea. *IEEE Trans. Biomed. Eng.* **2003**, *50*, 686–696. [CrossRef] [PubMed]
25. Correa, L.S.; Laciar, E.; Torres, A.; Jane, R. Performance evaluation of three methods for respiratory signal estimation from the electrocardiogram. In Proceedings of the 2008 30th Annual International Conference of the IEEE Engineering in Medicine and Biology Society, Vancouver, BC, Canada, 20–25 August 2008; IEEE: Piscataway, NJ, USA, 2008; pp. 4760–4763.
26. Widjaja, D.; Taelman, J.; Vandeput, S.; Braeken, M.A.; Otte, R.A.; Van den Bergh, B.R.; Van Huffel, S. ECG-derived respiration: Comparison and new measures for respiratory variability. In Proceedings of the 2010 Computing in Cardiology, Belfast, UK, 26–29 September 2010; IEEE: Piscataway, NJ, USA, 2010; pp. 149–152.
27. Campolo, M.; Labate, D.; La Foresta, F.; Morabito, F.C.; Lay-Ekuakille, A.; Vergallo, P. ECG-derived respiratory signal using empirical mode decomposition. In Proceedings of the 2011 IEEE International Symposium on Medical Measurements and Applications, Bari, Italy, 30–31 May 2011; IEEE: Piscataway, NJ, USA, 2011; pp. 399–403.
28. Bao, X.; Gall, N.; Deng, Y.; Kamavuako, E.N. Analysis of ECG and PCG Time Delay around Auscultation Sites. In Proceedings of the 13th International Joint Conference on Biomedical Engineering Systems and Technologies, Valletta, Malta, 24–26 February 2020; Volume 4, pp. 206–213.
29. Ricciardi, D.; Cavallari, I.; Creta, A.; Giovanni, G.D.; Calabrese, V.; Belardino, N.D.; Mega, S.; Colaiori, I.; Ragni, L.; Proscia, C.; et al. Impact of the high-frequency cutoff of bandpass filtering on ecg quality and clinical interpretation: A comparison between 40hz and 150hz cutoff in a surgical preoperative adult outpatient population. *J. Electrocardiol.* **2016**, *49*, 691–695. [CrossRef] [PubMed]
30. Laguna, P.; Jane, R.; Caminal, P. Adaptive filtering of ecg baseline wander. In Proceedings of the 1992 14th Annual International Conference of the IEEE Engineering in Medicine and Biology Society, Paris, France, 29 October–1 November 1992; Volume 2, pp. 508–509.
31. Pan, J.; Tompkins, W.J. A real-time qrs detection algorithm. *IEEE Trans. Biomed. Eng.* **1985**, *32*, 230–236. [CrossRef] [PubMed]
32. Charlton, P.H.; Birrenkott, D.A.; Bonnici, T.; Pimentel, M.A.; Johnson, A.E.; Alastruey, J.; Tarassenko, L.; Watkinson, P.J.; Beale, R.; Clifton, D.A. Breathing rate estimation from the electrocardiogram and photoplethysmogram: A review. *IEEE Rev. Biomed. Eng.* **2017**, *11*, 2–20. [CrossRef] [PubMed]
33. Mason, L. *Signal Processing Methods for Non-Invasive Respiration Monitoring*; University of Oxford: Oxford, UK, 2002.
34. Kitney, R.I.; Rompelman, O. *The Study of Heart-Rate Variability (No. 3)*; Oxford University Press: Oxford, UK, 1980.
35. Bao, X.; Howard, M.; Niazi, I.K.; Kamavuako, E.N. Comparison between Embroidered and Gel Electrodes on ECG-Derived Respiration Rate. In Proceedings of the 2020 42nd Annual International Conference of the IEEE Engineering in Medicine & Biology Society (EMBC), Montreal, QC, Canada, 20–24 July 2020; IEEE: Piscataway, NJ, USA, 2020; pp. 2622–2625.

Article

EMG-Free Monitorization of the Acoustic Startle Reflex with a Mobile Phone: Implications of Sound Parameters with Posture Related Responses

Christopher L. Gowen [1], Prashanna Khwaounjoo [1,2] and Yusuf O. Cakmak [1,2,3,4,*]

1 Department of Anatomy, School of Biomedical Sciences, University Of Otago, Po Box 56, Dunedin 9054, New Zealand; chris.gowen@postgrad.otago.ac.nz (C.L.G.); prash.khwaounjoo@otago.ac.nz (P.K.)
2 Medtech Core, Auckland 1010, New Zealand
3 Brain Health Research Centre, Dunedin 9054, New Zealand
4 Centre for Health Systems and Technology, Dunedin 9054, New Zealand
* Correspondence: yusuf.cakmak@otago.ac.nz; Tel.: +64-03-479-4030

Received: 4 September 2020; Accepted: 14 October 2020; Published: 22 October 2020

Abstract: (1) Background: Acute acoustic (sound) stimulus prompts a state of defensive motivation in which unconscious muscle responses are markedly enhanced in humans. The orbicularis oculi (OO) of the eye is an easily accessed muscle common for acoustic startle reaction/response/reflex (ASR) investigations and is the muscle of interest in this study. Although the ASR can provide insights about numerous clinical conditions, existing methodologies (Electromyogram, EMG) limit the usability of the method in real clinical conditions. (2) Objective: With EMG-free muscle recording in mind, our primary aim was to identify and investigate potential correlations in the responses of individual and cooperative OO muscles to various acoustic stimuli using a mobile and wire-free system. Our secondary aim was to investigate potential altered responses to high and also relatively low intensity acoustics at different frequencies in both sitting and standing positions through the use of biaural sound induction and video diagnostic techniques and software. (3) Methods: This study used a mobile-phone acoustic startle response monitoring system application to collect blink amplitude and velocity data on healthy males, aged 18–28 community cohorts during (n = 30) in both sitting and standing postures. The iPhone X application delivers specific sound parameters and detects blinking responses to acoustic stimulus (in millisecond resolution) to study the responses of the blinking reflex to acoustic sounds in standing and sitting positions by using multiple acoustic test sets of different frequencies and amplitudes introduced as acute sound stimuli (<0.5 s). The single acoustic battery of 15 pure-square wave sounds consisted of frequencies and amplitudes between 500, 1000, 2000, 3000, and 4000 Hz scales using 65, 90, and 105 dB (e.g., 3000 Hz_90 dB). (4) Results: Results show that there was a synchronization of amplitude and velocity between both eyes to all acoustic startles. Significant differences (p = 0.01) in blinking reaction time between sitting vs. standing at the high intensity (105 dB) 500 Hz acoustic test set was discovered. Interestingly, a highly significant difference (p < 0.001) in response times between test sets 500 Hz_105 dB and 4000 Hz_105 dB was identified. (5) Conclusions: To our knowledge, this is the first mobile phone-based acoustic battery used to detect and report significant ASR responses to specific frequencies and amplitudes of sound stimulus with corresponding sitting and standing conditions. The results from this experiment indicate the potential significance of using the specific frequency, amplitude, and postural conditions (as never before identified) which can open new horizons for ASR to be used for diagnosis and monitoring in numerous clinical and remote or isolated conditions.

Keywords: acoustic; startle; reaction; response; reflex; blink; mobile; sound

1. Introduction

Dysfunctional mental health affects nearly 300 million people globally with the World Health Organization defining mood and cognitive disorders as the largest contributors to human disability [1]. The burden of diseases revolving around mental health conditions is difficult to quantify given the complexity of standards of care and recording capabilities from 2nd and 3rd world nations as well as individual reporting/withholding. Information from Europe and the United States describe global costs comprising medication, physician visits, as well as hospitalization and indirect costs such as mortality, disability, and production losses accumulate to ~1.7 trillion USD [2]. Aside from these strains, additional socio-economic impact falls on the effects generated from mental health fraud and abuse. Condition masking, abuse of prescription medication as well as disability compensation have influenced patient reporting and very well may continue to without more objective and precise methods for accurate diagnosis.

The current means to examine mental health disorders are not as easily identifiable as symptoms, for example, as physical asymmetry in stroke, but usually rely on the blend of patient history, mental, and physical status examination, and laboratory and/or neuroimaging methods to detect impairments [3–6]. However, diagnosis of mood disorders where resources are constrained may solely rely on patient reporting and invite the feigning of symptoms [1]. Because of these limitations, many leading authorities on psychiatric diagnosis such as Allen Francis, have cautioned health care professionals about the diagnostic in-/deflation in both marginally symptomatic or healthy individuals while using current self-reporting practices [3,7,8].

While combining structured interviews with patient records, and laboratory and imaging review appears to produce more accurate primary and secondary diagnoses than routine clinical methods, there is still significant controversy as to what is considered the *gold standard* towards psychiatric diagnosis as well as what is the laboratory or neuroimaging test's *expected utility*, or the difference between benefit and cost [9].

The two major diagnostic manuals for mood disorders: The Diagnostic and Statistical Manual of Mental Disorders Fifth Edition (DSM-V) and the International Classification of Diseases provide classification systems for clinical identification which encourage self-reporting and questionnaire-literary responses [3,10]. However, these systems are objectively flawed in that their recommended methods (questionnaires) cannot control reporter/assessor bias. To assist with authenticity, biologic and physiologic surrogates of neural states have involved anomalous stress hormones, heart rate variability (HRV), blood pressure, and others have been employed to describe the nervous, cognitive, and physiologic symptoms of mental health [3,8]. Bearing in mind the development of more technological methods, the acoustic startle reflex (ASR) has also proved to be a promising approach in quantifying mental health [11].

The ASR is an aversive response which is enhanced during a fear state and is diminished in a pleasant emotional context [12]. The neuronal arcade responsible for the ASR comprises unconsciously regulated brainstem and cerebral structures where diverse conditions have been found to alter both response time as well as intensity of muscle reaction to sound [13–34]. To assess anomalous blink latencies and amplitudes of the ASR, the prominent blinking muscle, the orbicularis oculi (OO), is easily and commonly accessed using electromyogram (EMG) [34,35]. Using EMG involves specialized training and equipment (wired sensors) and have not been found to be common psychiatric practice even though "each psychiatrist has their own personal style" [3]. Although correlations of OO-ASR responses have been examined between sitting and supine conditions in a post-traumatic stress disorder population [15,23,36], sitting and standing postures were not found to be examined. Standing may not always be a possibility for some participants and these conditions may also reflect differential pathways to sound and reflex pathogenesis [31].

With EMG-free muscle recording in mind, our primary aim was to identify and investigate potential correlations in the responses of individual and cooperative OO muscles to various acoustic stimuli using a mobile and wire-free system. Our secondary aim was to investigate potential altered responses to high and also relatively low intensity acoustics at different frequencies in both sitting and standing positions through the use of biaural sound induction and video diagnostic techniques and software. Sitting and standing methods provide flexibility for the use of devices for people with disabilities. These aims may then clarify the use of the ASR to researchers, medical care providers, and scientists in using sounds and postures to differentiate populations, and/or subpopulate groups into distinct neurophysiologies. We hypothesize that we may find significant details of sound amplitudes and frequencies for use in future experimentation.

Purpose and Goals

The ASR has been investigated to a high degree using a number of tools [35]. However, an EMG free wireless ASR tool or system has yet to be developed. Additionally, using such a system to correspond responses of the left and right eye as well as a comparison between standing and sitting postures to a range of acoustic test sets have yet to be investigated. Within these contexts, the purpose of this pilot study was to develop acoustically repeatable parameters for use in ASR investigations and to subsequently develop an acoustic response spectrum. The goals of this study was to develop an inexpensive, mobile, and clinically relevant biomedical device through the use of an application (app) to deliver a specific acoustic test set and monitor the ASR responses of the eye muscles responsible for blinking.

The outcomes of the present study may provide a detailed profile of the startle reflex which has various clinical and therapeutic significances.

2. Methods

2.1. Ethics and Environment

All subjects gave their informed consent for inclusion before they participated in the study. The study was conducted in accordance with the Declaration of Helsinki, and the protocol was approved by the Human Ethics Committee of the University of Otago (Project identification code D18/407, 11.12.2018). ASR investigations were carried out in a testing environment <65 dB under interior-overhead lighting, and the environment remained at a stable temperature (21 °C).

2.2. Recruitment

Thirty male subjects (Otago University volunteers) between the ages of 18 and 28 (mean 24, standard deviation (SD) 3.7) were recruited and required ~15 min of ASR collection. With each participant, the subject was briefed of the proceedings, signed the consent form, and was alternated to either sitting (n = 15) or standing (n = 15) positions.

Inclusion criteria for the study required participants to be male, between the ages of 18 and 28, and in good health. The exclusion criteria were (a) medical history of neurological disease, and (b), having active stimulants and/or depressants in their system during testing time.

2.3. Hardware and Software

2.3.1. Mobile Sensing Platform Architecture

Due to the novelty of this method and testing equipment, we had to design and cooperate individual equipment and software in order to deliver pure and repeatable sound sets (variable in frequency and amplitude) as well as collect the blinking responses of the participants. To perform these tasks, the iPhone X and iPhone X insert earphones (Apple Incorporated, Cupertino, CA, USA) were used as the hardware platforms to both deliver the acoustic battery and collect the blinking amplitude (magnitude of the blink) and response time of each blinking reflex.

The Sound Stimulus App was created (in collaboration with CodeFluegel GmbH., Graz, Austria) to integrate the acoustic test sets from Table 1 using audio files derived from (https://www.nch.com.au/tonegen/index.html) into the 2018 iPhone X operating system (iOS11.4.1) and collect the ASR blinking data. As acoustic outputs may differ between different phones, operation platforms, speakers, and sound files, we utilized iPhone X insert earphones to limit possible cross compatibility limitations.

Table 1. ASR sound stimulus battery.

Test Set	Frequency	~Decibels	Delay to Startle (s)	Volume Scales (Phone)
1	500	65	5	2
2	1000	90	6	10
3	500	105	8	12
4	500	90	5	7
5	1000	65	8	4
6	2000	65	7	3
7	4000	90	7	10
8	2000	90	7	8
9	4000	105	5	14
10	3000	105	1	12
11	4000	65	3	3
12	1000	105	7	14
13	3000	65	8	3
14	2000	105	4	12
15	3000	90	9	12

2.3.2. Application Overview

In order to collect blinking data, the Sound Stimulus App identified the eyes of an individual and geometrically designated points along the eyelids to measure the movements between geometric anchors (P1–6) against others across time. This method enabled the ability to draw parameters describing the blink completeness as well as the reaction time of both eyes to each test set. The app uses a data computing Dlib library and the included default face landmarking model file [37]. Dlib is a modern C++ toolkit containing machine learning algorithms and tools for creating complex software in C++. This model provides 2D facial feature points when applied on a camera stream containing a human face. In Figure 1, we removed unwanted feature points and maintained only the eyes (6 points for each eye). The equation in Figure 1 provides an output for the eye size and hence acts as a blink marker:

$$EAR = \frac{(P^2\text{-}P^6)+(P^3\text{-}P^5)}{2(P^1\text{-}P^4)}$$

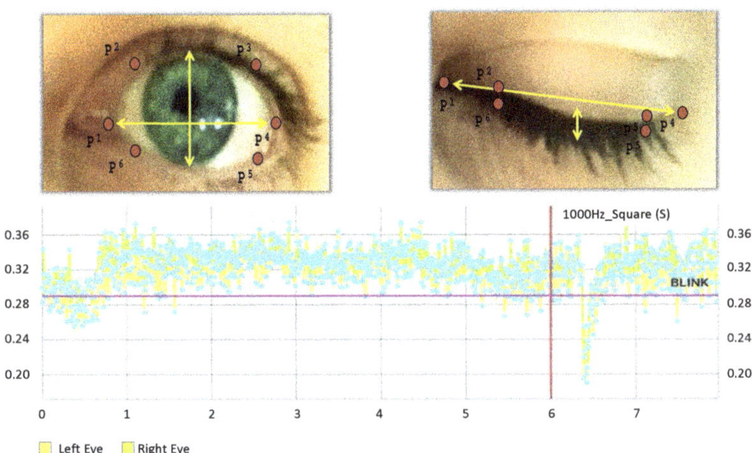

Figure 1. Blink–reflex detection using eye aspect ratio (EAR) across time and the geometric anchors (P1–6). The equation that provides an output for the eye size and hence acts as a blink marker based on Soukupová and Čech (2016).

The idea and formula were based on the work by Soukupová and Čech (2016), who developed a real-time algorithm to detect eye blinks in a video sequence from a standard camera [38]. We adapted this tracking method and integrated sound stimuli details within the Apple iPhone iOS.

This algorithm calculates the distances between vertical and horizontal eye feature points (one horizontal, two vertical lines) and computes the aspect ratio of acquired distances. The aspect ratio is approximately constant while the eye is opened and rapidly falls when the eye blinks. This change in EAR during a blink is used as the blink response amplitude.

2.3.3. ASR Sound Stimulus Battery

The use of this novel mobile ASR monitoring system enabled the introduction of a sound stimulus battery (Table 1) of 15 pure acoustic sounds at delayed intensities using amplitudes of 65, 90, 105 dB and frequency variables of 500, 1000, 2000, 3000 and 4000 Hz. Sound sets were relative to normal human hearing ranges (1–20,000 Hz) introduced high- (4000 Hz) and low- (500 Hz) pitches/frequencies from parameters adapted from previous studies [31,33,36]. The sound sets we developed included novel use of the 3000 and 4000 Hz frequencies as well as the 65 dB amplitude to explore more diversified sound ranges beyond those historically used in literature for ASR elicitation and muscle response monitoring.

To measure the dB output of our mobile ASR monitoring system, 500, 1000, 2000, 3000, and 4000 Hz sounds were selected at 100% (Sound Stimulus App specific) volume for each test set delivery. The side buttons on the iPhone X controls the volume of the speaker (volume scales) which we had to manually select for each test set to deliver either 65, 90, or 105 dB.

We used a Digitech professional sound level meter (SLM) and a sound level calibrator (Harman International Industries, Salt Lake, UT, USA) to monitor the loudness of the testing environment as well as authenticate the dB output from the insert earphones. To accomplish this, the SLM was configured to record the highest dB output, set to "C" weighting (for checking the low-frequency content the sound), and set to "Fast" for normal measurements (fast varying noise) sound recording prior to each use. We obtained laboratory acoustic background noise using the SLM throughout a workday and measured the highest dB readings in the laboratory with the door closed to be <65 dB with the SLM placed 1 m from the door 1 m high.

2.4. Experimental Protocol

For both postures, the mobile device was placed at eye level on an adjustable tripod ~30 cm away from the volunteer's face for optimum ASR measurements (Figure 2).

Figure 2. Seated volunteer with Sound Stimulus App, iPhone X, insert earphones, and noise reduction cups.

During this time, the app was set to record maximum frames per second (120 fps) for increased data collection. The participant inserted the earphones and placed over the ears further insulating (−28 dB) noise reduction cups (Work Force Maxi Muffs, Maxisafe, New South Wales, Australia) prior to the stimulus delivery.

The variables: Hz, dB, delay, and volume scale for each test were manually inputted for each delivery. Following input settings and initiation of the application, the camera begins recording. After establishing these stimulus parameters, the volunteer was instructed to remain still, and look at eye level with the iPhone X until the stimulus delivery and recordings were concluded.

The sound stimulus battery encompasses a set specific order of acoustic signals which were initially randomized to determine delay and sequence (Table 1). Each of the 15 acoustic test sets (delivered as 100 ms square-wave sounds) were delivered biaurally to the ears of the participant who reported as comfortable throughout the ~15 min of stimulus delivery and blink reflex collection. The blinking amplitude and reaction time data from the total 30 tests were converted from the app display (Figure 3) to our data tables for analysis.

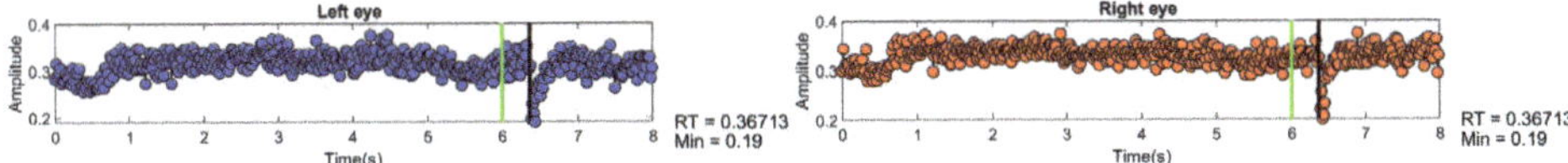

Figure 3. Display of eyelid geodynamics across collection time (~10 s). Left eye (blue) and right eye (red) response amplitudes and velocities. RT = Reaction time of blink reflex to acoustic stimuli, Green line = ASR sound stimulus, Black line = Blink reflex.

3. Statistical Analysis

To determine the significance of these data, each recording was transferred from the Sound Stimulus App to a laptop computer (2012 MacBook Pro, Apple Inc., Cupertino, CA, USA) and underwent statistical analysis and cross platform data interpretation using MATLAB (The Mathworks Inc., Natick, MA, USA), and Microsoft Office 365 computing (Microsoft Corp, Washington, DC, USA). SPSS (IBM Corp, New York, NY, USA) was utilized for data management as well. Pairwise comparisons using two tailed *t*-tests and a repeated measures one-way ANOVA for the responses for both left/right eyes and sitting/standing with Bonferroni adjusted and Tukey post-hoc test were conducted.

4. Results

4.1. Eye Synchronization

The use of our mobile acoustic-startle reflex monitoring system (MARS) allowed the collection of right and left eye responses for the acoustic battery (Table 1). The activity of the eyelid geometry from the sound-initiation onset or previous blink allowed us the ability to differentiate between open alert (yellow and green readings markers up to 6 s) and blink response or closed (Figure 1).

4.2. Blink Reaction Time and Response Amplitudes: Left and Right Eye, Sitting and Standing

Blink reaction times (RT) between both eyes showed no significant differences across the ASR sound stimulus battery. Whereas quickest blink reflexes for the standing and sitting postures occurred after 4000 Hz_105 dB and 4000 Hz_90 dB respectively (stand mean = 0.29 s, sit mean = 0.26 s; SD = 0.05, 0.06). The slowest RT between postures were 500 Hz_105 dB for standing (mean = 0.62 s, SD = 0.04), and 4000 Hz_65 dB for sitting (mean = 0.41 s; SD = 0.02). A comparison between sitting and standing blink reaction time identified a significant difference ($p < 0.05$) between these two postures post 500 Hz_105 dB stimuli (Figure 4). Additionally, left and right eye response amplitudes showed no significant differences. Averages of blinking amplitude for both eyes to the acoustic battery were significantly different ($p = 0.01$) between sitting and standing only with a stimulus of 1000 Hz_65 dB (Figure 4).

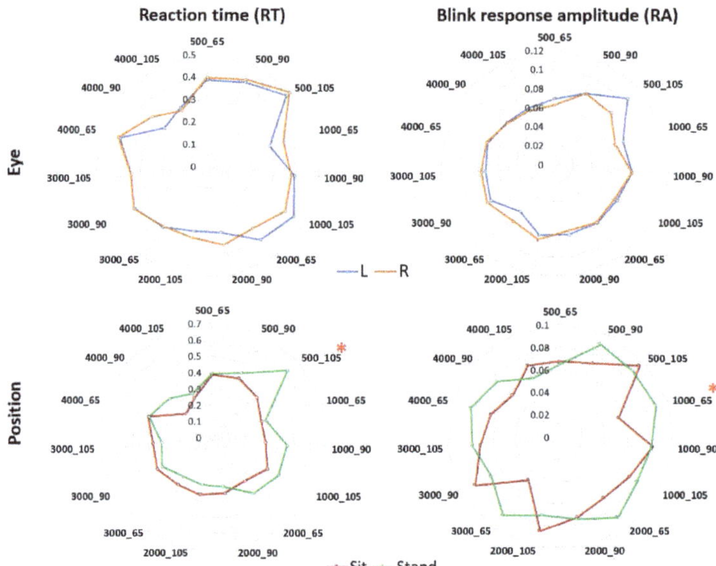

Figure 4. Radar plots showing the average response times (s) and amplitudes for left (L, blue) and right (R, orange) eyes and sitting (red) and standing (green) position/postures for the multiple stimuli. RT = Reaction time (s), RA = Blink response amplitude (given by change in EAR during blink). * Statistically significant $p < 0.05$.

4.3. Response Times to Stimulus Parameters

Combined pairwise comparisons at the different frequencies showed significant differences between the following test sets: 500 Hz_65 dB–500 Hz_105 dB, ($p = 0.018$); 4000_65 dB–4000 Hz_105 dB, ($p = 0.002$) and a highly significant difference ($p = 0.000022$) between test sets 500 Hz_105 dB and 4000 Hz_105 dB with the faster response at 4000 Hz_105 dB (Figure 5).

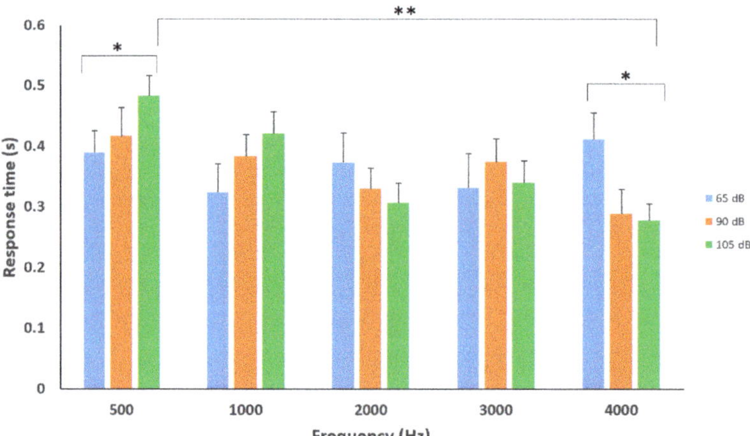

Figure 5. Average response times at varying frequencies and intensities. Significant amplitude-specific differences demonstrated within 500 Hz and 4000 Hz frequencies between 65 and 105 dB. Highly significant frequency-specific differences between 500 and 4000 Hz were also identified from 105 dB amplitudes. Statistically significant * $p < 0.05$, ** $p < 0.001$.

Additionally, we concluded the following significant differences ($p < 0.05$) between frequencies when analyzing the specific acoustic intensities of 65, 90 and 105 dB. At 65 dB: 500 Hz–1000 Hz and 1000 Hz–4000 Hz. At 90 dB: 500 Hz–4000 Hz and 1000 Hz–4000 Hz. At 105 dB: 500 Hz–2000 Hz, 500 Hz–3000 Hz, 500 Hz–4000 Hz, 1000 Hz–4000 Hz (Figures 6–8).

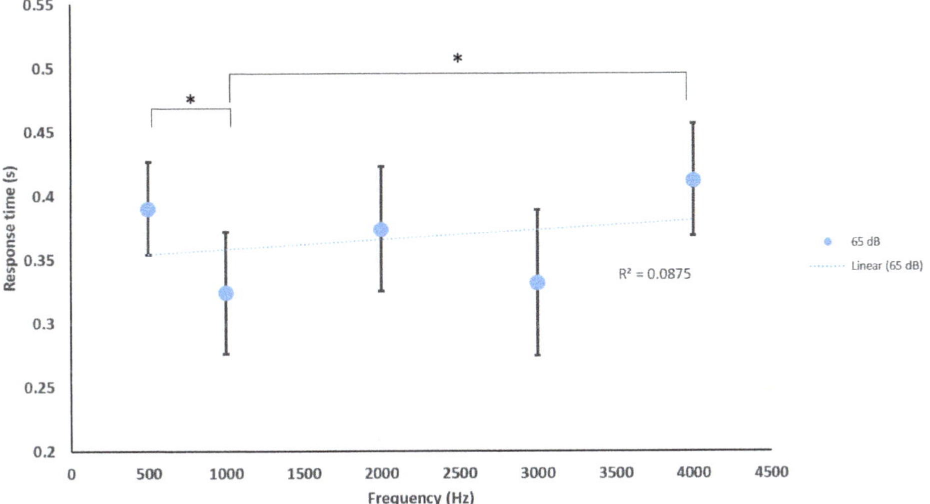

Figure 6. Average blink response times of sitting and standing to various frequencies at 65 dB. Significant differences were found between 500 and 1000 Hz and 1000 and 4000 Hz. R^2 = fit of the line to the data, minimal trend in RT that can be explained by the frequency ($R^2 = 0.087$). * Statistically significant $p < 0.05$.

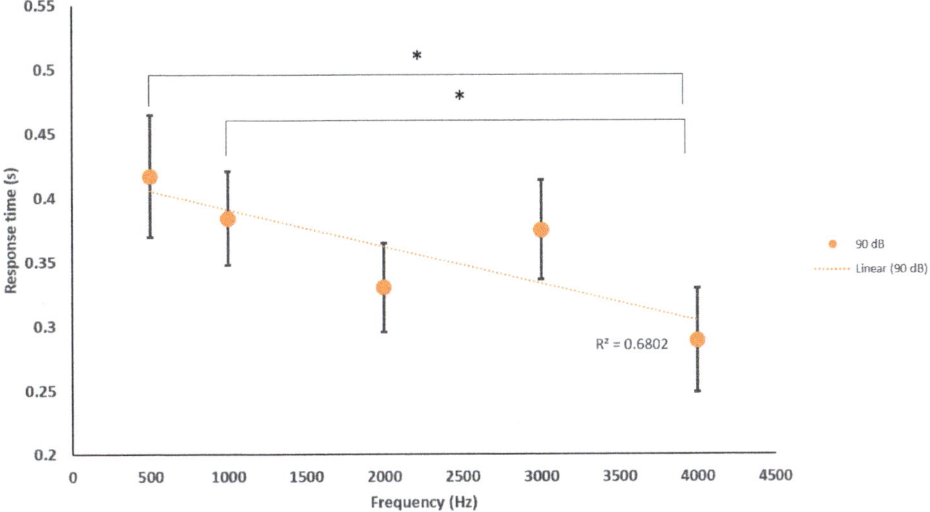

Figure 7. Average blink response times of sitting and standing to various frequencies at 90 dB. Significant differences were found between 500 and 4000 Hz and 1000 and 4000 Hz Indication of reduction in RT with increasing frequency, moderate linear trend ($R^2 = 0.680$). * Statistically significant $p < 0.05$.

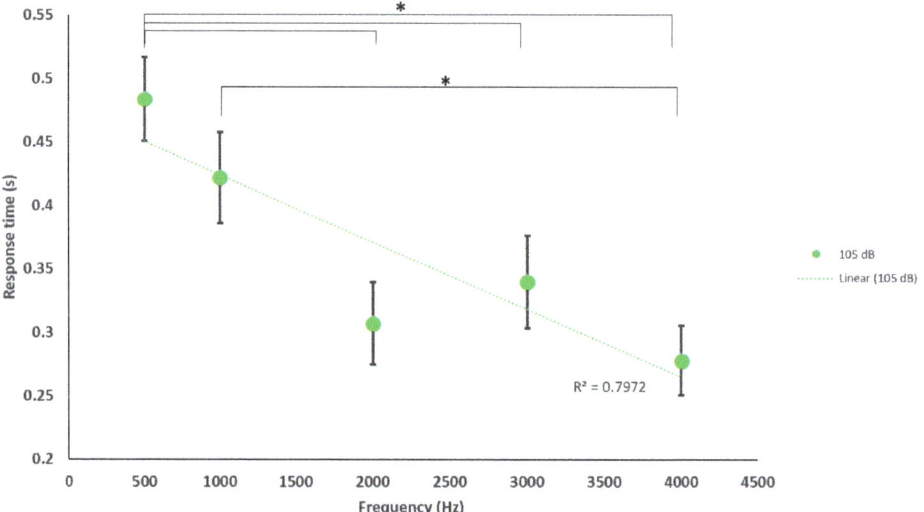

Figure 8. Average blink response times of sitting and standing to various frequencies at 105 dB. Significant differences were found between 500 and 2000, 3000, and 4000 Hz and between 1000 and 4000 Hz. Indication of reduction in RT with increasing frequency, moderate to strong linear trend ($R^2 = 0.797$). * Statistically significant $p < 0.05$.

5. Discussion

Using unique and historically used ASR sound sets, we were able to cue the ASR networks of healthy male participants and track the kinetics of the OO across time with a novel mobile acoustic-startle reflex monitoring system. Individual test set responses between left and right eyes were compared and showed no statistically significant differences between the reaction times or amplitudes between the eyes to any of the test sets within the acoustic battery (Figure 3). However, significant differences between the frequency response times and amplitudes after the delivery of specific sound sets were discovered. Additionally, postural conditions were found to alter the ASR network from unique acoustic stimuli.

In the standing and sitting analysis of the data, the 500 Hz_105 dB cued a faster blink reflex in the sitting posture yet the 1000 Hz_65 dB sound-set cued a stronger response for amplitude for standing ($p < 0.05$). These results were unexpected but may explain a sound set specific postural connectome in the ASR network similar to that described by [39,40], where evidence of fMRI anatomical segregation of auditory information relevant to recognition and localization is processed by distinct neuronal populations. However, these postures have yet to be evaluated using any sound stimuli while using fMRI. The combined overall quickest and slowest blink reflexes were identified at 4000 Hz_105 dB, and 500 Hz_105 dB respectively (Figure 5). These were both high- and low-pitched "loud" sounds which in non-human primates describe an evolutionary context with alertness and our relations to these sounds to the (potentially dangerous) environment [41,42]. Significant differences in blink reaction time and amplitude were determined from comparing sitting and standing postures after the delivery of the 500 Hz_105 dB and 1000 Hz_65 dB test sets respectively (Figure 4). Because these results show blink reaction time to be quicker at the 500 Hz_105 dB and a lower blink reflex amplitude at 1000 Hz_65 dB test sets in the sitting condition, these stimuli may be specified to investigations in subjects unable to stand or participants in unique environments (zero gravity or underwater environments) to address strength of response (over/underreaction) and the speed of ASR network (hyper-/hypoarousal). Furthermore, these sound sets may be expanded to additional frequency-decibel variants.

Between both postures, pairwise comparisons for the different frequencies concluded significant differences between the following test sets: 500 Hz_65 dB–500 Hz_105 dB (p = 0.01), 4000 Hz_65 dB–4000 Hz_105 dB (p = 0.02) and 500 Hz_105 dB-4000 Hz_105 dB (p = 0.000022). These unique sound sets may shift attitudes towards testing toward the higher and lower frequency (±4000 and 500 Hz) with different decibel ranges for future investigations, contrary to previous ASR literature using 1000 and 2000 Hz frequencies for induction. The differences in reactions to these sounds may be significant in terms of cueing the unconscious regions of reactive "survival" limbic, reticular, and autonomic systems for investigating neuronal operating bands in both healthy and disordered (hypo/hyperactive) states. Additionally, these unique sounds may establish further population-specific common ranges of activity such as those identified in children with autism spectrum disorders where hyperreactivity to weak acoustic stimuli and prolonged acoustic startle latency were found in a specific age and condition (autism) cohort [32]. Using varying frequencies and sound intensities i.e., low (500 Hz,65 dB) and high (4000 Hz, 105 dB) to test the ASR, may alter the level of pleasantness/arousal of the sound, which in the past has been associated with emotional and affectional contexts and in turn responses [43–46]. Due to these parameters causing either a very quick or slow blink reflex, investigators may utilize these tones in populations associated with neurotone hyper-/hypoarousal such as anxiety and depression [30,44], or may look at specific dysfunction (brain injury) in autonomic, reticular, limbic, or other networks intrinsic to sound processing and reflex outputs [47,48].

We found additional significant differences ($p < 0.05$) at 65 dB: 500 Hz–1000 Hz and 1000 Hz–4000 Hz. At 90 dB: 500 Hz–4000 Hz and 1000 Hz–4000 Hz. At 105 dB: 500 Hz–2000 Hz, 500 Hz–3000 Hz, 500 Hz–4000 Hz, and 1000 Hz–4000 Hz for our sound sets. These correlations are not yet understood but may represent connectome-specific tonotopy thresholds or transitions between specific frequency groups of stereocilia or ASR processes of deeper brain centers [48–50]. Nevertheless, using the ASR to describe more neurophysiologic spectrums of arousing and depressing activity within a reflex latency scale has applicability in defining neurological tone in patients/participants suffering from central nervous system trauma (traumatic brain injury, stroke) and neurological degenerative disorders (Alzheimer's and Parkinson's). For example, we know that we should see a normal range of blink, amplitude, and eye synchronicity responses from a young healthy male, but, if he were to suffer a concussion or brain injury, the ASR networks may display anomalous responses [51]. The ASR may be delayed, accelerated, or show no reflex at all during the peri/post recovery period and, during this time, the blink reflex and acoustic-processing networks may be conveniently monitored with a mobile phone before/during/after clinical examination for functional abnormalities. Further investigations incorporating the subject's emotional state, arousal or attention, and comfort level may better correlate more specific outputs within a psychiatric context. From our analyses, there is an indication that both higher and lower frequencies with high sound amplitudes may be suitable for future studies in fields of psychiatry, specifically, anxiety, depression, and post-traumatic stress disorder (PTSD).

In this study, we created a mobile acoustic startle response monitoring app and showed that the app produced certain frequency-decibel sound sets in order to record the ASR for blink amplitude and latency in healthy adults for sitting and standing postures. We were able to identify and investigate potential correlations in the responses of individual and cooperative OO muscles to various acoustic stimuli using a mobile and wire-free system. Additionally, we found that certain sound sets induced contrasting reaction times in and between both postures and identified specific pure sounds for future startle response investigations as well as established a baseline ASR spectrum of responses for healthy adults. These healthy parameters may further be contrasted against future ASR spectrums for disordered mood and/or neurological condition monitoring.

Additionally, our findings may also lend to the classification of and recording of neurodegenerative and abnormal neurosystem conditions (such as paraplegia, cerebral palsy, or multiple sclerosis), using unique sound sets where sitting or standing postures are altered or not always possible. The ability to remotely and conveniently monitor for ASR-spectrum deviations and establish individual normative ASR reaction time and intensity ambits may also amplify performance training/conditioning schemes or establish neurosystem measurement parameters for elite programs (Astronaut/Cosmonaut/Military special operations/forces). The means of wire-free monitoring the state of the nervous system using long established and rugged hardware (smart phone with protective case) opens horizons to remote monitoring in extreme environments.

The results from this investigation indicate the potential significance of using specific frequency, amplitude, and postural conditions in ASR studies in addition to diversifying the remote monitoring capabilities of biometric devices from smart devices (iPhone). Utilizing sound parameters with posture related responses while monitoring the acoustic startle reflex with a mobile phone may open new horizons in ASR monitoring across a multitude of populations in order to identify biometric parameters of healthy responses.

5.1. *Limitations*

This was a pilot study, hence the sample size (N) was relatively small; nevertheless, our investigations were able to show usability and effectiveness in determining differences in responses to acoustic battery and posture. Future studies will utilize a larger N. Although the acoustic batteries used to induce the ASR were more diverse than those used in previous literature, time and resources limited the use of additional frequency and amplitude investigational acoustic sets, notably the use of more numerous higher or lower frequency sounds (>500 Hz and <3000 Hz). Age, sex, sleep schedule, and emotional state have been shown to affect ASR to certain degrees: lower response magnitude in aging, pre-pulse inhibition anomalies in sleep deprivation, and hyper/hypoaroused systems in emotion. However, blink response latency to diverse acoustic stimuli in these states is largely undescribed. While limitations of this study were considerations within these behavioral, mood, sleep, sex, and stress hormone variables at the time of testing, our focus and resources remained on the ASR delivery and capture system (MARS) with utility considerations of sitting and standing. As such, to reduce the variability within this pilot study, we utilized a specific age and sex cohort.

5.2. *Comparison with Prior Work*

To the best our knowledge, we are the first group to produce a mobile phone based device for ASR monitoring; hence, other direct comparisons to such systems were difficult to find in the literature. In this context, limited comparisons with EMG based ASR studies [25–36] were made. The average blink response times of the present study at 2000 Hz in standing and sitting were close to and corresponded with literature utilizing similar intensities and frequencies (90–105 dB and 2000 Hz respectively) [11,31] in the context of electromechanical delay, response latencies, response duration, late responses [11,31] and also considering the EAR threshold based ASR methodology of the present study.

These ASR latencies have the potential to be used as a diagnostic or monitoring adjunct which uses left and right eye responses to determine the presence or severity of brain injury victims [51]. Alternate frequencies outside of 2000 Hz were not identified in scientific literature and are considered novel.

6. Conclusions

We found significant details in the responses of cooperative OO muscles to various acoustic stimuli and identified altered responses to high and low intensity acoustics at different frequencies in both sitting and standing postures. Results suggested substantial links between individual differences in frequency, amplitude, sitting, and standing. However, further research is needed to disentangle the specific nature of these associations to one another as well as identify arousal and comfort relationships to specific frequency and amplitude acoustic test sets. To our knowledge, this is the first mobile phone-based monitoring system used to detect and report significant ASR responses to a variety of frequencies and amplitudes in sitting and standing postures. Therefore, it should be kept in mind that further replications are needed to contrast the present techniques and results or address any other potential topics. By visually recording through the use of a smart phone app, we demonstrated that it can be possible to detect and monitor the ASR in healthy population through the use of a mobile device.

This opens new horizons for the ASR to be used for diagnosis and monitoring in numerous clinical conditions (e.g., stroke, traumatic brain injury, and mood disorders). The findings in the present study suggest that MARS is a simple and mobile methodology used to study the links between acute acoustic variables and their subsequent effects on the human blink response.

Author Contributions: Y.O.C. and C.L.G.: Designed the study protocol, Y.O.C., P.K., and C.L.G.: Designed and contributed to the development of the mobile app, C.L.G.: Collected the data, P.K.: Analyzed the data, C.L.G., P.K., and Y.O.C.: Interpreted the results, C.L.G., P.K., and Y.O.C.: Wrote the manuscript. All authors have read and agreed to the published version of the manuscript.

Funding: The sound stimulus application development in this research was funded by Inventram (Istanbul, Turkey).

Conflicts of Interest: YOC has pending and granted patents involving/related to mobile monitorization of ASR.

Abbreviations

ASR	Acoustic startle reflex
OO	Orbicularis oculi
EAR	Eye Aspect Ratio
EMG	Electromyogram
SD	Standard Deviation
DSM-V	Diagnostic and Statistical Manual of Mental Disorders Fifth Edition
Hz	Hertz
dB	Decibels
N	Sample size
SLM	Sound level meter
MARS	Mobile acoustic-startle reflex monitoring system
RT	Reaction time
P	Probability value

References

1. World Health Organization. Depression and Other Common Mental Disorders. In *Global Health Estimates*; World Health Organization: Geneva, Switzerland, 2017.
2. Trautmann, S.; Rehm, J.; Wittchen, H. The economic costs of mental disorders: Do our societies react appropriately to the burden of mental disorders? *EMBO Rep.* **2016**, *17*, 1245–1249. [CrossRef] [PubMed]
3. Hackett, P.T.; Cassem, H.N. *Handbook of General Hospital Psychiatry*, 2nd ed.; PSG Publishing Company, INC.: Littleton, MA, USA, 1987.
4. Rauch, S.L.; Whalen, P.J.; McInerney, S.C.; Macklin, M.L.; Lasko, N.B.; Orr, S.P.; Pitmman, R.K. Exagerated amygdala response to masked facial stimuli in posttraumatic stress disorder: A functional MRI study. *Biol. Psychiatry* **2000**, *47*, 769–776. [CrossRef]
5. Koutsouleris, N.; Meisenzahl, E.M.; Borgwardt, S.; Rieher-Rossler, A.; Frodi, T.; Kambeitz, J.; Kohler, Y.; Falkai, P.; Moller, H.J.; Reiser, M.; et al. Individualized differential diagnosis of schizophrenia and mood disorders using neuroanatomical biomarkers. *Brain J. Neurol.* **2015**, *7*, 2059–2073. [CrossRef] [PubMed]
6. Le-Niculescu, H.; Kurian, S.M.; Yehyawi, N.; Dike, C.; Patel, S.D.; Edenberg, H.J.; Tsuang, M.T.; Salomon, D.R.; Nurnberger, J.I., Jr.; Niculescu, A.B. Identifying blood biomarkers for mood disorders using convergent functional genomics. *Mol. Psychiatry* **2009**, *14*, 1143. [CrossRef]
7. Francis, A. The past, present, and future of psychiatric diagnosis. *World Psychiatry* **2013**, *12*, 111–112. [CrossRef]
8. American Psychiatric Association. Trauma- and Stressor-Related Disorders. In *Diagnostic and Statistical Manual of Mental Disorders*, 5th ed.; American Psychiatric Association: Washington, DC, USA, 2013.
9. Phillips, K.A.; First, M.B.; Pincus, H.A. (Eds.) *Advancing DSM: Dilemmas in Psychiatric Diagnosis*; American Psychiatric Association: Washington, DC, USA, 2003.
10. International Classification of Diseases. 6B40 Post Traumatic Stress Disorder, ICD-11 for Mortality and Morbidity Statistics (Version 04). Available online: https://icd.who.int/browse11/l-m/en#/http://id.who.int/icd/entity/2070699808 (accessed on 1 June 2019).

11. Kaviani, H.; Gray, J.A.; Checkley, S.A.; Raven, P.W.; Wilson, G.D.; Kumari, V. Affective modulation of the startle response in depression: Influence of the severity of depression, anhedonia, and anxiety. *J. Affect. Dis.* **2004**, *15*, 21–31. [CrossRef]
12. Lang, P.J.; Bradley, M.M.; Cuthbert, B.N. Emotion, attention, and the startle reflex. *Psychol. Rev.* **1990**, *3*, 377–395. [CrossRef]
13. Bakker, M.J.; Boer, F.; Benninga, M.A.; Koelman, J.H.T.M.; Tijssen, M.A.J. Increased Auditory Startle Reflex in Children with Functional Abdominal Pain. *J. Pediatrics* **2010**, *156*, 285–291. [CrossRef]
14. Benning, S.D. Postauricular and superior auricular reflex modulation during emotional pictures and sounds. *Psychophysiology* **2011**, *48*, 410–414. [CrossRef]
15. Benvenuti, S.M.; Bianchin, M.; Angrilli, A. Effects of simulated microgravity on brain plasticity: A startle reflex habituation study. *Physiol. Behav.* **2011**, *104*, 503–506. [CrossRef]
16. Blanch, A.; Balada, F.; Aluja, A. Habituation in acoustic startle reflex: Individual differences in personality. *Int. J. Psychophysiol.* **2014**, *91*, 232–239. [CrossRef] [PubMed]
17. Delwaide, P.J.; Shepens, B. Auditory startle (audio-spinal) reaction in normal man: EMG responses and H reflex changes in antagonistic lower limb muscles. *Electroencephalogr. Clin. Neurophysiol.* **1995**, *97*, 416–423. [CrossRef]
18. Grillon, C.; Morgan, C.A.; Southwich, S.M.; Davis, M.; Charney, D.S. Baseline startle amplitude and prepulse inhibition in Vietnam veterans with posttraumatic stress disorder. *Psychiatry Res.* **1995**, *64*, 169–178. [CrossRef]
19. Grillon, C.; Pellowski, M.; Merikangas, K.R.; Davis, M. Darkness facilitates the acoustic startle reflex in humans. *Biol. Psychiatry* **1997**, *42*, 453–460. [CrossRef]
20. Kiziltan, M.E.; Gunduz, A.; Apaydin, H.; Ertan, S.; Kiziltan, G. Auditory startle reflex and startle reflex to somatosensory inputs in generalized dystonia. *Clin. Neurophysiol.* **2015**, *126*, 1740–1745. [CrossRef]
21. Korn, C.W.; Staib, M.; Tzovara, A.; Castegnetti, G.; Bach, D.R. A pupil size response model to assess fear learning. *Psychophysiology* **2017**, *54*, 330–343. [CrossRef]
22. Kumari, V.; Gray, J.A. Smoking withdrawal, Nicotine dependence and prepulse inhibition of the acoustic startle reflex. *Psychopharmacology* **1999**, *141*, 11–15. [CrossRef]
23. Morgan, C.A.; Grillon, C.; Lubin, H.; Southwick, S.M. Startle Reflex Abnormalities in Women with Sexual Assault-Related Posttraumatic Stress Disorder. *Am. J. Psychiatry* **1997**, *154*, 1076–1080.
24. Muller, J.; Kofler, M.; Wenning, G.K.; Seppi, K.; Valls-Sole, S.; Poewe, W. Auditory startle response in cervical dystonia. *Mov. Disord.* **2003**, *18*, 1522–1526. [CrossRef]
25. Nonnekes, J.; Geel, K.V.; Oude Nijhuis, L.B.; Bloem, B.R.; Geurts, A.C.; Weerdesteyn, V. Loading enhances the occurrence of startle responses in leg muscles. *Neuroscience* **2013**, *240*, 186–190. [CrossRef]
26. Nonnekes, J.; Geurts, A.C.H.; Oude Nijhuis, L.B.; Geel, K.V.; Snijders, A.H.; Bloem, B.R.; Weerdesteyn, V. Reduced StartReact effect and freezing of gait in Parkinson's disease: Two of a kind? *J. Neurol.* **2014**, *261*, 943–950. [CrossRef]
27. O'beirne, G.A.; Patuzzi, R.B. Basic properties of the sound-evoked post-auricular muscle response (PAMR). *Hear. Res.* **1999**, *138*, 115–132. [CrossRef]
28. Oude Jijhuis, L.B.; Allum, J.H.J.; Valls-Sole, J.; Overeem, S.; Bloem, B.R. First Trial Postural Reactions to Unexpected Balance Disturbances: A comparison With the Acoustic Startle Reaction. *J. Neurophysiol.* **2010**, *104*, 2704–2712. [CrossRef] [PubMed]
29. Rigato, S.; Rieger, G.; Romei, V. Multisensory signaling enhances pupil dilation. *Sci. Rep.* **2015**, *6*, 26188. [CrossRef] [PubMed]
30. Samuels, E.R.; Hou, R.H.; Langley, R.W.; Szabadi, E.; Bradshaw, C.M. Modulation of the Acoustic Startle Response by the Level of Arousal: Comparison of Clonidine and Modafinil in Healthy Volunteers. *Neuropsychopharmacology* **2007**, *32*, 2405–2421. [CrossRef]
31. Siegelaar, S.E.; Olff, M.; Bour, L.J.; Veelo, D.; Zwinderman, A.H.; Bruggen, G.V.; Vries, G.J.D.; Raabe, S.; Cupido, C.; Koelman, J.H.T.M.; et al. The auditory startle response in post-traumatic stress disorder. *Exp. Brain Res.* **2006**, *174*, 1–6. [CrossRef]
32. Takahashi, H.; Nakahachi, T.; Komatsu, S.; Ogino, K.; Iida, Y.; Kamio, Y. Hyperreactivity to weak acoustic stimuli and prolonged acoustic startle latency in children with autism spectrum disorders. *Mol. Autism* **2014**, *5*, 23. [CrossRef]
33. Talaat, H.S.; Kabel, A.H.; Khalil, L.H.; Said, N.M. Post Auricular Muscle Response in Auditory Neuropathy. *J. Int. Adv. Otol.* **2010**, *6*, 360–364.

34. Blumenthal, T.D.; Cuthbert, B.N.; Filion, D.L.; Hackley, S.; Lipp, O.V.; van Boxtel, A. Committee report: Guidelines for human startle eyeblink electromyographic studies. *Psychophysiology* **2005**, *42*, 1–15. [CrossRef]
35. Yeomans, J.S.; Liang, L.; Scott, B.W.; Frankland, P.W. Tactile, acoustic and vestibular systems sum to elicit the startle reflex. *Neurosci. Biobehav. Rev.* **2001**, *26*, 1–11. [CrossRef]
36. Koch, J.; Flemming, J.; Zeffiro, T.; Rufer, M.; Orr, S.P.; Mueller-Pfeiffer, C. Effects of Posture and Stimulus Spectral Composition on Peripheral Physiological Responses to Loud Sounds. *PLoS ONE* **2016**, *11*, e0161237. [CrossRef] [PubMed]
37. Dlib C++ Library, Algorithm. Available online: http://dlib.net/algorithms.html#murmur_hash3 (accessed on 1 June 2019).
38. Soukupova, T.; Cech, J. Real-Time Eye Blink Detection using Facial Landmarks. In Proceedings of the 1st Computer Vision Winter Workshop, Rimske Toplice, Slovenia, 3–5 February 2016.
39. Maeder, P.P.; Meuli, R.A.; Bellmann, A.; Fornari, E.; Thiran, J.-P.; Pittet, A.; Clarke, S. Distinct Pathways Involved in Sound Recognition and Localization: A Human fMRI Study. *Neuroimage* **2001**, *14*, 802–816. [CrossRef] [PubMed]
40. Lewis, J.W.; Wightman, F.L.; Brefczynski, J.A.; Phinney, R.E.; Binder, J.R.; DeYoe, E.A. Human brain regions involved in recognizing environmental sounds. *Cereb. Cortex.* **2004**, *14*, 1008–1021. [CrossRef]
41. Aboitiz, F. A Brain for Speech. Evolutionary Continuity in Primate and Human Auditory-Vocal Processing. *Front. Neurosci.* **2018**, *12*, 174. [CrossRef]
42. Bocchi, C.; Severi, F.M.; Bruni, L.; Filardi, G.; Delia, A.; Boni, C.; Altomare, A.; Bellieni, C.V.; Petraglia, F. Ultrasound and Fetal Stress: Study of the Fetal Blink-Startle Reflex Evoked by Acoustic Stimuli. *Neonatal Pain* **2008**, *4*, 31–33. [CrossRef]
43. Quam, R.; Martinez, I.; Lorenzo, C.; Bonmati, A.; Rosa-Zurera, M.; Jarabo, P.; Arsuaga, J.L. Studying audition in fossil hominins: A new approach to the evolution of language? In *Psychology of Language*; Nova Science Publishers, Inc.: New York, NY, USA, 2012; pp. 1–37.
44. Fujisawa, T.; Cook, N.D. Identifying emotion in speech prosody using acoustical cues of harmony. In Proceedings of the Eighth International Conference on Spoken Language Processin INTERSPEECH, Jeju Island, Korea, 4–8 October 2004; pp. 1333–1336.
45. Russel, J.A. A circumplex model of affect. *J. Personal. Soc. Psychol.* **1980**, *39*, 1161–1178. [CrossRef]
46. Hagman, F. Emotional Response to Sound. Master's Thesis, Chalmers University of Technology, Goteborg, Sweden, 2010.
47. Wu, M.-F.; Suzuki, S.S.; Siegel, J.M. Anatomical distribution and response patterns of reticular neurons active in relation to acoustic startle. *Brain Res.* **1988**, *2*, 399–406.
48. Hall, D.A.; Plack, C.J. Pitch Processing Sites in the Human Auditory Brain. *Cereb. Cortex* **2008**, *19*, 576–585. [CrossRef]
49. Hackett, T.A. The Human Auditory System; Fundamental Organization and Clinical Disorders Chapter 2, Anatomic organization of the auditory cortex. *Handb. Clin. Neurol.* **2015**, *129*, 27–53.
50. Smith, S.W. Chapter 22—Audio Processing/Human Hearing. In *The Scientist and Engineer's Guide to Digital Signal Processing*; California Technical Publishing: San Diego, CA, USA, 1998.
51. Zivi, I.; Bertelli, E.; Bilotti, G.; Clemente, A.; Saltuari, L.; Frazzitta, G. Blink-associated contralateral eccentric saccades as a rare sign of unilateral brain injury. *Neurology* **2017**, *88*, 2. [CrossRef]

Publisher's Note: MDPI stays neutral with regard to jurisdictional claims in published maps and institutional affiliations.

Article

Enhancing Classification Performance of fNIRS-BCI by Identifying Cortically Active Channels Using the z-Score Method

Hammad Nazeer [1], Noman Naseer [1,*], Aakif Mehboob [2], Muhammad Jawad Khan [2,3], Rayyan Azam Khan [4], Umar Shahbaz Khan [5,6] and Yasar Ayaz [2,3]

1 Department of Mechatronics Engineering, Air University, Islamabad 44000, Pakistan; hammad@mail.au.edu.pk
2 School of Mechanical and Manufacturing Engineering, National University of Science and Technology, Islamabad 44000, Pakistan; aakifmehboob@live.com (A.M.); jawad.khan@smme.nust.edu.pk (M.J.K.); yasar@smme.nust.edu.pk (Y.A.)
3 National Centre of Artificial Intelligence (NCAI), Islamabad 44000, Pakistan
4 Department of Mechanical Engineering, University of Saskatchewan, Saskatoon, SK S7N5A9, Canada; rayyan.khan@usask.ca
5 Department of Mechatronics Engineering, National University of Sciences and Technology, H-12, Islamabad 44000, Pakistan; u.shahbaz@ceme.nust.edu.pk
6 National Centre of Robotics and Automation (NCRA), Rawalpindi 46000, Pakistan
* Correspondence: noman.naseer@mail.au.edu.pk

Received: 26 October 2020; Accepted: 3 December 2020; Published: 7 December 2020

Abstract: A state-of-the-art brain–computer interface (BCI) system includes brain signal acquisition, noise removal, channel selection, feature extraction, classification, and an application interface. In functional near-infrared spectroscopy-based BCI (fNIRS-BCI) channel selection may enhance classification performance by identifying suitable brain regions that contain brain activity. In this study, the z-score method for channel selection is proposed to improve fNIRS-BCI performance. The proposed method uses cross-correlation to match the similarity between desired and recorded brain activity signals, followed by forming a vector of each channel's correlation coefficients' maximum values. After that, the z-score is calculated for each value of that vector. A channel is selected based on a positive z-score value. The proposed method is applied to an open-access dataset containing mental arithmetic (MA) and motor imagery (MI) tasks for twenty-nine subjects. The proposed method is compared with the conventional *t*-value method and with no channel selected, i.e., using all channels. The z-score method yielded significantly improved ($p < 0.0167$) classification accuracies of 87.2 ± 7.0%, 88.4 ± 6.2%, and 88.1 ± 6.9% for left motor imagery (LMI) vs. rest, right motor imagery (RMI) vs. rest, and mental arithmetic (MA) vs. rest, respectively. The proposed method is also validated on an open-access database of 17 subjects, containing right-hand finger tapping (RFT), left-hand finger tapping (LFT), and dominant side foot tapping (FT) tasks.The study shows an enhanced performance of the z-score method over the *t*-value method as an advancement in efforts to improve state-of-the-art fNIRS-BCI systems' performance.

Keywords: functional near-infrared spectroscopy; brain–computer interface; z-score method; channel selection; region of interest; channel of interest

1. Introduction

Functional near-infrared spectroscopy (fNIRS) is a noninvasive optical imaging technique used to measure blood oxygenation changes as brain activity to develop a brain–computer interface (BCI) [1].

Among other noninvasive modalities used for BCIs like functional magnetic resonance imaging (fMRI) and electroencephalography (EEG); applications of fNIRS are increasing steadily in the BCI community [2–5]. fNIRS is a cheap, portable, and safe optical brain imaging technique used in state-of-the-art BCI systems to control and drive external devices using brain signals [1,6]. fNIRS is also used to analyse the brain at work and during complex everyday life situations [7–10]. fNIRS records brain activity as cortical blood oxygenation changes using two or more wavelengths of near-infrared lights ranging from 700 to 1000 nm. It measures the changes in oxy- and deoxy-hemoglobin (ΔHbO and ΔHbR) using the modified Beer–Lambert law [11,12]. The theoretical principles, advancements, and practical fNIRS applications have previously been described in detail [2–4,13]. fNIRS systems are portable, wearable, and overall user-friendly, making fNIRS a suitable choice for BCI. Recently, fNIRS has demonstrated successful and promising results in several BCI applications [14–18] and clinical applications [19,20] for brain imaging and brain signal acquisition purpose.

BCI systems have enhanced patients' quality of life in clinics, hospitals, daily life activities, and at work [21,22]. A state-of-the-art BCI system includes recording brain signal, noise reduction, channel selection, extracting features, classification, and an application interface [2]. Initially, a suitable brain imaging modality is used to record brain signals. In the second step, preprocessing is performed which consists of detrending, removing physiological and instrumental noises, and cortical activity-based channel selection. Different methods can be used to select brain activation channels. These methods include the t-value method, baseline correction method, source analysis of brain activation, and others. In the third step, appropriate features are extracted, followed by brain signals classification using suitable machine learning algorithms. Finally, the control unit generates control commands using discriminated brain signals to control external devices. Brain signals with high signal-to-noise ratio, brain-activation-based channel selection, and suitable machine learning algorithms are essential components of a state-of-the-art BCI system. Recent studies have been conducted to enhance fNIRS-BCI systems' performance by enhancing classification accuracy using different methods and techniques at every stage of the BCI system [2,14,16,23]. The studies showed efforts to improve classification accuracy by applying cortical-activity-based channel selection techniques, extracting novel features, determining optimal features, and optimal feature-combinations for fNIRS-BCI [24–26].

In BCI, selecting channels of interest (COI) or a region of interest (ROI) has manifold objectives: reducing processing time, reducing dimensionality, enhancing performance, and suitable brain region identification containing low noise signals. The selection of appropriate channels in EEG-based BCI has shown encouraging results [27]—Mainly filtering, wrapper, embedded, hybrid, and human-based techniques have been used for the purpose [27]. The sequential floating forward selection (SFFS) algorithm [28] and iterative relief based on distance from centre (IterRelCen) algorithm [29] were applied for channel selection of motor imagery (MI) tasks for EEG-BCI. Likewise, Pearson's correlation method was used for channel selection of three different EEG datasets of MI tasks for EEG-BCI [3]. Li et al. [30] implemented three different strategies for channel selection for stroke patients. Feng et al. [5] used the CSP-rank channel selection methods along with multiband signal decomposition filtering for selection of optimal channels. Similarly, an attention-based convolutional recurrent neural network (ACRNN) was used to extract more discriminative features from EEG signals and improve the accuracy of emotion recognition [31]. Jin et al. [32] applied the bi-spectrum-based channel selection algorithm on MI tasks for EEG-BCI. For EEG-based BCI, cross-correlation [33], probabilistic mapping methods [34], contrast-to-noise-ratio [35], and principal component analysis [36] have also been used for channel selection and ROI identification.

However, for fNIRS-BCI, very few channel selection methods and techniques have been found in the literature, which includes the t-value method [37–39], baseline correction method [40], and hardware-based approach, i.e., bundled-optode method [41]. The t-value method is used excessively by researchers for selecting ROI and COI. This statistical-based approach considers only those channels that give a positive t-value ($t > 0$) or greater than the critical value ($t > t_{crt}$) and $p < 0.05$, where the value of t_{crt} depends upon the degree of freedom (i.e., number of samples in the signal). The method includes a step-wise procedure of (a) generating a canonical hemodynamic response function (cHRF) using 2-gamma functions [42] or 3-gamma functions [43], (b) convolving cHRF with known stimulation interval (boxcar function) to get a modelled/desired hemodynamic response function (dHRF), (c) applying iteratively reweighted least squares algorithm to estimate parameters by using a general linear regression model with dHRF, and (d) final significance of the hypothesis is calculated through these estimated parameters. If the estimated parameters are positive, then specific stimulation is assumed active and vice versa. Sontosa et al. [44] described the method in detail for finding out the most significant stimulation through *t*-values. However baseline correction technique simply compares peak value of tasks with peak value of rest in brain signals. If peak value of task is greater than the peak value of rest, the channel is selected. In this paper, we propose a novel method, the z-score method that uses cross-correlation and z-scores for ROI/COI selection to enhance the fNIRS-BCI system's performance.

In the proposed methodology, conventional steps of data acquisition and reduction of noise are followed. In the second step, brain-activation-based channel selection is performed. cHRF is calculated using two-gamma functions, followed by dHRF estimation. Cross-correlation is applied to dHRF and each channel of averaged trial. The max value of correlation coefficients is selected for each channel and forms another vector of all channels' max values. The z-score is calculated for the vector of max values. If the z-score is greater than zero, then the channel is selected (z-score > 0). After that, features are calculated, and classification is performed. The proposed methodology is applied to an open-access dataset of left motor imagery (LMI), right motor imagery (RMI), and mental arithmetic (MA) in this study. All channels and the *t*-value-method-selected channels are used for verification, following the same classification steps. Results show that the classification accuracy achieved using the z-score method is significantly higher ($p < 0.0167$; Bonferroni correction applied) than the *t*-value method and by using all channels. For validation of the proposed method, it is also applied on another open-access database of 17 subjects having RFT, LFT, and dominant side FT tasks. The results also show better performance of the z-score method on the conventional *t*-value method, baseline correction method, and by using all channels.

2. Materials and Methods

2.1. Subjects/Participants

An open-access dataset of fNIRS single-trial classification for LMI vs. rest, RMI vs. rest, and MA vs. rest is used in this study [45]. The dataset contains brain signals of twenty-nine healthy subjects with mean age of 28.5 ± 3.7 years. There were 14 males and 15 females and none of them had any mental, neurological, or visual disorder. The experimental paradigm was explained in detail to subjects before taking the written consent. The experiments were conducted following the latest Declaration of Helsinki. The Ethics Committee approved this study for the Institute of Psychology and Ergonomics, Technical University of Berlin (approval number: SH_01_20150330).

2.2. Experimental Paradigm/Protocol

In the literature, researchers used mental arithmetic, visual tasks, letter padding, word generation, object rotation, motor imagery, motor execution, and music imagery as brain activities for data acquisition for fNIRS-BCI [22,40,46–49]. In this study, motor imagery of left- and right-hand and mental arithmetic were selected as the brain activities.

The subjects were seated on a comfortable chair facing a screen. They were asked to control their body movements and stay still as much as possible during data acquisition. The experiment contained three sessions of LMI, RMI, and MA tasks. Each session started with an initial rest of 60 s to set up the baseline followed by 20 repetitions of the selected tasks with 60 s of final rest at the end of the session. Each task started with 2 s of the visual introduction of the task. Then the subject was asked to perform a task for 10 s followed by rest for a period of 15–17 s. A short beep (250 ms) was played at the start and end of each task. Task instructions were displayed on the screen. During the rest period, the subjects were asked to relax—further details can be found in [45]. The experimental paradigm is shown in Figure 1.

2.2.1. Motor Imagery (MI)

For MI tasks, subjects were asked to perform kinaesthetic MI, i.e., to imagine the opening and closing of their hands as they were grabbing a ball. As all subjects were naive, visual instruction using a black arrow pointing left or right side was displayed on screen for 2 s. A short beep sound was played before the arrow disappeared, followed by a fixation cross during the task period. The subjects were told to imagine opening and closing of the hand at a self-paced frequency of 1 Hz. Again, a short beep sound was played with 'STOP' written and displayed on the screen to end the task period. The fixation cross was also displayed on the screen during the rest period. This pattern was repeated twenty times in a single session keeping a balanced count of 10 trials for each LMI and RMI.

2.2.2. Mental Arithmetic (MA)

For the MA task, subjects were instructed to perform the initial subtraction of a one-digit number from a three-digit number, e.g., 384-8, by displaying it on the screen for 2 s. They were asked to memorize the numbers shown on screen for subtraction. The screen changed to a black fixation cross for the task period with a short beep sound. During the task period of 10 s, the subjects were instructed to subtract the one-digit number from the result of the previous subtraction repeatedly. Followed by a 15–17 s rest period, subjects were allowed to relax, and a black fixation cross was also displayed on the screen. Just like the MI paradigm, task periods were ended by playing a short beep sound, and "STOP" written and displayed on the screen. Likewise, the MI paradigm, initial, and final rest of the 60 s, was included in the MA paradigm to set up a baseline.

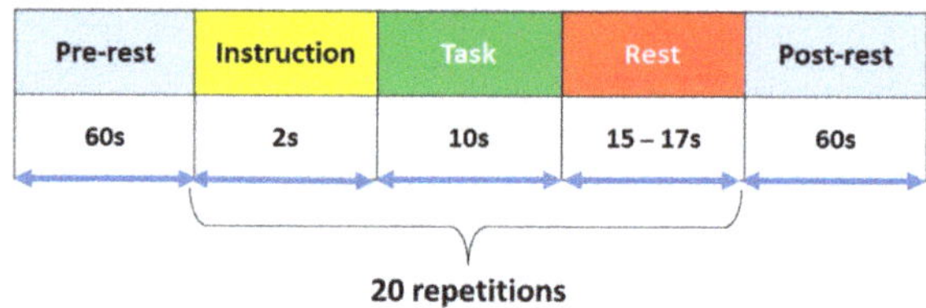

Figure 1. Experimental paradigm for data acquisition. After the initial 60-s rest, a single trial consisted of a 2-s visual instruction period, then 10 s left motor imagery (LMI), right motor imagery (RMI), motor imagery (MI), and mental arithmetic (MA) tasks followed by a 15–17 s rest.

2.3. *Experimental Setup/Optode Placement*

Fourteen emitters and sixteen detectors were used to record fNIRS signals with separation of 3 cm [50,51], resulting in thirty-six physiological channels. Nine channels were placed at the frontal cortex around Fp1, Fp2, and Fpz. Twelve channels were positioned at the motor cortex around C3 and C4 respectively. And three channels were placed at the visual cortex around Oz. Optodes were arranged according to the 10–20 international system as shown in Figure 2.

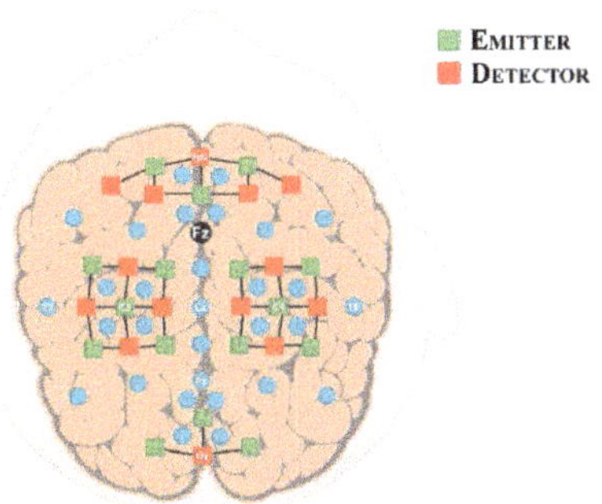

Figure 2. Optodes were placed at the frontal, motor, and visual cortex following the 10–20 international system [45]. Green and red squares represent emitters and detectors, respectively. Fourteen emitters and sixteen detectors were used to record functional near-infrared spectroscopy (fNIRS) signals with separation of 3 cm, resulting in a total of thirty-six.

2.4. Signal Acquisition

fNIRS data were measured by NIRScout (NIRx GmbH, Berlin, Germany). Additionally, an opaque cap was used over a stretchy fabric cap to block ambient light, and also firm contact was observed between the optodes and scalp. The sampling frequency was set to 12.5 Hz. The brain imaging system used two wavelengths, 760 and 850 nm. Following the literature [11], the modified Beer–Lambert law (MBLL) was applied to convert brain signals recorded into ΔHbO and ΔHbR.

$$\begin{bmatrix} \Delta HbO(t) \\ \Delta HbR(t) \end{bmatrix} = \frac{\begin{bmatrix} \varepsilon_{HbO}(\lambda_1) & \varepsilon_{HbR}(\lambda_1) \\ \varepsilon_{HbO}(\lambda_2) & \varepsilon_{HbR}(\lambda_2) \end{bmatrix}^{-1} \begin{bmatrix} \frac{\Delta ¥=(t,\lambda_1)}{\Delta ¥=(t,\lambda_2)} \end{bmatrix}}{d \times l} \tag{1}$$

where $\varepsilon_{HbO}(\lambda)$ and $\varepsilon_{HbR}(\lambda)$ are extinction coefficients of ΔHbO and ΔHbR in $\mu M^{-1} cm^{-1}$ respectively, d is the differential path-length factor in [mm], l is the distance between emitter and detector in [mm], and $\Delta ¥ = (t)$ is the absorbance difference of light source wavelength of λ_i (where i = 1,2).

2.5. Signal Processing

Various noises like instrumental, physiological, and experimental noises contained by acquired hemodynamic signals had to be removed before feature extraction and classification [49]. Following the instructions [45] about preprocessing, ΔHbO and ΔHbR data were band-pass filtered using a fourth-order Butterworth filter with a passband of 0.03–0.15 Hz to remove physiological noises. A Savitzky–Golay filter was applied for smoothing [2] in MATLAB® 2019b (The MathWorks, Inc., Natick, MA, USA). The averaged ΔHbO signals of all trials for channels 10, 12, and 22 for the MA, LMI, and RMI tasks after noise removal are shown in Figure 3 for an example subject.

2.6. Channel Selection/Channel of Interest/Region of Interest

In conventional BCI systems, either all channels are used, or channels are selected based on brain activation. In this study, the z-score method for COI/ROI is proposed and used for channel selection based on brain activation. The researchers have used the *t*-value method excessively for this purpose; therefore, it is included in the study. The conventional and proposed methodology is shown in Figure 4.

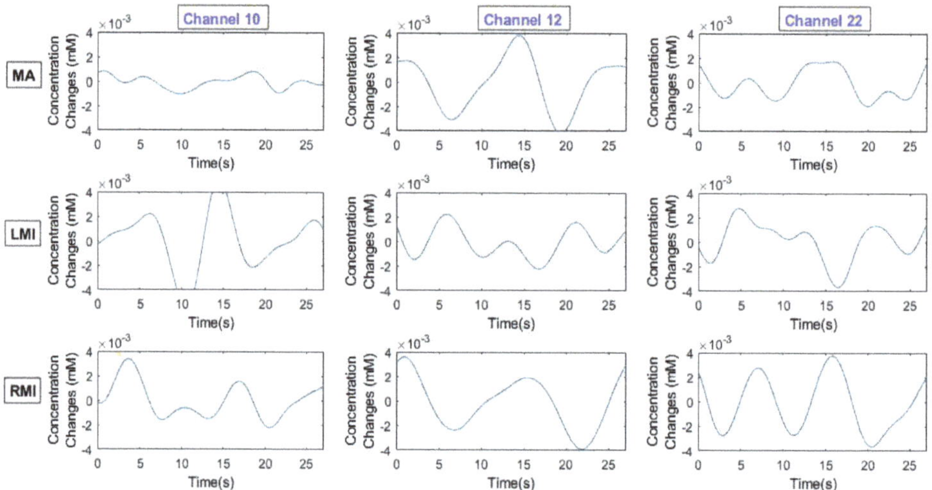

Figure 3. Averaged ΔHbO of all trials of channels 10, 12, and 22 for tasks and rest of the MA, LMI, and RMI.

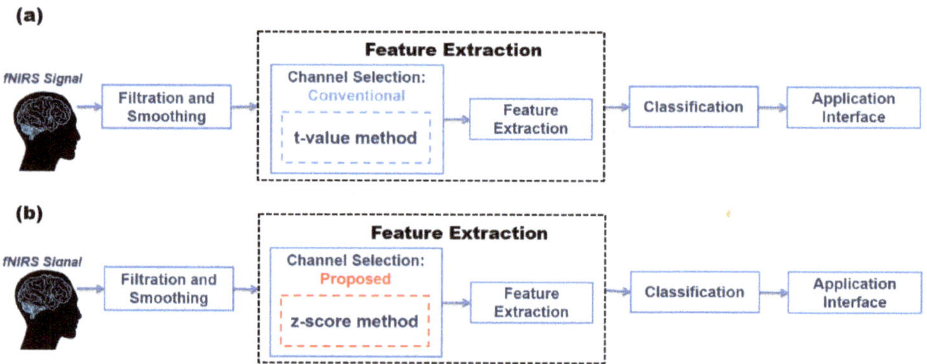

Figure 4. Methodology of (**a**) conventional and (**b**) proposed brain–computer interface (BCI) system.

2.6.1. *t*-value Method

The *t*-value method is an estimation-based channel selection or COI or ROI approach in which channels with a positive *t*-value are selected. Alternatively, a threshold value of 't' ($t > t_{crt}$) can also be set for the selection of active channels. In that case, the degree of freedom ($k - 1$) is used to determine the threshold value, where 'k' is the number of samples in an activity. The *t*-value method determines cortical activation through statistical estimation by fitting the linear regression model [44]. The estimation can be calculated by fitting dHRF, with measured hemodynamic response function resulted from cortical activation. It can be formulated as given below:

$$h_j^i(k) = \phi_j^i h_M(k) + \psi_j^i.1 + \varepsilon_j^i \tag{2}$$

The term on the left side of the equation i.e., $h_j^i(k) \in R^{k\times 1}$ is the measured response function in which 'k' is the number of samples in each stimulus, subscript 'j' denotes the stimulus number, and superscript 'i' represents the channel number. 'ϕ' is the unknown coefficient to be estimated, the coefficient 'ψ' is multiplied by column vectors of $1 \in R^{k\times 1}$ for correction of baseline drift in the signal, and $\varepsilon \in R^{k\times 1}$ is the error term in the linear regression method. The unknown coefficients 'ϕ' are estimated through *robustfit* function in MATLAB®. $h_M(k) \in R^{k\times 1}$ is calculated by convolution of cHRF $h_c(k)$ with boxcar function $s(k)$ $h_c(k)$ and can be modeled using two-gamma functions [42,43,52], as shown below:

$$h_c(k) = A\left[\frac{k^{\alpha_1-1}\beta_1^{\alpha_1}e^{-\beta_1 k}}{\Gamma(\alpha_1)} - c\frac{k^{\alpha_2-1}\beta_2^{\alpha_2}e^{-\beta_2 k}}{\Gamma(\alpha_2)}\right] \tag{3}$$

The parameter 'A' sets the amplitude, 'α_1' and 'α_2' set the peak and undershoot delays, respectively. In contrast, 'β_1' & 'β_2' set dispersions of the peak and undershoot respectively, 'c' is the ratio of the peak to the undershoot, and 'Γ' is the gamma distribution. $h_M(k)$ can be calculated using the formula

$$h_M(k) = \sum_{n=1}^{k} h_c(n)s(k-n) \tag{4}$$

And the boxcar function is:

$$s(k) = \begin{cases} 0, \; if \; k \in rest \\ 1, \; if \; k \in task \end{cases} \tag{5}$$

The boxcar function is a unit step function having a value of '0' for the rest period and '1' for the task period. After estimating the coefficient 'ϕ', its statistical significance is calculated by the ratio of the estimated coefficient and its standard error (SE). The said statistical significance is also called as '*t*-value'. Its positive or threshold value greater than the critical value shows that the channel is active or otherwise.

$$t_j^i = \frac{\phi_j^i}{SE\left(\phi_j^i\right)} \tag{6}$$

The above formula gives a *t*-value in *i*-th channel of *j*-th stimulus. In our case, active channels are considered which have *p*-value less than 0.05 and a *t*-value greater than 't_{crt}', which is 1.65 (degree of freedom is 299, i.e., $k - 1$). This method was initially used to measure the statistical significance of channel [44]. The step-by-step procedure of the *t*-value method is shown in Figure 5a.

2.6.2. z-Score Method

The z-score method uses cross-correlation as the mutual relationship between two signals and measures the strength of the relationship among the acquired signal and dHRF. Cross-correlation matches two signals temporally to find out the strength of similarity between each other, and mathematical expression is given in the equation below

$$r_{xy}(\tau) = \sum_{-\infty}^{\infty} x(t)y(t-\tau) \tag{7}$$

where τ is the time-lag between $x(t)$ and $y(t)$, and the value of r_{xy} denotes the difference between acquired signal $x(t)$ and modelled signal $y(t)$. This cross-correlation has been used earlier for finding the relationship of the potential dominant channel with its adjacent channels by observing delay in response between the channels [5]. In the current study, the dHRF signal is swept over the measured signal, and the integral of its product is found at each discrete position 't'. The maximum value of the integral product, i.e., the correlation coefficient, is selected for each channel showing the temporal similarity between two signals at that time instant 'τ'.

$$r^i = max\left(\sum_{t=-k}^{k} h^i(t)h_M^i(t-\tau)\right) \tag{8}$$

The average trial value of measured response is taken for each stimulus type (i.e., LMI, RMI, and MA), and afterward, cross-correlation is calculated with dHRF. Maximum strength of similarity occur when τ is selected for task vs. task intervals of measured and desired hemodynamic response function to overlap, it is also the highest value of cross-correlation coefficient. If τ is selected for rest vs. rest or rest vs. task or task vs. rest intervals of measured and desired hemodynamic response function to overlap, it will give lower values of cross-correlation coefficient. And if τ is selected for complete intervals i.e., calculate cross-correlation coefficient for complete time period, the highest value remains the same, as shown in Figure 6a. The maximum value of correlation coefficient 'r' is selected for each channel 'i', between measured hemodynamic response function $h(t) \in R^{k\times1}$ and dHRF $h_M(t) \in R^{k\times1}$, where 'k' is the number of samples in the signal. The vector r^i contains maximum values of cross-correlation coefficients for each channel. The magnitude of each maximum value varies for each channel and forms a new range as shown in Figure 6b. The z-score measures the distance of raw score from mean value i.e., how far from mean a data point is in population. In this study, z-score represents the channel activation in the form of matching and strength of similarity based on the value of the cross-correlation coefficient. A positive z-score represents higher strength and similarity and a negative z-score shows lower strength and no matching. The z-scores of the channel vary as the (max of) cross-correlation coefficient value varies with respect to task, as shown in Figure 6 for (c) LMI, (d) RMI, and (e) MA. Additionally, the z-score varies with subject. z-scores of vector containing maximum values of cross-correlation coefficients for each channel are then calculated using the formula.

$$z^i = \frac{\left(r^i - \bar{r}\right)}{\sigma_r} \tag{9}$$

where '$\bar{r}$' is the mean value of correlation coefficients and 'σ_r' is the standard deviation. Only those channels are selected which have a positive z-score (i.e., $z > 0$). The step-by-step procedure of channel selection using the z-score method is shown in Figure 5b.

Both the t-value method and z-score method are used to select cortical-activation-based channels.

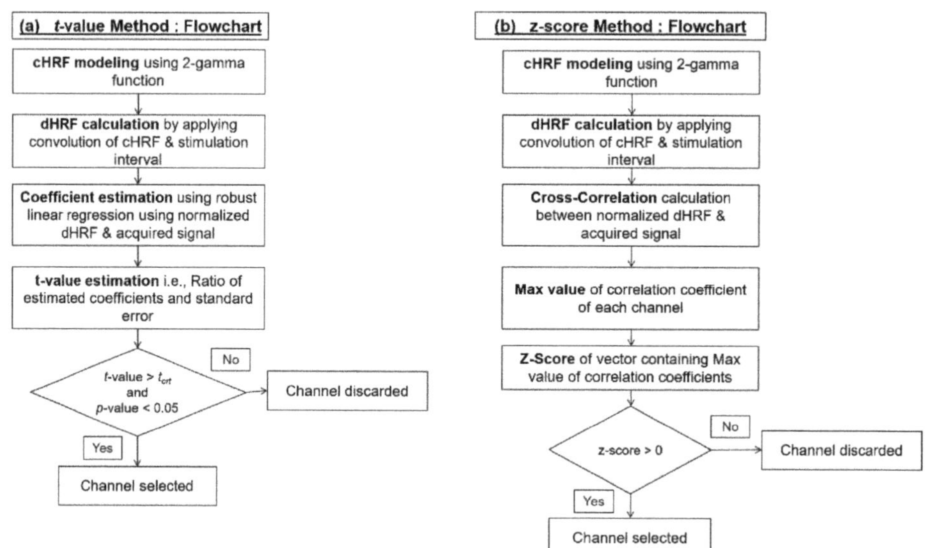

Figure 5. Step-by-step procedure of (**a**) t-value and (**b**) z-score method.

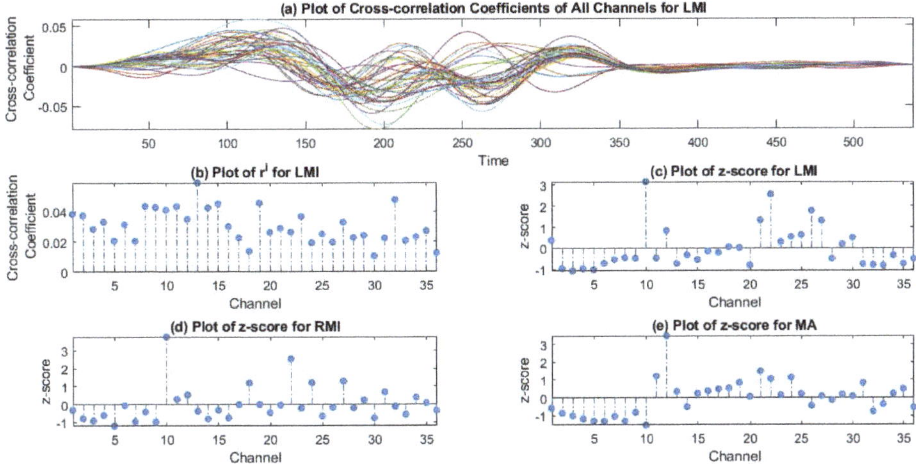

Figure 6. (**a**) Plot of cross-correlation coefficients of all channels for LMI for subject 28, (**b**) plot of r^i; vector containing maximum value of all channels for LMI for subject 28 and plot of z-score for (**c**) LMI, (**d**) RMI, and (**e**) MA task of subject 28.

2.7. Feature Extraction

2.7.1. Statistical Features

In fNIRS-BCI, statistical measures such as peak, mean, variance, kurtosis, skewness, and slope are extracted as features for classification [2,24,53–56]. However, mean and peak and mean, peak, and slope were optimal two- and three-feature combinations to achieve enhanced classification accuracies for the fNIRS-BCI system [57]. In this study, the mean, peak, and slope are used as features for the fNIRS-BCI problem classification. All features are calculated for ΔHbO spatio-temporally. Spatio-temporal features are calculated using a two-step procedure: (1) averaging all channels (spatial average) and (2) aggregating using a statistic across each task window (temporal statistic).

Mean is calculated as:

$$Mean = \mu = \frac{1}{n}\sum_{x=1}^{k} B_x \tag{10}$$

where B_x is the input signal such as $\Delta HbO(t)$ and n is the total number of observations, aggregated by *mean* function of MATLAB®.

The slope is aggregated using MATLAB® *polyfit* function, which fits the line to all input data points.

The peak is the maximum value of the signal, calculated using *max* function of MATLAB®.

2.7.2. Normalization

Features are normalized by rescaling using the following equation:

$$Y' = \frac{Y - min(Y)}{max(Y) - min(Y)} \tag{11}$$

where Y' is the normalized feature, and Y is the original feature values. This normalization has been applied to all features before classification. The final feature-matrix calculated is of size 20 × 3 for each task.

2.8. Linear Discriminant Analysis (LDA)

The linear discriminant analysis draws a hyper-plane in feature space to discriminate between classes. The hyper-plane is drawn based on minimizing the inter-class variance and maximizing the distance between classes mean. The optimal projection matrix to maximize Fisher's criterion is formulated as

$$J(X) = \frac{X^T S_B X}{X^T S_W X} \tag{12}$$

where S_W is with-in class scatter matrix and S_B is a between-class scatter matrix, defined as:

$$S_W = \sum_{i=1}^{n} \sum_{y_k \in class\ i} (y - \mu_i)(y - \mu_i)^T \tag{13}$$

$$S_B = \sum_{i=1}^{n} k_i(\mu_i - \mu)(\mu_i - \mu)^T \tag{14}$$

In the above equation μ_i denotes the sample mean of class i, and μ represents the total mean of all samples, n is the total number of classes, k_i designates the number of samples of class i, and k is the total number of samples. The largest eigenvalues contained by optimal vector X is calculated by Equation (12) as a generalized eigenvalue problem. To estimate classification performance leave-one-out cross-validation (LOOCV) is used. The dataset is divided into training and testing sets, to ensure separation of data for training and testing of classifier for each channel selection method and activity used. Due to a limited number of samples i.e., twenty, LOOCV is applied. There is one sample for testing and nineteen for training the classifier, repeated twenty times. In MATLAB® the following functions were used; *cvpartition* for data partition in folds, *classify* for classification, and *crossval* for cross-validation purposes.

3. Results

The analysis shows that both methods vary in channel selection and the total number of selected channels for a specific task. The t-value method measures the activation as statistical significance of signal of channel, while the z-score measures the activation as z-score of (max of) cross-correlation coefficients with respect to all channels population. The activation map is drawn for both methods using normalized t-values and z-scores. Figure 7 shows the cortical activation-map of the t-value and z-score method for MA, LMI and, RMI tasks. The figure shows the difference in measuring cortical activation by using both methods. It can be seen in Figure 7 that cortical activation does not occur in all channels for a specific activity, and also not in all channels of the designated region. For MA tasks the proposed method selected major channels from the prefrontal region (details can be found in Supplementary files) and the visual cortex (instructions were displayed on screen), however some other channels from the motor cortex also showed a positive z-score value. Similarly for LMI and RMI major channels are selected from the motor and prefrontal cortex (details can be found in Supplementary files) and the visual cortex (instructions were displayed on screen). The areas of activation found are similar to designated areas of neural activity [55]. The selection of extra channels may be due to human error, lack of concentration, multiple thinking during experiment, or induced neuronal activity. This also varies with respect to activity and subject. In Figure 7, the spatial difference of identifying cortical activation is found in both methods because of the fact that both methods apply different scientific and mathematical principles. The t-value method uses statistical significance by GLM and the z-score method applies signal matching through cross-correlation. A similar pattern of differences is found in all subjects. The proposed method is used to select channels having cortical activity during particular tasks. The number of chosen common channels between the t-value and z-score methods also varies for a specific task. For the MA task, the range of the total number of selected channels using the z-score method is 14–20 channels. The t-value method selected 6–36 channels, and the number of common channels between both methods varies from 2–20 channels. Similarly, for the LMI task, the chosen z-score method channels range from 14–20 channels, while the *t*-value has a range of 2–24 channels, and the number of common channels between both methods ranges from 1–20 channels. Likewise, for the RMI task, z-score-method-nominated channels range from 16–21 channels, whereas the *t*-value method selected 6–36 channels, and the amount of common channels ranges from 2–20. All three tasks of MA, LMI, and RMI were analyzed for all twenty-nine subjects. A plot of total number of channels selected by *t*-value, z-score method, and common channels for the MA, LMI, and RMI task is shown in Figure 8a–c respectively. Details of channels selected using the *t*-value and z-score method are available in Supplementary files for the MA, LMI, and RMI tasks, respectively.

After selecting channels, Spatio-temporal features of the mean, peak, and slope were extracted from nominated channels' data. In addition to the *t*-value method, all channel data were also used to compare the results. The feature scatters plot of the *t*-value-method-selected channels, the z-score-method-chosen channels, and all channels were drawn for each task. The feature scatter plot for MA, LMI, and RMI tasks for subject 28 is shown in Figures 9–11, respectively.

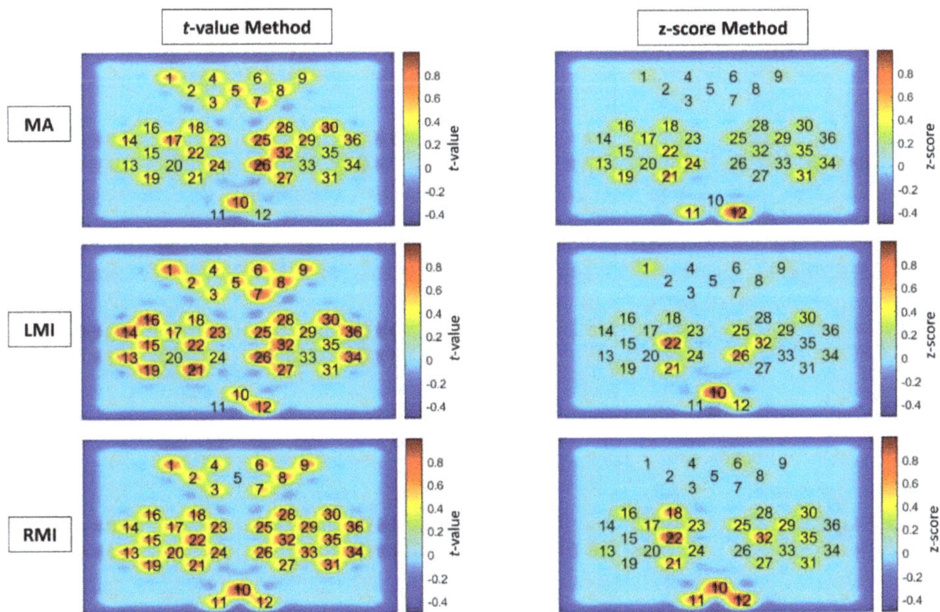

Figure 7. Activation map of *t*-value and z-score method for LMI, RMI, and MA tasks, for example subject 28.

The average classification accuracy obtained using channels selected by the z-score method is 87.2 ± 7.0%, 88.4 ± 6.2%, 88.1 ± 6.9% for LMI vs. rest, RMI vs. rest, and MA vs. rest; respectively. While the average classification accuracies yielded by using channels selected by the *t*-value method are 74.5 ± 9.3%, 70.3 ± 14.2%, and 73.9 ± 12.2% for LMI vs. rest, RMI vs. rest, and MA vs. rest; respectively. Likewise baseline-correction-technique-selected channels achieved classification accuracies of 79.3 ± 10.7%, 78.4 ± 13.3%, and 79.1 ± 18.1% for LMI vs. rest, RMI vs. rest, and MA vs. rest; respectively. However, all channels' data achieved classification accuracies of 77.7 ± 8.9%, 75.0 ± 10.8%, and 77.5 ± 9.6% for LMI vs. rest, RMI vs. rest, and MA vs. rest; respectively. Table 1 shows the detailed subject-wise classification accuracies using the z-score method, *t*-value method, and all channels' data for the MA, LMI, and RMI tasks. Figure 12 shows the bar chart for obtained accuracies using the z-score method, *t*-value method, and all channels for the MA, LMI, and RMI tasks. The better results obtained by the z-score method compared to conventional *t*-value and all channels' data are statistically verified by applying a two-tailed paired sample Student's *t*-test. For two comparisons, Bonferroni [58] correction was used to find the adjusted confidence interval level of 0.0167. Table 2 shows the *p*-values obtained for two comparisons for each task: the z-score method vs. *t*-value method and the z-score method vs. all channels. It can be seen that the z-score-method-selected channels' performance is significantly better (*p*-value < 0. 0167) than the *t*-value method and all channels for MA, LMI, and RMI fNIRS-BCI.

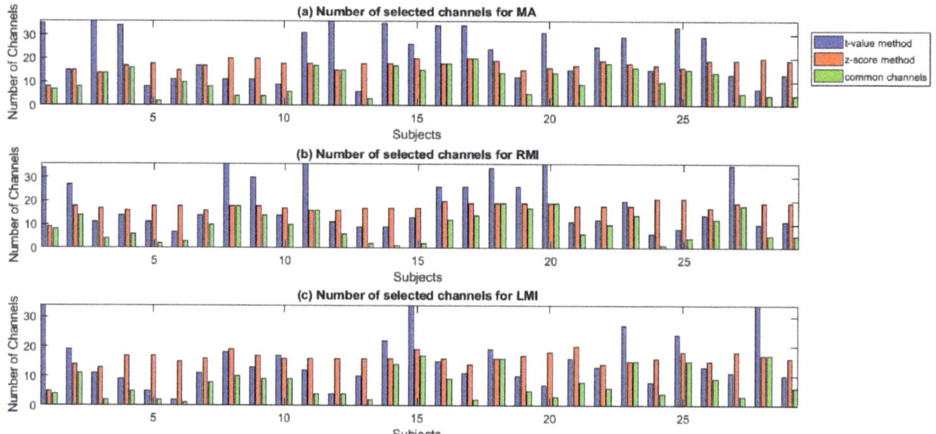

Figure 8. Total number of channels selected by *t*-value, z-score method, and common channels for (**a**) MA, (**b**) RMI, and (**c**) RMI task.

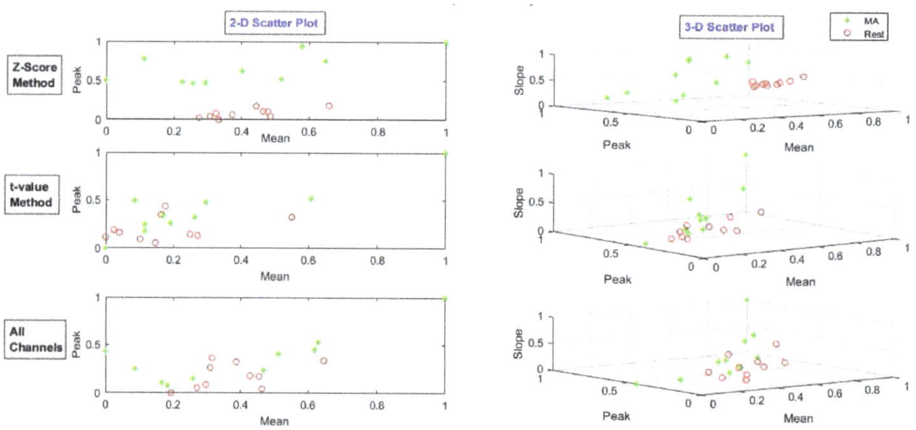

Figure 9. Feature scatter plot of the *t*-value-method-selected channels' data, z-score-method-selected channels' data, and all channels' data for MA task.

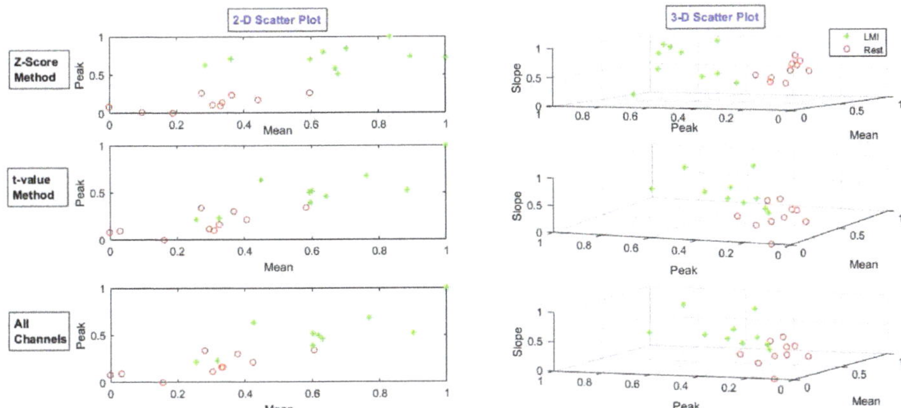

Figure 10. Feature scatter plot of the *t*-value-method-selected channels' data, z-score-method-selected channels' data, and all channels' data for LMI task.

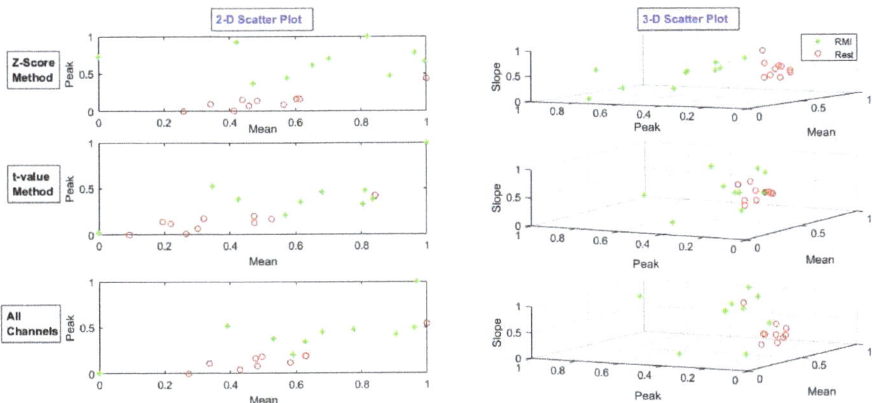

Figure 11. Feature scatter plot of the *t*-value-method-selected channels' data, z-score-method-selected channels' data, and all channels' data for RMI task.

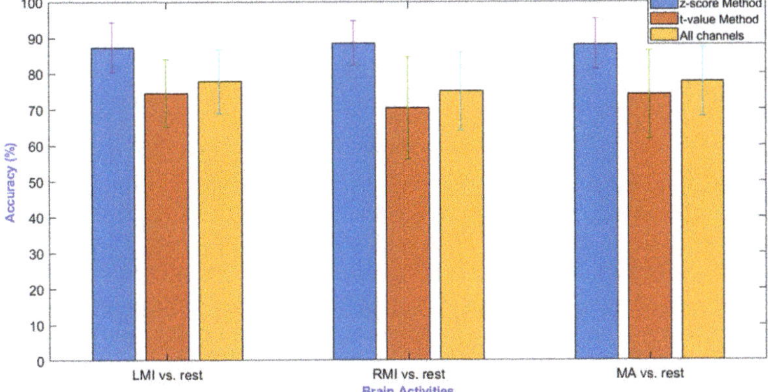

Figure 12. Average classification accuracies for the z-score method, *t*-value method, and all channels' data for MA, LMI, and RMI tasks.

Table 1. Subject-wise classification accuracies by using the z-score method, *t*-value method, and all channels' data for MA, LMI, and RMI tasks.

	MA			LMI			RMI		
	z-Score Method	*t*-Value Method	All Channels	z-Score Method	*t*-Value Method	All Channels	z-Score Method	*t*-Value Method	All Channels
					(%)				
Sub 1	75	85	90	90	70	55	90	70	75
Sub 2	90	70	80	85	80	90	85	65	75
Sub 3	80	70	70	95	75	65	90	80	75
Sub 4	95	80	80	85	70	75	85	80	65
Sub 5	90	65	55	90	55	80	85	75	65
Sub 6	90	45	55	95	80	85	90	60	65
Sub 7	90	70	80	80	65	65	90	55	85
Sub 8	90	60	90	90	80	85	85	85	85
Sub 9	90	80	60	75	65	70	100	90	85
Sub 10	85	80	85	80	70	75	85	80	70
Sub 11	95	90	85	85	70	70	95	70	70
Sub 12	90	75	75	95	80	80	75	65	65
Sub 13	95	45	70	90	70	70	90	90	80
Sub 14	90	70	70	95	85	90	95	65	65
Sub 15	95	70	75	85	85	85	90	75	90
Sub 16	75	75	70	90	80	75	90	60	80
Sub 17	85	80	80	95	80	85	95	80	85
Sub 18	85	90	90	90	70	85	90	90	90
Sub 19	85	80	85	70	85	85	80	85	80
Sub 20	95	95	85	95	50	80	95	60	60
Sub 21	75	80	80	85	75	80	85	60	90
Sub 22	85	75	85	80	60	70	95	80	75
Sub 23	80	75	80	95	80	80	80	65	70
Sub 24	95	60	80	80	70	85	90	35	75
Sub 25	100	90	80	95	85	90	75	45	50
Sub 26	90	85	90	80	80	70	95	70	75
Sub 27	80	65	75	85	75	65	95	85	85
Sub 28	100	60	70	95	85	85	90	75	90
Sub 29	85	80	80	80	85	80	80	45	55
Average	**88.1**	**74.0**	**77.6**	**87.2**	**74.5**	**77.8**	**88.4**	**70.3**	**75.0**

Table 2. Statistical significance of the z-score method.

	Bonferroni Correction Applied ($p < 0.0167$)		
	LMI vs. Rest	RMI vs. Rest	MA vs. Rest
z-score method vs. *t*-value method	2.21×10^{-7}	5.34×10^{-8}	1.47×10^{-6}
z-score method vs. all channels	3.50×10^{-5}	3.54×10^{-7}	1.37×10^{-5}

4. Validation

The validation of proposed z-score method has also been performed on a publicly available fNIRS dataset. Left- and right-hand finger tapping and dominant foot tapping tasks were included in the dataset for 17 subjects [59]. Each of the aforementioned activities consists of 25 trial data. Six conventional spatio-temporal features that include signal mean, peak, slope, skewness, kurtosis, and variance of ΔHbO signal are extracted. LDA classifier is applied on the said extracted features for a two-class fNIRS-BCI problem.

The average classification accuracies yielded by using the selected channels through the proposed z-score-method-based selected channels are 72 ± 8%, 66 ± 9.8%, and 67.41 ± 9.5% for RFT, LFT, and FT tasks respectively. In comparison, the *t*-value-method-based selected channels resulted in average classification accuracies of 54.47 ± 10%, 54.12 ± 13.8%, and 56.24 ± 9.3% for RFT, LFT, and FT tasks respectively. Nonetheless, average classification accuracies obtained for RFT, LFT, and FT tasks by selecting all channels remained 54.71 ± 10.3%, 54.47 ± 14.2%, and 54.12 ± 11.1% respectively. However, the classification accuracies resulted using baseline-correction-selected channels for RFT, LFT, and FT tasks were 52.82 ± 7.4%, 51.06 ± 9.5%, and 55.53 ± 12.2% respectively. The results achieved using z-score method have significantly (p-value < 0.0167) better performance as compared to the results of the t-value method, the baseline correction technique and using all channels for two-class fNIRS-BCI problem.

5. Discussion

In the present study, the authors proposed a new method of selecting cortical-activation-based channels to increase fNIRS-BCI performance, especially in terms of classification accuracy and COI/ROI. In the literature, recent studies have also focussed on enhancing classification accuracies of fNIRS-BCI systems by optimal classification technique [60], optimal feature selection [24,54], optimal feature-combination [57], general linear model [25], vector-based phase analysis [26,61,62], t-value method [22,37,41,63], cross-correlation [33], and dominant channel selection [64]. Accurate and reliable fNIRS-BCI performance may lead to producing applications in neurorobotics, rehabilitation, clinical BCI for monitoring and analysis, and neuroergonomics [10,65–67].

In previous studies, bundled optode configurations have been used to precisely identify active regions of the brain [41] in spatially resolved spectroscopy. Santosa et al. [44] first applied the t-value method to select hemodynamic responses with positive t-values. Another study detected ROI against different sound stimuli by placing optodes on both right and left hemispheres [63]. It is worth mentioning here that different channels against different subjects and stimuli were obtained. The baseline correction method was used in hybrid-BCI to select active channels by calculating and comparing the maximum values during rest and corresponding task stages [40]. The cross-correlation method was used to identify potentially dominant channels in both hemispheres for pain-related cortical activations. First visual inspection was used to identify potential dominant channels followed by calculating the delay of response—The adjacent active channels were selected [33]. In the present study, a novel method for cortical-activity-based channel selection is proposed and validated for three different brain activation fNIRS-BCIs. The results have shown improved classification performance as compared to previous methods. The proposed method is compared to the conventional excessively used t-value method and without channel selection using all channels' data. The proposed method uses cross-correlation to match and measure strength of similarity between dHRF and recorded brain signals. Followed by forming a vector of each channel correlation coefficients' max values, and the z-score is aggregated for each value of that vector. On the basis of z-score value (z-score > 0) a channel is selected. This study shows the improvement in classification accuracies of three activity-based fNIRS-BCIs using the z-score method compared to the t-value method for channel selection and without channel selection, i.e., using all channels' data. The classification accuracies are improved from 77.7 ± 8.9% to 87.2 ± 7.0% for MA vs. rest, 77.6 ± 9.6% to 88.1 ± 7.0% for LMI vs. rest, and 75.0 ± 10.9% to 88.4 ± 6.3% for RMI vs. rest. The channel selection results also verify that activation of channels is not uniform among different subjects due to variation in brain sizes. Similarly, a specific task is associated to a certain brain region—that is why identification of correct COI/ROI is extremely important.

The performance improvement may be since the z-score-method-selected channels represent brain activity more informatively and specifically than using the t-value method. The t-value method finds out the statistical significance by fitting the actual response to the estimated coefficients' desired response. The t-value is the ratio of weighting coefficients to its standard error, and its positive ($t > 0$) or threshold value ($t > t_{crt}$ and p-value < 0.05) decides whether the channel activity is significant or not. The t-value method was first used by Santosa et al. to measure the statistical significance of a channel [44]. Later studies used this method to select channels based on cortical activity in terms of a channel's statistical significance. However, the z-score method measures the strength of similarity as cross-correlation coefficient between the desired response and measured response. The maximum value of the cross-correlation coefficient represents the extent of similarity and matching between desired and measured responses. Furthermore, the z-score value is the measure of distance from the mean value in a data, and a positive z-score value decides whether the raw score is at the right side of bell's curve within a population as the channel's cortical activation. The proposed method selects a channel from a designated region with respect to activity along with a few other channels. The extra channels are selected show a positive z-score because of human error, which includes lack of concentration towards the experiment, multiple thoughts during the experiment, and arbitrary neuronal activity. Cortical-activity-based channels possess intrinsic brain activation information, which plays an essential role in improving the fNIRS-BCI system's classification accuracy. The z-score method can be used to identify cortical-activity-based brain activation regions, subregions, and channels to analyze, perform, and develop state-of-the-art fNIRS-BCI applications, including prosthetics, exoskeleton controls, and communications with stroke or locked-in syndrome patients.

This study has few limitations, including the fact that it applies to a single activity at a time as specific task is associated with a particular brain region. Furthermore, activation of subject-based specific channels occurs due to different brain sizes. The z-score method selects major chunks of channel from a designated region and a few other channels from other regions as well; this occurrence was also found in the *t*-value method. Further improvement can be made to reduce the selection of undesignated channels by only performing analysis on subregions. Moreover, the study includes only an LDA-based classification model because of its low computational cost and high-speed performance. LDA is a commonly used classifier for the fNIRS-BCI system [60]. Other machine learning algorithms may also be used for analysis and may perform better [55].

6. Conclusions

The aim of this study was to improve classification accuracy for a fNIRS-BCI system by selecting cortical-activity-based channels. The z-score method selects cortical-activity-based channels based on cross-correlation coefficients and z-score value ($z > 0$). Average classification accuracies achieved for MA vs. rest, LMI vs. rest, and RMI vs. rest by using the proposed method are significantly ($p < 0.0167$) higher than the *t*-value method and without channel selection, i.e., using all channels for classification. The results show enhanced performance for the proposed method over conventional methods as an advancement in efforts to identify cortically active channels/regions and to improve the classification performance of state-of-the-art fNIRS-BCI systems.

Supplementary Materials: The following are available online at http://www.mdpi.com/1424-8220/20/23/6995/s1, Table S1: Details of the selected channels using *t*-value and z-score method for MA, LMI and, RMI task.

Author Contributions: H.N. conceived this study and was involved in the data processing, and writing of the manuscript. A.M. was involved in the data analysis. U.S.K. and R.A.K. were involved in rechecking of results, and revision. M.J.K. and Y.A. helped in revision of the manuscript. N.N. was involved in the writing of the manuscript and supervised the research. All authors have read and agreed to the published version of the manuscript.

Funding: This research received no external funding.

Acknowledgments: We would like to thank Shin et al. (Berlin Institute of Technology, 10587 Berlin, Germany) for making this dataset open access to the public and letting researchers apply new methods and techniques to contribute to the advancement of state-of-the-art fNIRS-BCI.

Conflicts of Interest: The authors declare that the research was conducted in the absence of any commercial or financial relationships that could be construed as a potential conflict of interest.

References

1. Wolpaw, J.R.; Birbaumer, N.; McFarland, D.J.; Pfurtscheller, G.; Vaughan, T.M. Brain-computer interfaces for communication and control. *Clin. Neurophysiol.* **2002**, *113*, 767–791. [CrossRef]
2. Naseer, N.; Hong, K.-S. fNIRS-based brain-computer interfaces: A review. *Front. Hum. Neurosci.* **2015**, *9*, 1–15. [CrossRef]
3. Curtin, A.; Tong, S.; Sun, J.; Wang, J.; Onaral, B.; Ayaz, H. A systematic review of integrated functional near-infrared spectroscopy (fNIRS) and transcranial magnetic stimulation (TMS) studies. *Front. Neurosci.* **2019**, *13*. [CrossRef] [PubMed]

4. Pinti, P.; Aichelburg, C.; Gilbert, S.; Hamilton, A.; Hirsch, J.; Burgess, P.; Tachtsidis, I. A Review on the use of wearable functional near-infrared spectroscopy in naturalistic environments. *Jpn. Psychol. Res.* **2018**, *60*, 347–373. [CrossRef]
5. Basura, G.J.; Hu, X.-S.; Juan, J.S.; Tessier, A.-M.; Kovelman, I. Human central auditory plasticity: A review of functional near-infrared spectroscopy (fNIRS) to measure cochlear implant performance and tinnitus perception. *Laryngoscope Investig. Otolaryngol.* **2018**, *3*, 463–472. [CrossRef] [PubMed]
6. Ramadan, R.A.; Vasilakos, A.V. Brain computer interface: Control signals review. *Neurocomputing* **2017**, *223*, 26–44. [CrossRef]
7. Liu, Y.; Piazza, E.A.; Simony, E.; Shewokis, P.A.; Onaral, B.; Hasson, U.; Ayaz, H. Measuring speaker–listener neural coupling with functional near infrared spectroscopy. *Sci. Rep.* **2017**, *7*, 43293. [CrossRef]
8. Gateau, T.; Durantin, G.; Lancelot, F.; Scannella, S.; Dehais, F. Real-time state estimation in a flight simulator using fNIRS. *PLoS ONE* **2015**, *10*, e0121279. [CrossRef]
9. McKendrick, R.; Parasuraman, R.; Ayaz, H. Wearable functional near infrared spectroscopy (fNIRS) and transcranial direct current stimulation (tDCS): Expanding vistas for neurocognitive augmentation. *Front. Syst. Neurosci.* **2015**, *9*, 27. [CrossRef]
10. Ayaz, H.; Dehais, F. *Neuroergonomics: The Brain at Work and in Everyday Life*, 1st ed.; Academic Press: Cambridge, MA, USA; Elsevier: Amsterdam, The Netherlands, 2019.
11. Delpy, D.T.; Cope, M.; van der Zee, P.; Arridge, S.; Wray, S.; Wyatt, J. Estimation of optical pathlength through tissue from direct time of flight measurement. *Phys. Med. Biol.* **1988**, *33*, 1433–1442. [CrossRef]
12. Villringer, A.; Planck, J.; Hock, C.; Schleinkofer, L.; Dirnagl, U. Near infrared spectroscopy (NIRS): A new tool to study hemodynamic changes during activation of brain function in human adults. *Neurosci. Lett.* **1993**, *154*, 101–104. [CrossRef]
13. Kopton, I.M.; Kenning, P. Near-infrared spectroscopy (NIRS) as a new tool for neuroeconomic research. *Front. Hum. Neurosci.* **2014**, *8*, 1–13. [CrossRef] [PubMed]
14. Naseer, N.; Hong, K.S. Classification of functional near-infrared spectroscopy signals corresponding to the right- and left-wrist motor imagery for development of a brain-computer interface. *Neurosci. Lett.* **2013**, *553*, 84–89. [CrossRef] [PubMed]
15. Naseer, N.; Hong, M.J.; Hong, K.S. Online binary decision decoding using functional near-infrared spectroscopy for the development of brain-computer interface. *Exp. Brain Res.* **2014**, *232*, 555–564. [CrossRef]
16. Hong, K.S.; Naseer, N.; Kim, Y.H. Classification of prefrontal and motor cortex signals for three-class fNIRS-BCI. *Neurosci. Lett.* **2015**, *587*, 87–92. [CrossRef]
17. Naseer, N.; Hong, K. Functional near-infrared spectroscopy based brain activity classification for development of a brain-computer interface. In Proceedings of the 2012 International Conference of Robotics and Artificial Intelligence, Rawalpindi, Pakistan, 22–23 October 2012; pp. 174–178.
18. Naseer, N.; Hong, K.; Bhutta, M.R.; Khan, M.J. Improving classification accuracy of covert yes/no response decoding using support vector machines: An fNIRS study. In Proceedings of the 2014 International Conference on Robotics and Emerging Allied Technologies in Engineering (iCREATE), Islamabad, Pakistan, 22–24 April 2014; pp. 6–9.
19. Gramann, K.; Fairclough, S.H.; Zander, T.O.; Ayaz, H. Editorial: Trends in neuroergonomics. *Front. Hum. Neurosci.* **2017**, *11*. [CrossRef]
20. Franceschini, M.A.; Joseph, D.; Huppert, T.; Diamond, S.; Boas, D. Diffuse optical imaging of the whole head. *J. Biomed. Opt.* **2006**, *11*, 54007. [CrossRef]
21. Khan, R.; Naseer, N.; Nazeer, H.; Khan, M.N. Control of a prosthetic leg based on walking intentions for gait rehabilitation: An fNIRS study. *Front. Hum. Neurosci.* **2018**, *12*. [CrossRef]
22. Asgher, U.; Ahmad, R.; Naseer, N.; Ayaz, Y.; Khan, M.J.; Amjad, M.K. Assessment and classification of mental workload in the prefrontal cortex (PFC) using fixed-value modified beer-lambert law. *IEEE Access* **2019**, *7*, 143250–143262. [CrossRef]
23. Bosworth, A.; Russell, M.; Jacob, R.J.K. Update of fNIRS as an input to brain-computer interfaces: A review of research from the tufts human-computer interaction laboratory. *Photonics* **2019**, *6*, 90. [CrossRef]
24. Noori, F.M.; Naseer, N.; Qureshi, N.K.; Nazeer, H.; Khan, R.A. Optimal feature selection from fNIRS signals using genetic algorithms for BCI. *Neurosci. Lett.* **2017**, *647*, 61–66. [CrossRef]

25. Qureshi, N.K.; Naseer, N.; Noori, F.M.; Nazeer, H.; Khan, R.A.; Saleem, S. Enhancing classification performance of functional near-infrared spectroscopy-brain-computer interface using adaptive estimation of general linear model coefficients. *Front. Neurorobot.* **2017**, *11*. [CrossRef] [PubMed]
26. Nazeer, H.; Naseer, N.; Khan, R.A.; Noori, F.M.; Qureshi, N.K.; Khan, U.S.; Khan, M.J. Enhancing classification accuracy of fNIRS-BCI using features acquired from vector-based phase analysis. *J. Neural Eng.* **2020**, *17*, 56025. [CrossRef]
27. Alotaiby, T.; El-Samie, F.E.A.; Alshebeili, S.A.; Ahmad, I. A review of channel selection algorithms for EEG signal processing. *EURASIP J. Adv. Signal. Process.* **2015**, *2015*, 66. [CrossRef]
28. Qiu, Z.; Jin, J.; Lam, H.-K.; Zhang, Y.; Wang, X.; Cichocki, A. Improved SFFS method for channel selection in motor imagery based BCI. *Neurocomputing* **2016**, *207*, 519–527. [CrossRef]
29. Shan, H.; Xu, H.; Zhu, S.; He, B. A novel channel selection method for optimal classification in different motor imagery BCI paradigms. *Biomed. Eng. Online* **2015**, *14*, 93. [CrossRef] [PubMed]
30. Li, C.; Jia, T.; Xu, Q.; Ji, L.; Pan, Y. Brain-computer interface channel-selection strategy based on analysis of event-related desynchronization topography in stroke patients. *J. Healthc. Eng.* **2019**, *2019*, 3817124. [CrossRef]
31. Tao, W.; Li, C.; Song, R.; Cheng, J.; Liu, Y.; Wan, F.; Chen, X. EEG-based emotion recognition via channel-wise attention and self attention. *IEEE Trans. Affect. Comput.* **2020**, *70*, 1. [CrossRef]
32. Jin, J.; Liu, C.; Daly, I.; Miao, Y.; Li, S.; Wang, X.; Cichocki, A. Bispectrum-based channel selection for motor imagery based brain-computer interfacing. *IEEE Trans. Neural Syst. Rehabil. Eng.* **2020**, *28*, 2153–2163. [CrossRef]
33. Rojas, R.F.; Huang, X.; Ou, K.L.; Lopez-Aparicio, J. Cross correlation analysis of multi-channel near infrared spectroscopy. *Comput. Sci. Inf. Technol.* **2016**, *6*, 23–33. [CrossRef]
34. Verner, M.; Herrmann, M.J.; Troche, S.J.; Roebers, C.M.; Rammsayer, T.H. Cortical oxygen consumption in mental arithmetic as a function of task difficulty: A near-infrared spectroscopy approach. *Front. Hum. Neurosci.* **2013**, *7*, 217. [CrossRef] [PubMed]
35. Song, X.; Pogue, B.W.; Jiang, S.; Doyley, M.M.; Dehghani, H.; Tosteson, T.D.; Paulsen, K.D. Automated region detection based on the contrast-to-noise ratio in near-infrared tomography. *Appl. Opt.* **2004**, *43*, 1053–1062. [CrossRef] [PubMed]
36. Rivera, M.M.; Arentoft, A.; Kubo, G.K.; D'Aquila, E.; Scheiner, D.; Pizzirusso, M.; Sandoval, T.C.; Gollan, T.H. Neuropsychological, cognitive, and theoretical considerations for evaluation of bilingual individuals. *Neuropsychol. Rev.* **2008**, *18*, 255–268. [CrossRef]
37. Al-Shargie, F.; Kiguchi, M.; Badruddin, N.; Dass, S.C.; Hani, A.F.M.; Tang, T.B. Mental stress assessment using simultaneous measurement of EEG and fNIRS. *Biomed. Opt. Express* **2016**, *7*, 3882. [CrossRef] [PubMed]
38. Li, R.; Potter, T.; Huang, W.; Zhang, Y. Enhancing performance of a hybrid EEG-fNIRS system using channel selection and early temporal features. *Front. Hum. Neurosci.* **2017**, *11*, 1–11. [CrossRef]
39. Hu, X.-S.; Hong, K.-S.; Ge, S.S.; Jeong, M.-Y. Kalman estimator- and general linear model-based on-line brain activation mapping by near-infrared spectroscopy. *Biomed. Eng. Online* **2010**, *9*, 82. [CrossRef] [PubMed]
40. Hong, K.S.; Khan, M.J. Hybrid brain-computer interface techniques for improved classification accuracy and increased number of commands: A review. *Front. Neurorobot.* **2017**, *11*. [CrossRef]
41. Nguyen, H.-D.; Hong, K.-S. Bundled-optode implementation for 3D imaging in functional near-infrared spectroscopy. *Biomed. Opt. Express* **2016**, *7*, 3491. [CrossRef]
42. Lindquist, M.A.; Meng Loh, J.; Atlas, L.Y.; Wager, T.D. Modeling the hemodynamic response function in fMRI: Efficiency, bias and mis-modeling. *Neuroimage* **2009**, *45*, S187–S198. [CrossRef]
43. Shan, Z.Y.; Wright, M.J.; Thompson, P.M.; McMahon, K.L.; Blokland, G.G.A.M.; de Zubicaray, G.I.; Martin, N.G.; Vinkhuyzen, A.A.E.; Reutens, D.C. Modeling of the hemodynamic responses in block design fMRI studies. *J. Cereb. Blood Flow Metab. J. Int. Soc. Cereb. Blood Flow Metab.* **2014**, *34*, 316–324. [CrossRef]
44. Santosa, H.; Hong, M.J.; Hong, K.S. Lateralization of music processing with noises in the auditory cortex: An fNIRS study. *Front. Behav. Neurosci.* **2014**, *8*, 1–9. [CrossRef] [PubMed]
45. Shin, J.; Von Luhmann, A.; Blankertz, B.; Kim, D.W.; Jeong, J.; Hwang, H.J.; Muller, K.R. Open access dataset for EEG+NIRS single-trial classification. *IEEE Trans. Neural Syst. Rehabil. Eng.* **2017**, *25*, 1735–1745. [CrossRef]

46. Abdalmalak, A.; Milej, D.; Norton, L.; Debicki, D.B.; Gofton, T.; Diop, M.; Owen, A.M.; St. Lawrence, K. Single-session communication with a locked-in patient by functional near-infrared spectroscopy. *Neurophotonics* **2017**, *4*, 1. [CrossRef]
47. Shin, J.; Kwon, J.; Choi, J.; Im, C.H. Performance enhancement of a brain-computer interface using high-density multi-distance NIRS. *Sci. Rep.* **2017**, *7*, 1–10. [CrossRef]
48. Trakoolwilaiwan, T.; Behboodi, B.; Lee, J.; Kim, K.; Choi, J.-W. P10 Convolutional neural network for high-accuracy functional near- infrared spectroscopy in a brain—computer interface: Three-class classification of rest, right-, and left-hand motor execution. *Neurophotonics* **2017**, *5*. [CrossRef] [PubMed]
49. Khan, R.A.; Naseer, N.; Saleem, S.; Qureshi, N.K.; Noori, F.M.; Khan, M.J. Cortical tasks-based optimal filter selection: An fNIRS study. *J. Healthc. Eng.* **2020**, *2020*, 1–15. [CrossRef]
50. Gratton, G.; Brumback, C.R.; Gordon, B.A.; Pearson, M.A.; Low, K.A.; Fabiani, M. Effects of measurement method, wavelength, and source-detector distance on the fast optical signal. *Neuroimage* **2006**, *32*, 1576–1590. [CrossRef] [PubMed]
51. Zhang, Q.; Brown, E.N.; Strangman, G.E. Adaptive filtering to reduce global interference in evoked brain activity detection: A human subject case study. *J. Biomed. Opt.* **2007**, *12*, 064009. [CrossRef]
52. Abdelnour, A.F.; Huppert, T. Real-time imaging of human brain function by near-infrared spectroscopy using an adaptive general linear model. *Neuroimage* **2009**, *46*, 133–143. [CrossRef]
53. Nicolas-Alonso, L.F.; Gomez-Gil, J. Brain computer interfaces, a review. *Sensors* **2012**, *12*, 1211–1279. [CrossRef]
54. Noori, F.M.; Qureshi, N.K.; Khan, R.A.; Naseer, N. Feature selection based on modified genetic algorithm for optimization of functional near-infrared spectroscopy (fNIRS) signals for BCI. In Proceedings of the 2016 2nd International Conference on Robotics and Artificial Intelligence (ICRAI), Rawalpindi, Pakistan, 1–2 November 2016; pp. 50–53.
55. Hong, K.S.; Khan, M.J.; Hong, M.J. Feature extraction and classification methods for hybrid fNIRS-EEG brain-computer interfaces. *Front. Hum. Neurosci.* **2018**, *12*, 1–25. [CrossRef] [PubMed]
56. Khan, R.A.; Naseer, N.; Qureshi, N.K.; Noori, F.M.; Nazeer, H.; Khan, M.U. FNIRS-based neurorobotic interface for gait rehabilitation. *J. Neuroeng. Rehabil.* **2018**, *15*, 7. [CrossRef] [PubMed]
57. Naseer, N.; Noori, F.M.; Qureshi, N.K.; Hong, K.S. Determining optimal feature-combination for LDA classification of functional near-infrared spectroscopy signals in brain-computer interface application. *Front. Hum. Neurosci.* **2016**, *10*, 1–10. [CrossRef]
58. Navidi, W.C. *Statistics for Engineers and Scientists*; McGraw-Hill Higher Education: New York, NY, USA, 2008.
59. Park, S.; Park, J.; Shin, J.; Jeong, J. Open Access fNIRS Dataset for Classification of the Unilateral Finger- and Foot-Tapping. Available online: https://figshare.com/articles/Open_access_fNIRS_dataset_for_classification_of_the_unilateral_finger-_and_foot-tapping/9783755/1 (accessed on 15 June 2020).
60. Naseer, N.; Qureshi, N.K.; Noori, F.M.; Hong, K. Analysis of different classification techniques for two-class functional near-infrared spectroscopy-based brain-computer interface. *Comput. Intell. Neurosci.* **2016**, *2016*, 1–11. [CrossRef] [PubMed]
61. Zafar, A.; Hong, K.S. Neuronal activation detection using vector phase analysis with dual threshold circles: A functional near-infrared spectroscopy study. *Int. J. Neural Syst.* **2018**, *28*. [CrossRef]
62. Zafar, A.; Ghafoor, U.; Yaqub, M.A.; Hong, K.-S. Initial-dip-based classification for fNIRS-BCI. In *Neural Imaging and Sensing 2019*; Luo, Q., Ding, J., Fu, L., Eds.; SPIE: San Francisco, CA, USA, 2019; p. 60.
63. Hong, K.S.; Santosa, H. Decoding four different sound-categories in the auditory cortex using functional near-infrared spectroscopy. *Hear. Res.* **2016**, *333*, 157–166. [CrossRef]
64. Petrantonakis, P.C.; Kompatsiaris, I. Single-trial NIRS data classification for brain-computer interfaces using graph signal processing. *IEEE Trans. Neural Syst. Rehabil. Eng.* **2018**, *26*, 1700–1709. [CrossRef]
65. Curtin, A.; Ayaz, H. The age of neuroergonomics: Towards ubiquitous and continuous measurement of brain function with fNIRS. *Jpn. Psychol. Res.* **2018**, *60*, 374–386. [CrossRef]
66. Gateau, T.; Ayaz, H.; Dehais, F. In silico vs. over the clouds: On-the-fly mental state estimation of aircraft pilots, using a functional near infrared spectroscopy based passive-BCI. *Front. Hum. Neurosci.* **2018**, *12*, 187. [CrossRef]
67. Khan, R.A.; Naseer, N.; Khan, M.J. *Neuroergonomics*; Elsevier: Amsterdam, The Netherlands, 2019; pp. 79–85.

Publisher's Note: MDPI stays neutral with regard to jurisdictional claims in published maps and institutional affiliations.

Article

Decoding Attempted Hand Movements in Stroke Patients Using Surface Electromyography

Mads Jochumsen [1,*], Imran Khan Niazi [1,2,3], Muhammad Zia ur Rehman [4], Imran Amjad [2,4], Muhammad Shafique [4], Syed Omer Gilani [5] and Asim Waris [5]

1 Department of Health Science and Technology, Aalborg University, 9220 Aalborg Øst, Denmark; imrankn@hst.aau.dk
2 Centre for Chiropractic Research, New Zealand College of Chiropractic, Auckland 1060, New Zealand; imran.amjad@nzchiro.co.nz
3 Health and Rehabilitation Research Institute, AUT University, Auckland 1010, New Zealand
4 Faculty of Rehabilitation and Allied Sciences & Faculty of Engineering and Applied Sciences, Riphah International University, Islamabad 44000, Pakistan; ziaur.rehman@riphah.edu.pk (M.Z.u.R.); muhammad.shafique@riphah.edu.pk (M.S.)
5 Department of Biomedical Engineering & Sciences, School of Mechanical & Manufacturing Engineering, National University of Sciences and Technology (NUST), Islamabad 44000, Pakistan; omer@smme.nust.edu.pk (S.O.G.); asim.waris@smme.nust.edu.pk (A.W.)
* Correspondence: mj@hst.aau.dk

Received: 30 October 2020; Accepted: 25 November 2020; Published: 26 November 2020

Abstract: Brain- and muscle-triggered exoskeletons have been proposed as a means for motor training after a stroke. With the possibility of performing different movement types with an exoskeleton, it is possible to introduce task variability in training. It is difficult to decode different movement types simultaneously from brain activity, but it may be possible from residual muscle activity that many patients have or quickly regain. This study investigates whether nine different motion classes of the hand and forearm could be decoded from forearm EMG in 15 stroke patients. This study also evaluates the test-retest reliability of a classical, but simple, classifier (linear discriminant analysis) and advanced, but more computationally intensive, classifiers (autoencoders and convolutional neural networks). Moreover, the association between the level of motor impairment and classification accuracy was tested. Three channels of surface EMG were recorded during the following motion classes: Hand Close, Hand Open, Wrist Extension, Wrist Flexion, Supination, Pronation, Lateral Grasp, Pinch Grasp, and Rest. Six repetitions of each motion class were performed on two different days. Hudgins time-domain features were extracted and classified using linear discriminant analysis and autoencoders, and raw EMG was classified with convolutional neural networks. On average, 79 ± 12% and 80 ± 12% (autoencoders) of the movements were correctly classified for days 1 and 2, respectively, with an intraclass correlation coefficient of 0.88. No association was found between the level of motor impairment and classification accuracy (Spearman correlation: 0.24). It was shown that nine motion classes could be decoded from residual EMG, with autoencoders being the best classification approach, and that the results were reliable across days; this may have implications for the development of EMG-controlled exoskeletons for training in the patient's home.

Keywords: stroke; EMG; brain-computer interface; myoelectric control; pattern recognition

1. Introduction

A stroke is a cardiovascular disease affecting millions of people each year, where approximately 80% of the survivors are left with motor disabilities, such as paresis or paralysis [1,2]. Even after rehabilitation, around 50% of the patients are left with disabilities such that they require assistance with some of their activities of daily living [3,4]. Since a stroke is heterogeneous, there is no effective treatment that works for all [2]. There seems to be a consensus that principles of motor learning are relevant to stroke recovery [5], and induction of neural plasticity, which is the underlying factor of motor learning [6]. New patient-driven technologies have emerged where motor learning principles, such as repetition and attention, are incorporated in training, examples of such technologies are muscle- and brain-triggered exoskeletons or robots [7–10]. With the advances in the design and production of exoskeletons and rehabilitation robots, it is possible to perform different motions, which can be used to introduce task variability in training that can maximize the retention and generalization of the relearned movements [5]. To use such devices, however, it is necessary to detect the movement intention of various movement types to pair the motor commands with relevant afferent feedback. In patients with paralysis and no detectable electromyography (EMG) activity, it is necessary to use a Brain-Computer Interface where movement intentions are detected through electroencephalography (EEG). It is possible to classify movement intentions from idle activity with accuracies up to 80% in stroke patients [11,12], but the accuracies decrease when different movement types are classified [11–13]. If EMG activity is preserved or regained [14], it is possible to classify various movement types with high accuracy [15], even in patients with severe impairments [16,17]. It has been shown that neuroplasticity can be introduced using both EEG- and EMG-triggered electrical stimulation for providing afferent feedback [18], but it may be advantageous to use EMG if different movement types need to be classified. It has been shown in different studies that different movement types can be classified from EMG activity from the muscles in the affected limb. These movements include finger movements [14], various functional hand movements, such as open/close [15,19–22] and grasps [23], wrist extension [16], elbow and shoulder movements [24], and reaching [25]. Some of the techniques that have been used for decoding the attempted movements from the EMG are amplitude thresholds of the EMG signal envelope and proportional control [14,20], and pattern recognition approaches using, e.g., Hudgins time-domain features [15], autoregressive coefficients [22], empirical mode decomposition [26], and wavelets [27]. The performance of the decoding algorithms spans a wide range of 38–100%. Generally, the highest accuracies were associated with binary tasks, such as detecting a movement versus no-movement, and the performance decreases when including more motion classes. Some results have also shown how the classification accuracy is affected by the severity of the stroke [21]. As outlined, several studies have investigated the possibility of decoding attempted movements from stroke patients using surface EMG. These studies have primarily been single-session studies; therefore, there is a need for reliability studies to see if the decoding results are reproducible over time. Moreover, in previous work, EMG electrodes have been positioned for each patient individually to account, e.g., for muscle weakness and spasticity [22], or several EMG electrodes have been used to capture the activity from several muscles [15]. Therefore, the aim of this study was to investigate if different hand and forearm movements can be classified using a simple electrode setup placed on the same three muscles across heterogeneous stroke participants over two different days using a simple pattern recognition (linear discriminant analysis) approach with a low computational complexity which implements low-cost embedded systems. For comparison purposes, the simple classification approach using linear discriminant analysis was compared to autoencoders and convolutional neural networks that have been shown previously to improve the classification performance [28], but are more computationally intensive. In addition, the reliability of the pattern recognition approach was evaluated over two days. Besides the reliability analysis, it was investigated if it was possible to use the EMG recorded on one day to classify the EMG on the other day. Lastly, it was investigated if there was an association between the level of motor impairment and classification accuracy. Such an association has been reported previously, but more evidence is important, especially in a heterogeneous condition, such as a stroke.

2. Materials and Methods

2.1. Participants

Sixteen stroke patients (one female; 53 ± 8 years old) were recruited for this study (see the patient demographics in Table 1) from Railway General Hospital in Rawalpindi, Pakistan. One patient dropped out during the data collection. All patients provided their informed consent prior to participation. The procedures were approved by the local ethical committee (Riphah/RCRS/REC/00651). All procedures were in accordance with the Declaration of Helsinki. The Fugl-Meyer Assessment was performed to indicate the motor impairment of the patients, (motor score). The motor part of the Fugl-Meyer Assessment consists of two scores (100 points in total), one for the upper (66 points) and lower extremities (34 points). In this study, the score for the upper extremities is of interest. It covers the functionality of shoulder, elbow, wrist, and finger movements, as well as grasping various objects [29].

Table 1. Patient demographics. Upper limb (UL), and lower limb (LL). The maximum score is 66 and 34 for UL and LL, respectively.

Patient	Months Since Injury	Affected Side	Type of Injury	Fugl-Meyer [UL/LL/Total]
1	24	Left	Ischemic	[55/22/77]
2	17	Right	Ischemic	[36/34/70]
3	18	Right	Ischemic	[23/28/51]
4	32	Left	Ischemic	[46/32/78]
5	36	Left	Ischemic	[26/18/44]
6	5	Right	Ischemic	[65/31/96]
7	38	Right	Ischemic	[17/22/39]
8	2	Left	Ischemic	[59/31/90]
9	38	Right	Ischemic	[55/30/85]
10	6	Left	Ischemic	[51/23/74]
11	3	Right	Ischemic	[56/24/80]
12	5	Left	Hemorrhagic	[44/20/64]
13	66	Right	Hemorrhagic	[28/18/46]
14	19	Left	Ischemic	[50/21/71]
15	70	Left	Hemorrhagic	[36/33/69]

2.2. Recordings—Surface EMG

Six surface EMG electrodes (Ambu Neuroline 720 surface electrodes, REF 72000-S/25, Ambu, Ballerup, Denmark) were placed on the forearm on Extensor Carpi Radialis, Flexor Carpi Radialis, and Flexor Carpi Ulnaris. Two electrodes were placed on each muscle two cm apart and used in a bipolar configuration to obtain a single channel. The signals were referenced to a moist wristband. The signals were amplified with a gain of 10,000 (OT Bioelettronica, Torino, Italy) and sampled with 2048 Hz.

2.3. Experimental Setup

The experiment consisted of two recording sessions performed on two different days. The same experimental procedure was followed in both sessions. The recordings were performed in a seated position. Initially, the EMG electrodes were placed on the forearm on the most affected side, and the signal quality was checked (the electrode positions were marked on the forearm to ensure the same placement of the electrodes on day two). The participants were instructed how to perform the motions, and during the recording of the signals, they were visually cued (a picture of the specific motion was shown). A digital trigger was sent to the amplifier to synchronize the visual cue with the EMG recordings at the beginning of the recording. The following motion classes were performed: Hand Close, Hand Open, Wrist Extension, Wrist Flexion, Supination, Pronation, Lateral Grasp, Pinch Grasp, and Rest. Each motion class consisted of six repetitions of attempted movement, and the participant was asked to maintain the contraction for six seconds. Between each movement, there was a break of six seconds. All repetitions of the motion class were completed before moving to the next motion class. The order of motion classes was randomized.

2.4. Data Analysis

2.4.1. Pre-Processing and Feature Extraction

The EMG was bandpass filtered between 20–500 Hz, and a Notch filter from 48–52 Hz was applied using a 2nd order Butterworth filter with zero phase shift. The onsets of the EMG activity were visually inspected to avoid a potential delay between the cue and onset of the movements, such that the movement onsets were correctly identified for further analysis. Each of the 6-s repetitions of the motion class were extracted, and the first and last second were removed from the analysis, which resulted in epochs of 4-s duration for each repetition of the motion class. Following the pre-processing, four features were extracted: Mean absolute value, waveform length, zero crossing, and slope sign changes [30]. The features were extracted from a 200-millisecond data window with no overlap to obtain more data for classification [31]. The same analysis was performed on the data from the two separate recording sessions. An example of the filtered and rectified EMG for each motion class is shown in Figure 1.

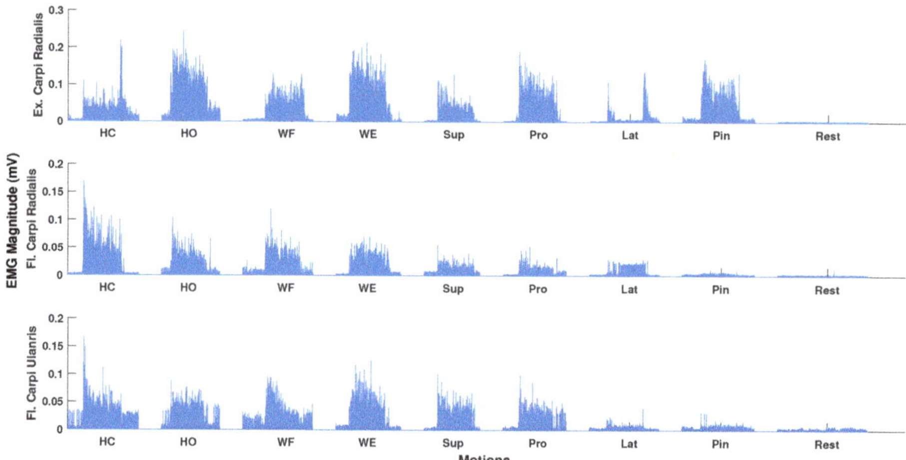

Figure 1. Rectified (only for visualization) and bandpass filtered surface EMG for the nine different motion classes for a single repetition and a single participant. Hand Close (HC), Hand Open (HO), Wrist Flexion (WF), Wrist Extension (WE), Supination (Sup), Pronation (Pro), Lateral Grasp (Lat), and Pin (Pinch Grasp). Flexor (Fl.), Extensor (Ex.). Clear EMG activity can be seen for most motion classes except the Lateral Grasp.

2.4.2. Classification

The classification was performed in two different ways: (1) Within-session calibration; and (2) between-session calibration. For the within-session calibration, 80% of the data windows were randomly selected for training, and 20% of the data windows were used for testing. The classifiers were trained on data windows from each subject individually and on the same randomly selected data windows to fairly compares classifiers. In the within-session calibration, the classifier was trained and tested on the recordings from the same day. In the between-session calibration, the classifier was trained on data from one day and tested on the other day. Moreover, confusion matrices were obtained.

Three classifiers were tested, two of them used features as input, while the third used bandpass filtered data windows as input. The features were classified using a linear discriminant analysis classifier (LDA) and autoencoders (AE), and the filtered data windows were classified with a convolutional neural network (CNN). The LDA is a linear classifier that can separate multiple classes using a linear combination of the input features [32], while AE is an artificial neural network. In this study, the default MATLAB implementation of the LDA was used where all classes have the same covariance matrix. In the implementation of the LDA in this study, all motion classes were included leading to a classification problem with nine classes. The AE network consisted of two layers, each with hidden units of 12 (length of the feature vector) and a softmax layer. Optimized parameters from previous work were used [33]. The 200-millisecond data windows of EMG were classified using a CNN. The CNN consisted of an input layer (200-millisecond data window), four convolutional layers, each with Relu and pooling layers, a fully connected layer, and a softmax layer. The architecture of the network was optimized randomly, and the network was trained using Adam optimizer with default values except for L2R (10×10^{-6}) and initial learning rate (5×10^{-3}) with a 'piecewise' learning rate schedule having a drop rate factor of 0.1 and drop period of 4. Maxepochs were set to 20, and a mini-batch size of 16 and 32 were used for within- and between-session analyses, respectively. All data processing and analyses were performed in MATLAB 2020a (MathWorks®). The computational time of the different classifiers was estimated on the training and test data. The classification was performed on a computer with 8 GB RAM, a core i5 processor, and a 64-bit operating system.

2.5. Statistics

All statistical analyses were performed in IBM® SPSS®. The test-retest reliability of the classification accuracies in the within- and between-session calibration was assessed using a two-way mixed-effect model with absolute agreement. The test was repeated three times for LDA, AE, and CNN, respectively. The mean classification accuracy was calculated across the two days for the within- and between-session calibration, and a two-way repeated-measures analysis of variance (ANOVA) was performed with "Calibration" (2 levels: Within-, and between-session calibration) and "Classifier" as factors (3 levels: LDA, AE, and CNN). Six Friedman tests were performed (the assumption of normality was violated) on the diagonal values in the confusion matrices (mean across the two days) with "Motion Class" as the factor (9 levels: Hand Close, Hand Open, Wrist Extension, Wrist Flexion, Supination, Pronation, Lateral Grasp, Pinch Grasp, and Rest) for the three classifiers in the within- and between-session calibration. Significant tests were followed up with a posthoc test using Bonferroni correction. Lastly, the Spearman correlation coefficient was calculated between the upper limb Fugl-Meyer score and the average classification accuracy for the within-session calibration (mean across the two days). Significant tests in all analyses were assumed when $p < 0.05$.

3. Results

The average classification accuracies across participants obtained in the within-session calibration were approximately 70% for the LDA and CNN on both days, while 80% of the motions were correctly classified with AE (see Figure 2). The average classification accuracies in the between-session calibration were approximately 30% for day 1 and 2, respectively, with slightly lower accuracies for AE. The results of the test-retest analysis are presented in Table 2. Good agreement was obtained for all classification and calibration scenarios except for the CNN in the between-session calibration, where the moderate agreement was obtained [34].

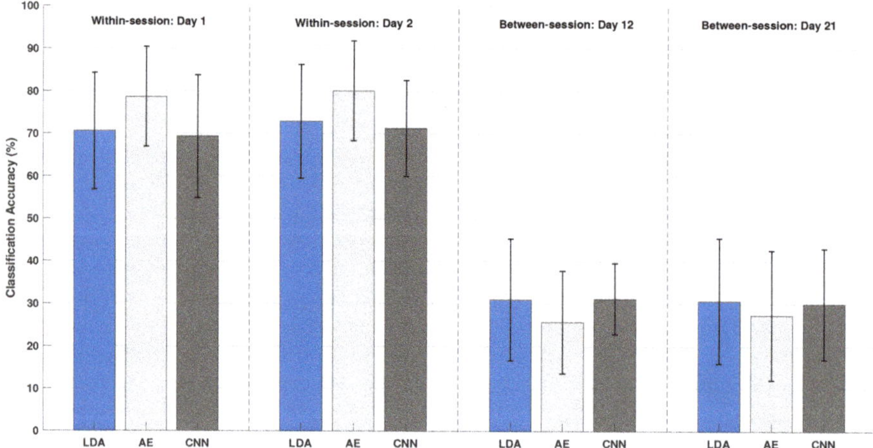

Figure 2. Overall classification accuracy for all motion types. The results are presented as mean ± standard deviation across participants. "Day12" indicates training on data from day 1 and testing on data from day 2. "Day21" indicates training on data from day 2 and testing on data from day 1. LDA (linear discriminant analysis), AE (autoencoders), and CNN (convolutional neural network).

Table 2. Intraclass correlation coefficients for the different calibration scenarios for the three classifiers. The intraclass correlation coefficient and 95% confidence intervals are reported.

	Within-Session	Between-Session
Linear discriminant analysis	0.84 [0.54:0.95]	0.88 [0.63:0.96]
Autoencoders	0.88 [0.63:96]	0.87 [0.62:0.96]
Convolutional neural network	0.86 [0.58:0.95]	0.69 [0.06:0.90]

The mean classification accuracy was calculated across the two days, and a two-way repeated-measures ANOVA revealed a significant interaction between Calibration and Classifier ($F_{(2,28)} = 27.05$; $p < 0.001$; $\eta^2 = 0.66$). This was followed up with two one-way repeated-measures ANOVA tests for the within- and between-session calibration. For the within-session calibration, there was a significant difference between the classifiers ($F_{(1.2,17.1)} = 17.27$; $p < 0.001$; $\eta^2 = 0.55$), and the posthoc analysis revealed higher classification accuracies for AE compared to the LDA and CNN. For the between-session calibration, there was no difference between the classifiers ($F_{(2,28)} = 3.10$; $p = 0.061$; $\eta^2 = 0.18$).

The confusion matrices (mean across the two days and across participants) for the within-session and between-session calibration are shown in Tables 3–8, respectively. For the within-session calibration, the highest numbers were on the diagonal for all motion classes, with Rest being the easiest to discriminate (92–95%). The other motion classes were in the range of 55–83%. A significant difference between the motion classes was found for the LDA ($\chi^2_{(8)} = 44.13$; $p < 0.001$), AE ($\chi^2_{(8)} = 39.63$; $p < 0.001$) and CNN ($\chi^2_{(8)} = 51.14$; $p < 0.001$). For the LDA, the posthoc analyses revealed that Rest had higher classification accuracies compared to the other classes except Wrist Extension and Hand Open. Wrist Extension and Hand Open had higher classification accuracies than Lateral Grasp. For AE, the classification accuracies for the Rest class were higher than Lateral Grasp, Supination, Pinch Grasp, and Pronation. For CNN, Rest had higher classification accuracies compared to the other classes except Wrist Extension and Wrist Flexion. Wrist Extension and Wrist Flexion had higher classification accuracies than Lateral Grasp.

Table 3. Confusion matrix based on within-session calibration (the mean across the two days have been calculated) using linear discriminant analysis. All values are in percent and presented as the mean across participants. HC (Hand Close), HO (Hand Open), WE (Wrist Extension), WF (Wrist Flexion), Sup (Supination), Pro (Pronation), Lat (Lateral Grasp), and Pin (Pinch Grasp).

	HC	**HO**	**WF**	**WE**	**Sup**	**Pro**	**Lat**	**Pin**	**Rest**
HC	72	5	2	2	4	2	10	4	2
HO	4	77	6	4	2	3	1	4	1
WF	4	9	71	4	3	4	4	3	1
WE	2	4	4	75	6	3	2	5	1
Sup	2	2	2	7	66	8	7	5	3
Pro	1	2	2	2	12	70	4	7	3
Lat	9	2	1	2	9	8	56	9	6
Pin	2	4	1	5	7	3	6	69	5
Rest	0	1	0	0	2	2	2	3	92

Table 4. Confusion matrix based on within-session calibration (the mean across the two days have been calculated) using autoencoders. All values are in percent and presented as the mean across participants. HC (Hand Close), HO (Hand Open), WE (Wrist Extension), WF (Wrist Flexion), Sup (Supination), Pro (Pronation), Lat (Lateral Grasp), and Pin (Pinch Grasp).

	HC	**HO**	**WF**	**WE**	**Sup**	**Pro**	**Lat**	**Pin**	**Rest**
HC	82	3	2	2	2	1	8	3	1
HO	3	83	5	2	2	2	2	2	0
WF	2	7	80	3	2	2	3	3	1
WE	1	4	4	79	5	2	2	3	0
Sup	2	2	2	7	73	7	6	3	2
Pro	1	2	2	3	9	76	3	6	2
Lat	7	1	4	2	4	6	70	7	2
Pin	2	2	2	3	4	3	6	77	3
Rest	0	0	1	0	1	1	2	2	94

Table 5. Confusion matrix based on within-session calibration (the mean across the two days have been calculated) using a CNN. All values are in percent and presented as the mean across participants. HC (Hand Close), HO (Hand Open), WE (Wrist Extension), WF (Wrist Flexion), Sup (Supination), Pro (Pronation), Lat (Lateral Grasp), and Pin (Pinch Grasp).

	HC	**HO**	**WF**	**WE**	**Sup**	**Pro**	**Lat**	**Pin**	**Rest**
HC	70	5	4	2	3	1	13	3	1
HO	5	69	8	4	5	4	2	3	0
WF	2	8	73	6	3	4	3	3	0
WE	1	4	4	76	4	4	4	5	1
Sup	2	4	3	7	61	9	6	8	2
Pro	1	3	3	4	11	68	4	7	2
Lat	13	3	4	2	6	6	55	10	4
Pin	2	4	2	5	7	7	6	68	2
Rest	0	0	0	0	0	2	2	2	95

Table 6. Confusion matrix based on between-session calibration (the mean across the two days have been calculated) using linear discriminant analysis. All values are in percent and presented as the mean across participants. HC (Hand Close), HO (Hand Open), WE (Wrist Extension), WF (Wrist Flexion), Sup (Supination), Pro (Pronation), Lat (Lateral Grasp), and Pin (Pinch Grasp).

	HC	HO	WF	WE	Sup	Pro	Lat	Pin	Rest
HC	41	6	5	6	13	6	18	5	1
HO	16	30	13	6	14	7	9	7	1
WF	16	9	39	6	12	6	10	3	2
WE	15	7	6	42	7	3	10	10	2
Sup	17	9	10	4	21	10	17	11	3
Pro	12	6	7	4	15	23	12	15	7
Lat	35	8	4	5	14	7	17	6	5
Pin	19	8	7	6	15	12	6	24	6
Rest	13	0	4	1	14	9	5	11	43

Table 7. Confusion matrix based on between-session calibration (the mean across the two days have been calculated) using autoencoders. All values are in percent and presented as the mean across participants. HC (Hand Close), HO (Hand Open), WE (Wrist Extension), WF (Wrist Flexion), Sup (Supination), Pro (Pronation), Lat (Lateral Grasp), and Pin (Pinch Grasp).

	HC	HO	WF	WE	Sup	Pro	Lat	Pin	Rest
HC	29	9	16	6	8	7	22	6	1
HO	9	28	15	10	12	6	12	9	0
WF	16	9	40	8	13	7	6	3	1
WE	11	7	8	41	7	8	11	9	0
Sup	14	9	12	8	22	12	21	3	2
Pro	12	10	11	8	13	28	9	7	4
Lat	24	10	14	9	11	9	16	7	2
Pin	17	8	13	15	12	13	7	14	4
Rest	13	2	7	10	13	13	14	8	22

Table 8. Confusion matrix based on between-session calibration (the mean across the two days have been calculated) using a CNN. All values are in percent and presented as the mean across participants. HC (Hand Close), HO (Hand Open), WE (Wrist Extension), WF (Wrist Flexion), Sup (Supination), Pro (Pronation), Lat (Lateral Grasp), and Pin (Pinch Grasp).

	HC	HO	WF	WE	Sup	Pro	Lat	Pin	Rest
HC	30	8	19	5	14	4	14	7	1
HO	16	21	17	8	9	8	13	7	1
WF	8	11	49	7	10	4	8	3	2
WE	8	10	13	43	8	2	8	9	1
Sup	13	12	13	7	22	11	11	6	7
Pro	7	10	12	7	15	19	12	14	7
Lat	19	11	17	5	9	9	15	10	7
Pin	10	10	11	15	15	12	8	13	6
Rest	1	0	7	0	12	4	7	4	66

For the between-session calibration, the highest numbers were on the diagonal for most motion classes except Lateral and Pinch Grasps. Rest was the motion class with the highest accuracies (22–66%). The other motion classes were in the range of 13–43%. A significant difference between the motion classes was found for the LDA ($\chi^2_{(8)} = 23.02$; $p = 0.003$), AE ($\chi^2_{(8)} = 20.52$; $p = 0.009$), and CNN ($\chi^2_{(8)} = 29.72$; $p < 0.001$). For the LDA, the posthoc analyses revealed no difference between the classes, which is due to the conservative nature of the Bonferroni correction. For AE, the classification accuracies for the Wrist Extension and Wrist Flexion classes were higher than the Pinch Grasp. For CNN, Rest had higher classification accuracies compared to the Lateral Grasp and Pinch Grasp. Wrist Extension had higher classification accuracies than the Pinch Grasp. There was a considerable standard deviation across the participants (see Figure 2), which may be attributed to the amplitude differences between movement and Rest for the different participants (see Figure 3).

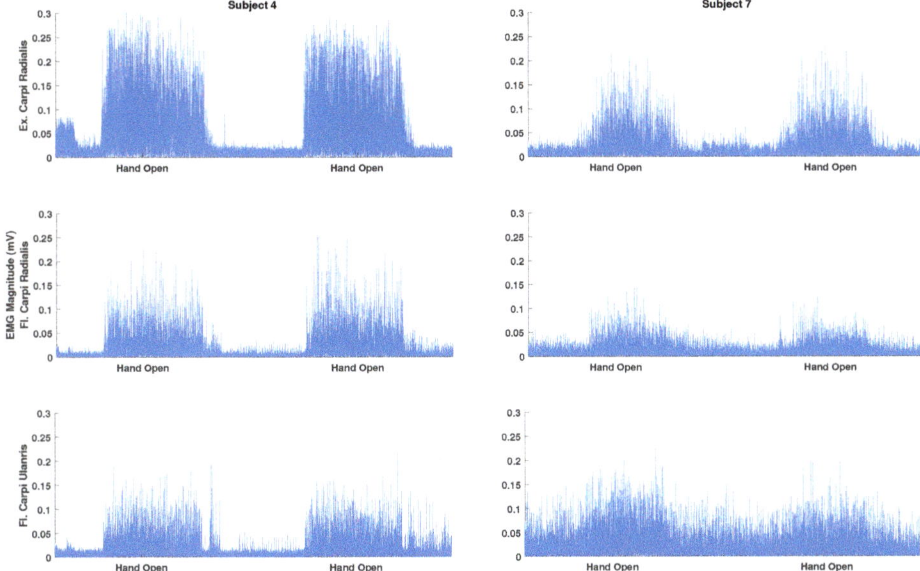

Figure 3. Rectified (only for visualization) and bandpass filtered surface EMG for the Hand Open motion class for the subject with the highest (subject 4) and lowest (subject 7) classification accuracy. The highest and lowest overall classification accuracies were 91% and 54% (classified with linear discriminant analysis), respectively. The amplitude of the EMG for the motions performed by the best subject is higher compared to the worst subject. Moreover, there is a smaller EMG amplitude for the resting state between the movements for the best subject.

The Spearman correlation was calculated between the upper limb Fugl-Meyer score and the classification accuracy across all motion classes for the within-session calibration. The results are presented in Table 9. There was no association between the functional score and the classification accuracies.

Table 9. Correlation analysis between the classification accuracies (mean across days) for the within-session calibration and the functional score (upper limb Fugl-Meyer score).

	Correlation Coefficients	p-Value
Linear discriminant analysis	0.29	0.30
Autoencoders	0.24	0.38
Convolutional neural network	0.37	0.18

The results of the computational time are presented in Table 10. The LDA was the fastest classifier to train, followed by the AE. For testing, the LDA and AE were faster than the CNN, but it only took 0.22 and 0.27 s to classify all test data with a CNN.

Table 10. The computational time of the training and test data for within- and between-session calibration. In the within-session scenario, the training data consisted of 828 data windows, and the test data consisted of 198 test windows. For the between-session scenario, the training and test data consisted of 1026 data windows.

Classifier	Training (Seconds)	Test (Seconds)
Linear discriminant analysis (within-session)	0.010	0.010
Autoencoders (within-session)	12.16	0.015
Convolutional neural network (within-session)	47.68	0.22
Linear discriminant analysis (between-session)	0.018	0.018
Autoencoders (between-session)	13.22	0.016
Convolutional neural network (between-session)	58.77	0.27

4. Discussion

The aim of this study was to decode attempted movements in stroke patients. Generally, it was possible to decode nine different motion classes of the hand/forearm with accuracies of 79 ± 12% and 80 ± 12% for day 1 and 2, respectively (using AE). There was a strong intraclass correlation between the classification accuracies, but there was no association between the classification accuracies and the upper limb Fugl-Meyer score. The classification accuracies obtained with AE were significantly higher than the LDA and CNN in the within-session calibration, but not significantly different for between-session calibration.

The findings in this study are in agreement with other studies that have found that EMG of attempted movements can be decoded from stroke patients with motor impairments [15–17,21,22,24]. Similar or slightly lower classification accuracies are obtained, although the studies differ in terms of methodology. The number of channels in this study (three bipolar channels) is low compared to other studies [15,26], which is likely to account for some of the differences in the classification accuracies. Moreover, a generalized approach was used where the electrodes were placed on the same three muscles on the contrary to other findings where electrodes have been positioned based on the impairment of the individual patient [22]. By increasing the number of channels, it is possible to record activity from more muscles; especially the chronic stroke patients may have developed coping strategies to perform the movements using altered activation patterns. For the different grasp motions, it would have been beneficial to place EMG electrodes closer to or on the hand. Different classifiers were tested, and it was shown that the classification performance could be significantly improved using AE, which is in agreement with previous findings [28]. It was also shown that classification accuracies comparable to a LDA can be obtained for a CNN without the need for extracting features, but this indicates that it may not be needed to use a CNN over LDA with features. Generally, Hudgins time-domain features [15,22,24,27,30] have been used, but to improve the classification accuracies further, other feature types could be added to the feature vector, such as spectral information, wavelets, autoregressive coefficients, and entropy [15,27,35].

In future studies, the most optimal or general electrode setup and feature types could be investigated, e.g., with and without constraints on computational power and energy consumption of the system. For a low-end implementation, Raspberry Pi could be used, while a Parallella Baseline System (PBS) could be used when more computational power is needed with a constraint of still being energy efficient [36]. These approaches should be validated using online control and with afferent feedback from an exoskeleton. Lastly, the impairment of the patients would probably affect the classification accuracies as well. There was no association between the level of motor impairment and classification accuracy in this study, although that was expected, as it has been shown previously that classification accuracies decrease as a function of the severity of the injury [21]. The limited sample size in this study could explain that no association was found between the classification accuracies and motor impairment. Another explanation could be that the machine learning approach was able to pick up movement patterns from participants with a low Fugl-Meyer score leading to reasonably high classification accuracy, or that the Fugl-Meyer score not only reflects the tasks performed in this study, but also reflexes and shoulder and elbow movements. Participants that have reduced hand and wrist movement may have functional reflexes and elbow and shoulder movements. Lastly, the movements and assessment method in this study differed from those performed in Reference [21]. In Reference [21], more functional hand movement tasks were performed, while more wrist movement tasks where performed in the current study, which may be easier to perform for participants with a low Fugl-Meyer score compared to functional hand movement tasks leading to higher classification accuracy, and hence, a lower correlation coefficient. In addition, in Reference [21], they used the Stage of Hand component of the Chedoke-McMaster Stroke Assessment scale, which may be more sensitive to hand movements compared to the total upper limb Fugl-Meyer score. However, these are speculations that need to be tested in a future study.

The motion class that was easiest to discriminate was the Rest condition, which was significantly different from the other classes except Wrist Extension and Wrist Flexion in the within-session calibration. This indicates that the patients did not suffer much from spasticity, which would reduce the ability to discriminate between the motion classes and the usability of EMG to control an exoskeleton/rehabilitation robot or functional electrical stimulation. However, a recent study has found that spasticity can be reduced by utilizing a myoelectric computer interface [37]. The motion class that was most difficult to discriminate was the Lateral Grasp. This is probably because it resembles the motion class Hand Close, 7–13% of the data from the Lateral Grasp motion class was classified as Hand Close. It has also been previously reported that the motions that resembled each other were more difficult to classify [21]. In a rehabilitation scenario, the motion classes that are difficult to classify could be performed in separate training sessions to improve the system performance. The test-retest reliability of the classification of the motion classes was good for both within- and between-session calibration, but the 95%-percent confidence intervals were wide. In addition, there was a large standard deviation of the accuracies for the different participants. This could potentially be explained by various factors, such as the patient's level of fatigue or that, the recruitment patterns of some motion classes differ slightly, and that the muscles from which the EMG was recorded were affected differently in the participants. Moreover, the amplitude of the EMG is related to the force the muscles can produce (see Figure 3). The classification accuracies may also be affected by the signal quality, which could be reduced over time if the impedance of the electrodes changed, due to, for example, sweating. The classification accuracies associated with the between-session calibration were significantly lower compared to the within-session calibration. However, if more days were included, it could potentially have improved the performance [28], but only 2–3 repetitions of each motion class are needed to perform within-session calibration [31], so it would be possible to quickly calibrate the EMG decoder each day. Another aspect that needs to be considered if an EMG-controlled exoskeleton/rehabilitation robot can be used by the patients in their own homes is if they can place the recording electrodes accurately on the muscles. An alternative is to use technology, such as the Thalmic Myo armband, which is easy to don and doff. It utilizes several dry electrodes around the forearm, and the activity from multiple muscles can be recorded. Previously it has been shown that comparable classification accuracies can be obtained using such a setup compared to state-of-the-art wet electrodes and expensive amplifiers [38].

5. Conclusions

In conclusion, it is possible to decode various motion classes of the hand and forearm in stroke patients using both a simple setup with few electrodes and a simple pattern recognition approach and a deep learning approach with and without feature extraction. These findings were consistent across days where the test-retest reliability was good. No association was found between the classification accuracies and the level of impairment. The Rest, Hand Open and Close, and Wrist Extension and Flexion were the classes that were easiest to classify, and the Lateral Grasp was the most difficult to classify. The best classification was obtained using AE. However, more patients with varying degrees of impairment should be included in future studies to validate these findings. Moreover, other feature types and electrode setups (number and location) should be investigated as well to improve the classification accuracy further. These approaches should be validated in online studies where afferent feedback is provided from an exoskeleton or rehabilitation robot.

Author Contributions: Conceptualization, M.J., I.K.N., I.A., M.S., S.O.G. and A.W.; Data curation, I.K.N., I.A., M.S., S.O.G. and A.W.; Formal analysis, M.J., M.Z.u.R.; Funding acquisition, M.J.; Methodology, M.J., M.Z.u.R.; Writing—original draft, M.J.; Writing—review & editing, M.J., I.K.N., M.Z.u.R., I.A., M.S., S.O.G. and A.W. All authors have read and agreed to the published version of the manuscript.

Funding: This work was funded by VELUX FONDEN (project no. 22357).

Acknowledgments: The Authors would like to thank Afaq Ahmed and Ahmed Saad Ullah for assistance in the data collection.

Conflicts of Interest: The authors declare no conflict of interest.

References

1. WHO MONICA Project Principal Investigators. The world health organization monica project (monitoring trends and determinants in cardiovascular disease): A major international collaboration. *J. Clin. Epidemiol.* **1988**, *41*, 105–114. [CrossRef]
2. Langhorne, P.; Coupar, F.; Pollock, A. Motor recovery after stroke: A systematic review. *Lancet Neurol.* **2009**, *8*, 741–754. [CrossRef]
3. Jørgensen, H.S. The Copenhagen Stroke Study experience. *J. Stroke Cerebrovasc. Dis.* **1996**, *6*, 5–16. [CrossRef]
4. Schaechter, J.D. Motor rehabilitation and brain plasticity after hemiparetic stroke. *Prog. Neurobiol.* **2004**, *73*, 61–72. [CrossRef]
5. Krakauer, J.W. Motor learning: Its relevance to stroke recovery and neurorehabilitation. *Curr. Opin. Neurol.* **2006**, *19*, 84–90. [CrossRef]
6. Pascual-Leone, A.; Nguyet, D.; Cohen, L.G.; Brasil-Neto, J.P.; Cammarota, A.; Hallett, M. Modulation of muscle responses evoked by transcranial magnetic stimulation during the acquisition of new fine motor skills. *J. Neurophysiol.* **1995**, *74*, 1037–1045. [CrossRef]
7. Xu, R.; Jiang, N.; Mrachacz-Kersting, N.; Lin, C.; Prieto, G.A.; Moreno, J.C.; Pons, J.L.; Dremstrup, K.; Farina, D. A Closed-Loop Brain-Computer Interface Triggering an Active Ankle-Foot Orthosis for Inducing Cortical Neural Plasticity. *IEEE Trans. Biomed. Eng.* **2014**, *20*, 2092–2101.
8. Ramos-Murguialday, A.; Broetz, D.; Rea, M.; Läer, L.; Yilmaz, Ö.; Msc, F.L.B.; Liberati, G.; Curado, M.R.; Garcia-Cossio, E.; Vyziotis, A.; et al. Brain–machine interface in chronic stroke rehabilitation: A controlled study. *Ann. Neurol.* **2013**, *74*, 100–108. [CrossRef]
9. Monte-Silva, K.; Piscitelli, D.; Norouzi-Gheidari, N.; Batalla, M.A.P.; Archambault, P.; Levin, M.F. Electromyogram-related neuromuscular electrical stimulation for restoring wrist and hand movement in Poststroke hemiplegia: A systematic review and meta-analysis. *Neurorehabilit. Neural Repair* **2019**, *33*, 96–111. [CrossRef]
10. Ambrosini, E.; Russold, M.; Gfoehler, M.; Puchinger, M.; Weber, M.; Becker, S.; Krakow, K.; Immick, N.; Augsten, A.; Rossini, M.; et al. A Hybrid Robotic System for Arm Training of Stroke Survivors: Concept and First Evaluation. *IEEE Trans. Biomed. Eng.* **2019**, *66*, 3290–3300. [CrossRef]
11. Jochumsen, M.; Niazi, I.K.; Mrachacz-Kersting, N.; Jiang, N.; Farina, D.; Dremstrup, K. Comparison of spatial filters and features for the detection and classification of movement-related cortical potentials in healthy individuals and stroke patients. *J. Neural Eng.* **2015**, *12*, 56003. [CrossRef] [PubMed]
12. Jochumsen, M.; Niazi, I.K.; Taylor, D.; Farina, D.; Dremstrup, K. Detecting and classifying movement-related cortical potentials associated with hand movements in healthy subjects and stroke patients from single-electrode, single-trial EEG. *J. Neural Eng.* **2015**, *12*, 56013. [CrossRef]
13. Jochumsen, M.; Niazi, I.K.; Dremstrup, K.; Kamavuako, E.N. Detecting and classifying three different hand movement types through electroencephalography recordings for neurorehabilitation. *Med. Biol. Eng. Comput.* **2015**, *54*, 1491–1501. [CrossRef] [PubMed]

14. Kawakami, M.; Fujiwara, T.; Ushiba, J.; Nishimoto, A.; Abe, K.; Honaga, K.; Nishimura, A.; Mizuno, K.; Kodama, M.; Masakado, Y.; et al. A new therapeutic application of brain-machine interface (BMI) training followed by hybrid assistive neuromuscular dynamic stimulation (HANDS) therapy for patients with severe hemiparetic stroke: A proof of concept study. *Restor. Neurol. Neurosci.* **2016**, *34*, 789–797. [CrossRef]
15. Zhang, X.; Zhou, N.P. High-density myoelectric pattern recognition toward improved stroke rehabilitation. *IEEE Trans. Biomed. Eng.* **2012**, *59*, 1649–1657. [CrossRef]
16. Balasubramanian, S.; Garcia-Cossio, E.; Birbaumer, N.; Burdet, E.; Ramos-Murguialday, A. Is EMG a Viable Alternative to BCI for Detecting Movement Intention in Severe Stroke? *IEEE Trans. Biomed. Eng.* **2018**, *65*, 2790–2797. [CrossRef]
17. Ramos-Murguialday, A.; García-Cossio, E.; Walter, A.; Cho, W.; Broetz, D.; Bogdan, M.; Cohen, L.G.; Birbaumer, N. Decoding upper limb residual muscle activity in severe chronic stroke. *Ann. Clin. Transl. Neurol.* **2015**, *2*, 1–11. [CrossRef] [PubMed]
18. Jochumsen, M.; Navid, M.S.; Rashid, U.; Haavik, H.; Niazi, I.K. EMG-versus EEG-Triggered Electrical Stimulation for Inducing Corticospinal Plasticity. *IEEE Trans. Neural Syst. Rehabil. Eng.* **2019**, *27*, 1901–1908. [CrossRef]
19. Camona, C.; Wilkins, K.B.; Drogos, J.; Sullivan, J.E.; Dewald, J.P.A.; Yao, J. Improving hand function of severely impaired chronic hemiparetic stroke individuals using task-specific training with the ReIn-Hand system: A case series. *Front. Neurol.* **2018**, *9*, 923. [CrossRef]
20. Huang, Y.; Lai, W.P.; Qian, Q.; Hu, X.; Tam, E.W.C.; Zheng, Y.-P. Translation of robot-assisted rehabilitation to clinical service: A comparison of the rehabilitation effectiveness of EMG-driven robot hand assisted upper limb training in practical clinical service and in clinical trial with laboratory configuration for chronic stroke. *Biomed. Eng. Online* **2018**, *17*, 91.
21. Lee, S.W.; Wilson, K.M.; Lock, B.A.; Kamper, D.G. Subject-specific myoelectric pattern classification of functional hand movements for stroke survivors. *IEEE Trans. Neural Syst. Rehabil. Eng.* **2010**, *19*, 558–566. [CrossRef] [PubMed]
22. Lu, Z.; Tong, R.K.; Zhang, X.; Li, S.; Zhou, P. Myoelectric pattern recognition for controlling a robotic hand: A feasibility study in stroke. *IEEE Trans. Biomed. Eng.* **2018**, *66*, 365–372. [CrossRef] [PubMed]
23. Gandolla, M.; Ferrante, S.; Ferrigno, G.; Baldassini, D.; Molteni, F.; Guanziroli, E.; Cottini, M.C.; Seneci, C.; Pedrocchi, A. Artificial neural network EMG classifier for functional hand grasp movements prediction. *J. Int. Med. Res.* **2017**, *45*, 1831–1847. [CrossRef] [PubMed]
24. Kopke, J.V.; Hargrove, L.J.; Ellis, M.D. Applying LDA-based pattern recognition to predict isometric shoulder and elbow torque generation in individuals with chronic stroke with moderate to severe motor impairment. *J. Neuroeng. Rehabil.* **2019**, *16*, 35. [CrossRef] [PubMed]
25. Subramaniam, S.; Varghese, R.; Bhatt, T. Influence of chronic stroke on functional arm reaching: Quantifying deficits in the ipsilesional upper extremity. *Rehabil. Res. Pr.* **2019**, *2019*, 1–10.
26. Zhang, X.; Zhou, P. Myoelectric pattern identification of stroke survivors using multivariate empirical mode decomposition. *J. Healthc. Eng.* **2014**, *5*, 261–274. [CrossRef] [PubMed]
27. Wang, D.; Zhang, X.; Gao, X.; Chen, X.; Zhou, P. Wavelet packet feature assessment for high-density myoelectric pattern recognition and channel selection toward stroke rehabilitation. *Front. Neurol.* **2016**, *7*, 197. [CrossRef]
28. Rehman, M.Z.U.; Waris, A.; Gilani, S.O.; Jochumsen, M.; Niazi, I.K.; Jamil, M.; Farina, D.; Kamavuako, E.N. Multiday EMG-based classification of hand motions with deep learning techniques. *Sensors* **2018**, *18*, 2497. [CrossRef]
29. Gladstone, D.J.; Danells, C.J.; Black, S.E. The fugl-meyer assessment of motor recovery after stroke: A critical review of its measurement properties. *Neurorehabil. Neural Repair* **2002**, *16*, 232–240. [CrossRef]
30. Hudgins, B.; Parker, P.; Scott, R.N. A new strategy for multifunction myoelectric control. *IEEE Trans. Biomed. Eng.* **1993**, *40*, 82–94. [CrossRef]
31. Jochumsen, M.; Waris, A.; Kamavuako, E.N. The effect of arm position on classification of hand gestures with intramuscular EMG. *Biomed. Signal Process. Control.* **2018**, *43*, 1–8. [CrossRef]
32. Duda, R.O.; Hart, P.E.; Stork, D.G. *Pattern Classification*; John Wiley & Sons: Hoboken, NJ, USA, 2012.
33. Rehman, M.Z.U.; Gilani, S.O.; Waris, A.; Niazi, I.K.; Slabaugh, G.; Farina, D.; Kamavuako, E.N. Stacked sparse autoencoders for EMG-based classification of hand motions: A comparative multi day analyses between surface and intramuscular EMG. *Appl. Sci.* **2018**, *8*, 1126. [CrossRef]

34. Koo, T.K.; Li, M.Y. A guideline of selecting and reporting intraclass correlation coefficients for reliability research. *J. Chiropr. Med.* **2016**, *15*, 155–163. [CrossRef] [PubMed]
35. Zhou, P.; Barkhaus, P.E.; Zhang, X.; Rymer, W.Z. Characterizing the complexity of spontaneous motor unit patterns of amyotrophic lateral sclerosis using approximate entropy. *J. Neural Eng.* **2011**, *8*, 66010. [CrossRef] [PubMed]
36. Hussain, T.; Haider, A.; Taleb-Ahmed, A. A heterogeneous multi-core based biomedical application processing system and programming toolkit. *J. Signal Process. Syst.* **2019**, *91*, 963–978. [CrossRef]
37. Mugler, E.M.; Tomic, G.; Singh, A.; Hameed, S.; Lindberg, E.W.; Gaide, J.; Alqadi, M.; Robinson, E.; Dalzotto, K.; Limoli, C.; et al. Myoelectric Computer Interface Training for Reducing Co-Activation and Enhancing Arm Movement in Chronic Stroke Survivors: A Randomized Trial. *Neurorehabilit. Neural Repair* **2019**, *33*, 284–295. [CrossRef]
38. Pizzolato, S.; Tagliapietra, L.; Cognolato, M.; Reggiani, M.; Müller, H.; Atzori, M. Comparison of six electromyography acquisition setups on hand movement classification tasks. *PLoS ONE* **2017**, *12*, e0186132. [CrossRef]

Article

Eye and Voice-Controlled Human Machine Interface System for Wheelchairs Using Image Gradient Approach

Saba Anwer [1], Asim Waris [1,*], Hajrah Sultan [1], Shahid Ikramullah Butt [1], Muhammad Hamza Zafar [2], Moaz Sarwar [3], Imran Khan Niazi [4,5,6], Muhammad Shafique [7] and Amit N. Pujari [8,9]

1 School of Mechanical and Manufacturing Engineering, National University of Sciences and Technology, Islamabad 45200, Pakistan; sanwer.bmes19smme@student.nust.edu.pk (S.A.); hsultan.bmes19smme@student.nust.edu.pk (H.S.); drshahid@smme.nust.edu.pk (S.I.B.)
2 Department of Electrical Engineering, University of Engineering and Technology Lahore-FSD Campus, Faisalabad 38000, Pakistan; Hamzazafar214@gmail.com
3 Department of Computer Sciences, Government College University, Faisalabad 38000, Pakistan; moazejaaz@gmail.com
4 Center of Chiropractic Research, New Zealand College of Chiropractic, Auckland 0600, New Zealand; imran.niazi@nzchiro.co.nz
5 Department of Health Science and Technology, Center for Sensory-Motor Interaction, Aalborg University, 9000 Alborg, Denmark
6 Faculty of Health and Environmental Sciences, Health and Rehabilitation Research Institute, AUT University, Auckland 0627, New Zealand
7 Head of Department, Biomedical Engineering, Riphah International University, Islamabad 45710, Pakistan; muhammad.shafique@riphah.edu.pk
8 School of Engineering and Technology, University of Hertfordshire, Hatfield AL10 9AB, UK; amit.pujari@ieee.org
9 School of Engineering, University of Aberdeen, Aberdeen AB24 3FX, UK
* Correspondence: asim.waris@smme.nust.edu.pk

Received: 26 August 2020; Accepted: 17 September 2020; Published: 26 September 2020

Abstract: Rehabilitative mobility aids are being used extensively for physically impaired people. Efforts are being made to develop human machine interfaces (HMIs), manipulating the biosignals to better control the electromechanical mobility aids, especially the wheelchairs. Creating precise control commands such as move forward, left, right, backward and stop, via biosignals, in an appropriate HMI is the actual challenge, as the people with a high level of disability (quadriplegia and paralysis, etc.) are unable to drive conventional wheelchairs. Therefore, a novel system driven by optical signals addressing the needs of such a physically impaired population is introduced in this paper. The present system is divided into two parts: the first part comprises of detection of eyeball movements together with the processing of the optical signal, and the second part encompasses the mechanical assembly module, i.e., control of the wheelchair through motor driving circuitry. A web camera is used to capture real-time images. The processor used is Raspberry-Pi with Linux operating system. In order to make the system more congenial and reliable, the voice-controlled mode is incorporated in the wheelchair. To appraise the system's performance, a basic wheelchair skill test (WST) is carried out. Basic skills like movement on plain and rough surfaces in forward, reverse direction and turning capability were analyzed for easier comparison with other existing wheelchair setups on the bases of controlling mechanisms, compatibility, design models, and usability in diverse conditions. System successfully operates with average response time of 3 s for eye and 3.4 s for voice control mode.

Keywords: human machine interface (HMI); rehabilitation; wheelchair; quadriplegia; Raspberry Pi; image gradient; AMR voice; Open-CV; image processing

1. Introduction

Significant strides made in the fields of rehabilitation, artificial intelligence (AI) (especially around the implementation of complex algorithms for analysis and interpretation of human cognition), and human machine interfaces (HMIs), have opened a new evolutionary pathway for the development of smart mobility aids [1]. People who accidentally lose their lower limbs or suffer from conditions such as quadriplegia or stroke, resulting in paralysis, and muscle stiffness are unable to make use of conventional wheelchairs [2]. Researchers across the world are engaged in developing medical devices/rehabilitation aids for physically challenged populations, such as quadriplegics, to enable them to carry out their daily work without or with minimal assistance from caregivers and nurses, etc. [3]. They are thus increasing the self-esteem and functional capabilities of such patients with the ultimate goal of improving the patients' quality of life. Biosignals recorded through electroencephalography (EEG), electromyography (EMG), and electrooculography (EOG), etc. [4,5], have been exploited by researchers in developing smart, responsive and real-time rehabilitative control systems.

More recently, eye gesture control-based systems have gained significant attention due to the fact that even in the most seriously physically challenged population, such as quadriplegics, eye movements are still intact; the main operating mechanism of the eye-controlled based systems [6]. Therefore, keeping in view this fact and the need to advance the adoption of independent mobile rehabilitation technology, such as smart wheelchairs and walkers, a distinctive eyeball movement-based technique for controlling a wheelchair is presented, leading to increased patient comfort.

The main purpose is to design an autonomous system that requires minimal manual assistance and thereby provides wheelchair users with a sense of confidence, competence, and independence. As such, the presented system should be easy to use for a paralyzed individual with a severe lower limb disability. Additionally, the system also incorporated voice-controlled technology [7]. The system is low-cost, easily manageable, scalable, and designed with the user's comfort in mind.

Algorithms currently used for face detection and feature extraction in eye controlled systems include Hough circle (feature extraction based approach used to detect circular objects in image processing) [8] and active infrared illumination (uses IR sensors to detect the eye movement and emit IR radiation ranging from 700–1000 nm in the electromagnetic spectrum) [9]. Although these methods accurately localize eyeball position and are relatively simple compared to other existing techniques, e.g., Haar cascade [10], they do have some drawbacks. For example, the Hough circle technique (CHT), when applied to discrete images, demands a large storage capacity, as well as computing power [11] and an active infrared illumination technique, which can cause irreversible damage to eyes, resulting in the loss of efficiency of a working body organ, thus worsening the situation [12]. Moreover, these algorithms sometimes fail during complex situations, for example, the low resolution of images and low contrast conditions.

Although the use of biosignals, i.e., EEG (electroencephalogram) and EMG (electromyography) is widely accepted to develop HMIs, for example, EMG based physiotherapy devices [13] and EEG based diagnostic medical equipment [14], however, these systems come with wearable technology which is not practical or comfortable in the case of mobility assistive mechanization. In EEG and EMG based systems, a user also must be in contact with electrodes while using the device/wheelchair, thus making the system cumbersome and uncomfortable. Further, electrodes (for EEG, EMG) are susceptible to a range of issues. For example, signals may be contaminated by a variety of noises at the electrode–skin interface, which can lead to contamination of the acquired biosignals [15,16]. The electrode issues include (1) motion artifacts, which occur when a force impulse travels through muscle causing an unwanted movement at the skin–electrode interface; (2) inherent noise in electrical components, which cannot be completely removed; (3) ambient noise occurring due to electromagnetic radiation as the human body is persistently exposed to this radiation; and (4) power line interference and other disruptions like baseline shifts (due to the excessive motion of cable, baseline shows a significant shift from the actual position). All these noises attenuate the desired signal causing undesired results [17].

A variety of control techniques have been used for creating assistive HMIs. However, all these systems had their own limitations in terms of operating efficiency. For example, head motion controlled and hand gesture operated wheelchairs [18,19], both of these controlling modes (head and hand) include the use of flex sensor and accelerometer. Sometimes a sensor's efficiency is greatly affected by environmental factors (temperature, dust and humidity). Once contaminated, this can cause controllability issues. Moreover, an accelerometer has a fixed operating range limiting its application, thus obstructing the way actual acceleration is read.

Although distinct eye motion-controlled wheelchairs were developed claiming to assist disabled individuals, they are limited in their functioning capability and comfort level; as these systems eyeball movements are processed using software, such as MATLAB, and computing devices (laptop, etc.) are required to be carried all the time, which occupy substantial space, making the system cumbersome and expensive [20,21].

Considering the limitations of existing systems discussed above, a system that tries to overcome these limitations and ensures patient safety and comfort, as well as be scalable and highly functional, is presented here. The operating system is installed in Raspberry Pi, and the language used for processing the eyeball movements in real-time captured images is C++, using the Open computer vision (Open CV) library [22]. As the introduced system is compact in design, it is relatively easy to install in a wheelchair. Along with the eyeball control option, this system can also be controlled via voice commands, increasing the system's adaptability and usability.

2. System Model

Components of the system model and their interaction with each other are shown as a block diagram (Figure 1). The webcam is fixed to a vertical pole precisely in front of the user's eye, and this webcam is connected to Raspberry Pi so that it can continuously capture images of the user's eyeball movement and respond. Raspberry Pi installed with Open CV has an image processing capability and generates an actuation signal. Raspberry Pi is coupled with the motor driving circuitry, which is responsible for directing the wheelchair according to a given command.

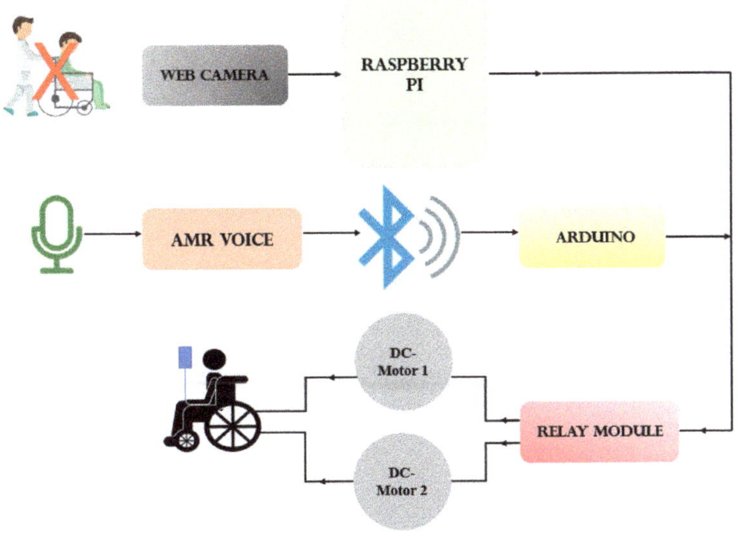

Figure 1. Block diagram of the system.

For the voice control mode, an audio signal is fed to the Arduino via a Bluetooth module (HC-05 BT). Arduino is programmed to process these voice commands and generate the required drive signal. Arduino is wired with a switching circuit (four-channel relay module) responsible for driving the motors in the designated direction.

Figure 1 depicts the complete mechanical control mechanism of the system, including all the major components, i.e., Raspberry Pi, webcam, power supply, microphone, Arduino, DC-motors, motor driving circuitry, and Bluetooth device. The system works using real-time data acquisition, with Raspberry Pi as the main controller used for eyeball tracking. Raspberry Pi is a low-cost single-board embedded processor, which thus reduces the complexity of the system and is suitable for real-time applications. In the present system, a distance of approximately 1–1.5 feet is maintained between the user's eye and the webcam. To keep the costs down, and for recognition accuracy and processing speed, a 1080P webcam is used. Figure 2 describes the process flow of the eye control mechanism. First, the webcam captures real-time images of the eyeball and then identifies whether the eye is open or closed. If closed, then images are recaptured and analyzed again to identify the direction of the eyeball. Once the eyeball direction is confirmed, the signal is processed, and an actuator signal is generated, which is then fed to the motor driving circuitry of the wheelchair.

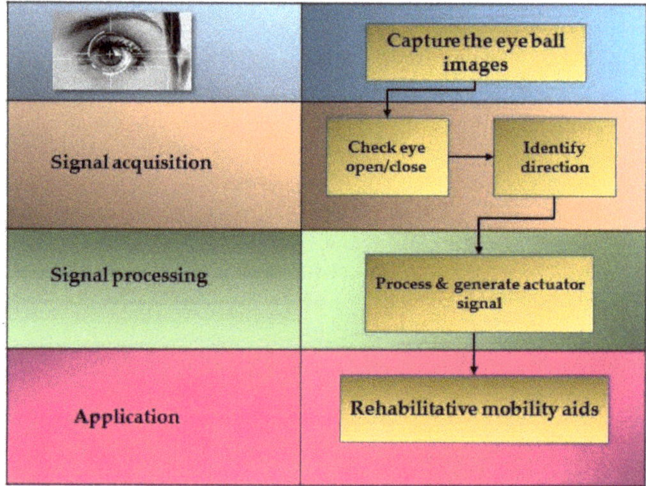

Figure 2. General process flow block diagram.

3. Methodology

Figure 3 illustrates the complete functional flow chart of the eye control system. The system begins with capturing images through the webcam. After capturing a real-time image, the system detects the face and then extracts eye images.

A complete flow chart is defined using state, condition, and decision boxes. State boxes are denoted by rectangular shapes with round corners, decision boxes are diamond-shaped, and condition boxes are rectangular with sharp corners. State boxes represent the status of the system (i.e., moving or not), decision boxes describe direction (i.e., left, right or forward), and condition boxes give information about the system's working condition (i.e., face detection or driving the wheelchair).

Paths indicate the process flow. For example, after initialization, the system detects the face, and then it checks whether the eye is open or closed, as shown in Figure 3. After this, the eye pupil's position is identified, i.e., whether the user is looking forward, left, or right. After the eyeball position is identified, this image is processed. Raspberry Pi then generates an actuator signal, which is fed to a switching circuit (relay) to drive the motors accordingly.

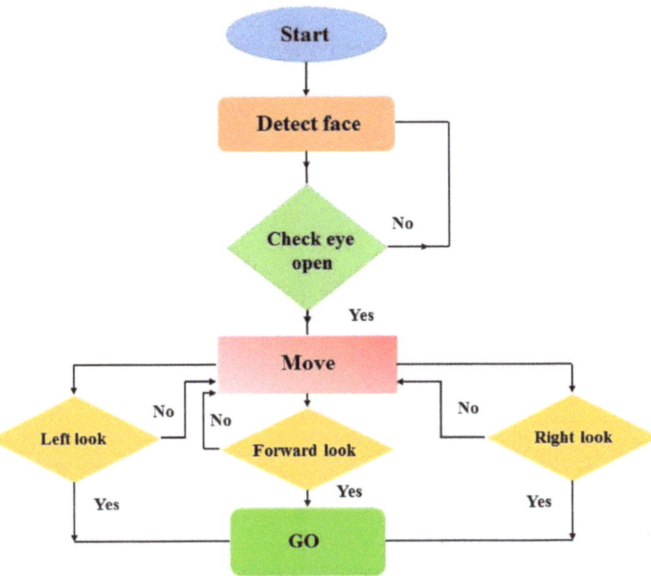

Figure 3. Functional flow chart of eye control system.

4. Algorithm

The most challenging task is to locate the eyeball movement. This task is accomplished using the image gradient approach described below. As mentioned earlier, various other techniques have been used to locate eyeballs. Although they provide accurate results and are used in commercially available eye-tracking and face recognition systems, they are not easy or practical to use as they come with head-mounted or wearable technology.

In the present system, a feature-based approach is applied that can accurately locate the eye centers using a webcam, even in low-resolution videos and images [23]. A simple and easy approach is applied, which defines the center of circular objects as the location where the intersection of multiple image gradients occurs.

Eye Center Localization by Gradient Vectors

By considering a vector field comprising of image gradients, geometrically, the eyeball center can be located. A fast iterative scheme is achieved by using a mathematical formula [23]. The formula describes a relation between the conceivable center and all the image gradients directed towards it.

Suppose n is a possible center, and G_k is the gradient vector. If the position of this gradient vector G_k is X_k then the direction of displacement vector D_k should be the same as gradient vector G_k (Figure 4).

If the vector field of image gradients is used, then this vector field can be exploited by calculating the dot products between displacement D_k and gradient vectors G_k, by using the (1). Center n of a circular object in an image with pixel positions $X_{k,}$ (where $k \in \{1, \ldots \ldots, \mathrm{N}\}$) is measured by (1).

$$n = \frac{1}{N}\sum_{i=1}^{N}(D_k{}^{\wedge}T{\cdot}G_k)^2 \tag{1}$$

$$D_k = \frac{x_k - n}{\|x_k - n\|^{\wedge}2} \tag{2}$$

$$\forall k\text{: } \| Gk \|2 = 1 \quad (3)$$

To get an equal weight for all the pixel positions, displacement vectors (D_k) are scaled to unit length (2). Robustness to the linear variations in luminous conditions can also be improved by scaling gradient vectors G_k to unit length (3).

Calculations can be simplified by considering only the gradient vectors (G_K). Partial derivatives are computed to get the image gradients (4).

$$G_k = \frac{\delta_P(x_k, y_k)}{\delta X_k}, \frac{\delta_P(x_k, y_k)}{\delta Y_k} \quad (4)$$

When gradients are computed, it is possible that images have extra structures, i.e., hairs, spectacles, and eyebrows. These structures are responsible for gradients that do not possess the same direction as image gradients of the eye. Due to these structures, eye center computation may become difficult. To overcome this difficulty, the threshold is applied to the objective function. This threshold is based on the maximum value, which eliminates all other remaining entities associated with the desired image boundary. After this, a maximum of prevailing entities is computed, and its position is taken as the eye center (Figure 5). This threshold does not have any negative impact on estimating the eye center. In the present system, the threshold is taken as 85% (0.85) of the overall maximum.

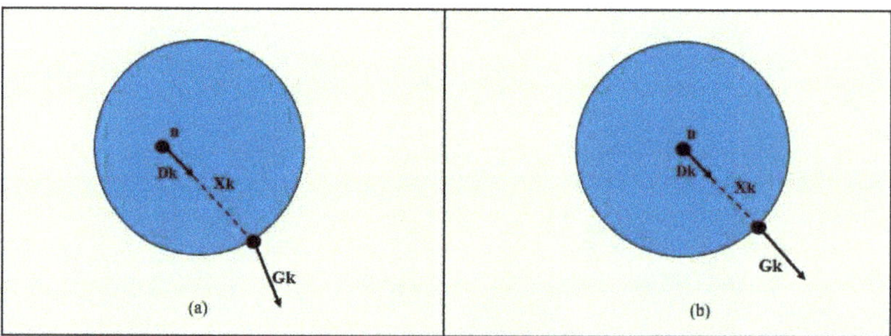

Figure 4. A contrived example is having a dark-colored circle against a light background, similar to an iris and sclera. In (**a**) D_k (displacement vector) and G_k (gradient vector) do not have same direction but in (**b**) the orientations are same.

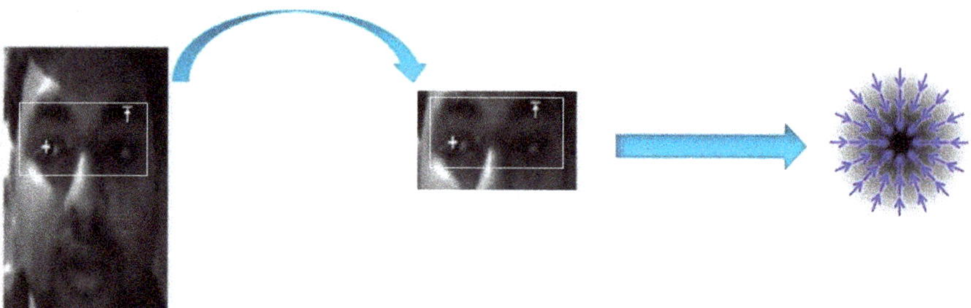

Figure 5. Eye core located (**left**) indicated with white mark in the presented system, using image gradients symbolically directed towards the center (**Right**).

5. Eye Control Mode

During eye control mode, webcam captures the real-time images and sends them to Raspberry Pi, Raspberry Pi processes them and generates an actuator signal. Raspberry Pi is further wired with a

4-channel relay module (switching circuit) responsible for driving the two DC-motors, thus making an efficient, reliable, and easily functional system.

In the present system, Raspberry Pi model 3B is used as the main controller. Its working resembles a CPU. It has its own RAM, ROM, internet port, 36 connection pins, four supply pins, four USB (universal serial bus) ports and one memory card holder up to 32 GB [24]. Raspberry Pi is Raspbian supportive hardware. Raspbian runs on the Linux operating system. Processing to track eye is done in Open CV 3.0 [17] (open computer vision) library. This library is commonly used for image processing. Open CV is installed under a BSD license, which is free for both commercial and academic purposes. Open CV has C++, C, and Java interfaces and is the most suitable platform for real-time applications.

Integrating Open CV, Linux operating system, and C++ language with hardware has improved the present system's constancy compared to other existing systems (mentioned earlier) and has reduced processing latency. These features have also improved the system's compatibility and level of convenience for the user. Figure 6 shows the results of detecting the eye center for right, left and forward commands, respectively. Through turning the wheelchair and then looking forward will move it in reverse direction safely as this system has been designed for an extreme disability (quadriplegic patient).

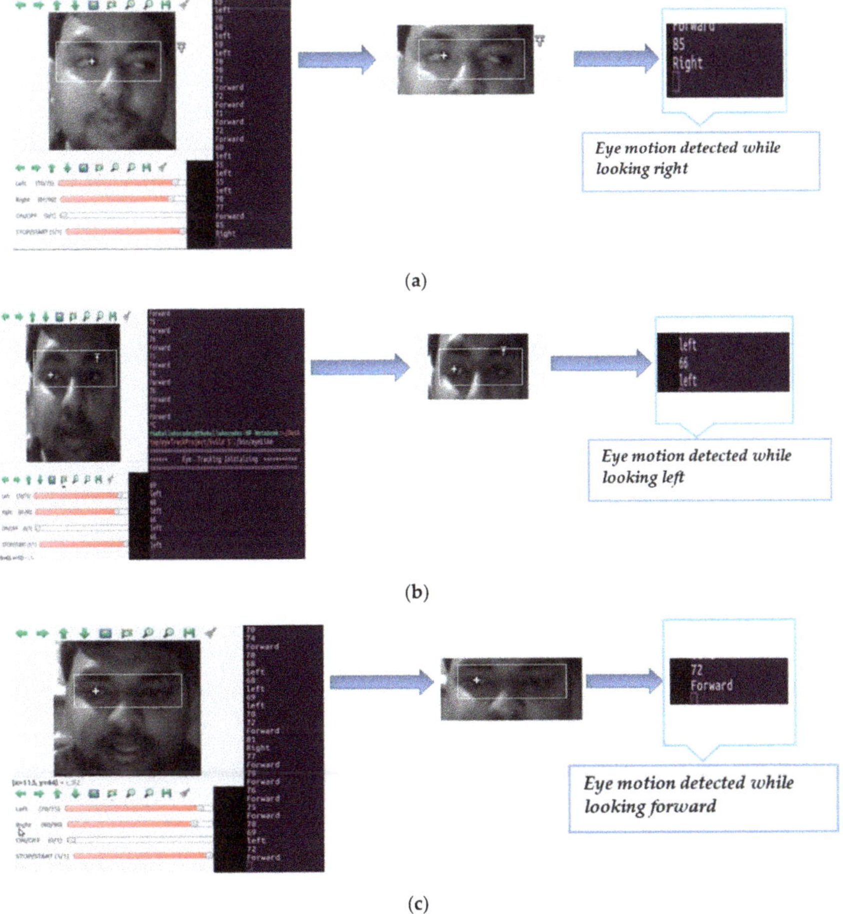

Figure 6. Designed system's results for eye control mode (**a**) right, (**b**) left, and (**c**) forward.

6. Vocal Control Mode

For voice control mode on the wheelchair, the adaptive multi-rate AMR voice app is used with four commands ON (for forwarding), left, right and stop. Arduino UNO is used as the main controller. Other main components of voice control mode are microphone, HC-05 Bluetooth module, and AMR voice recognition. The block diagram in Figure 7 depicts the functionality flow between all components of voice control mode.

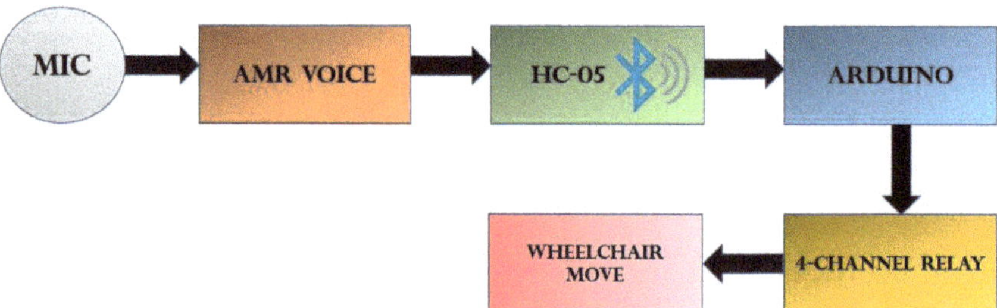

Figure 7. Block diagram for voice control mode.

For voice control, the microphone is connected with Arduino via the HC-05 Bluetooth module. Arduino transfers actuator signal to motor driving circuitry, which is a four-channel relay module. Two DC-motors are powered by two chargeable 12-V dry batteries (Figure 7).

7. Mechanical Assembly

The most important aspect during the mechanical assembly of the wheelchair is the proper selection of frames. This is necessary for the successful installation of all essential components in the wheelchair. During this phase, the patient's comfort and compatibility are the top priorities. Moreover, a wheelchair's material must be resistant to corrosion for its long-lasting utilization, and its posture should be such that the patient's weight is evenly distributed to avoid pain and pressure sores [25].

In the present wheelchair design (Figures 8 and 9), all the required components (batteries, motors, controllers and relays) are placed appropriately on a designated platform (length = 12 inches, width = 9 inches, thickness = 5 inches) welded to the wheelchair underneath the seat. A low weight frame is used so that a wheelchair can easily be propelled. Total weight of wheelchair is 17.35 kg (frame = 14.5 kg, other components = 2.85 kg). Further, the camera is located in such a position so that the user can easily gaze into the camera while remaining in the comfort zone, thus avoiding any tiredness.

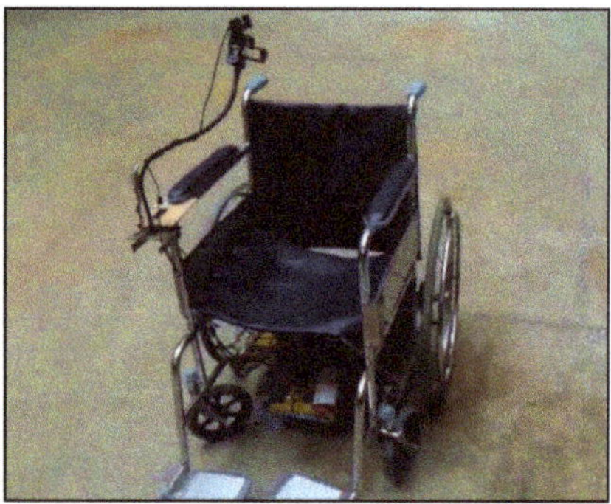

Figure 8. Wheelchair model with all essential components installed appropriately.

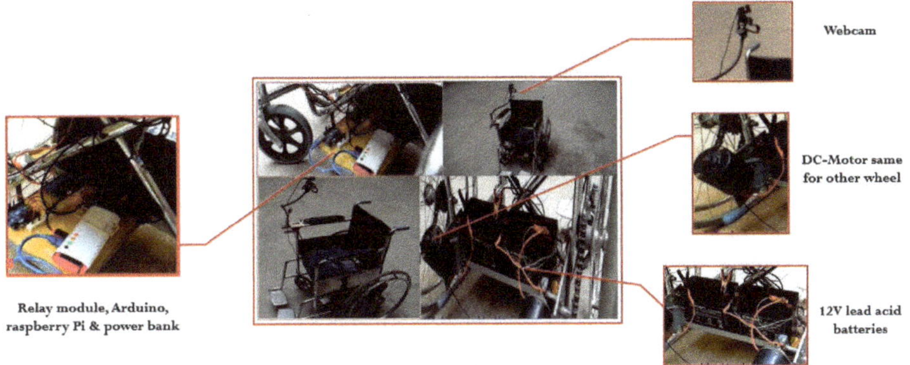

Figure 9. Complete mechanical assembly of wheelchair.

Two 12-V permanent magnet DC-motors are used in this system with 95 W output power. Further, the system is incorporated with two 12-V, rechargeable, lead-acid batteries for its appropriate functioning. The gear ratio used was 1:20.

8. Basic Skills Performance Test for Wheelchair

Wheelchair technology is diverse in nature. Therefore, a basic wheelchair skill test (WST) was carried out to analyze the system's efficiency and response time. This testing approach has advantages of being easy to manage, requires minimum testing equipment, is inexpensive, and has adequate measurement properties to quantify the performance of wheelchair movements. Results of WST can provide crucial data about the test subject's performance. For example, whether the subjects were able to accomplish the assigned movement task successfully up to the marked distance and the corresponding response time of the system. Thus, the results of WST are representative of the range of movement of the wheelchair that may be required to be performed regularly by the disabled.

The most suitable term used for an individual selected as the object of testing is 'subject' because he/she may be a researcher, caretaker, user, or health care student. However, it is necessary that the test subject meets the same criteria as specified for the wheelchair user. For example, he/she must remain within the designated space and operate a wheelchair as will be operated by a disabled individual.

Therefore, in order to assess the mobility and working proficiency of the wheelchair, a simple test was conducted. The main objective was to ensure and record a chair's maneuverability, performance, and user compatibility. Basic skills for which the presented setup of the wheelchair was tested included (Table 1).

Table 1. Basic skills, the designed wheel chair is capable of.

Basic Attributes
Relief from pressure
Forward move
Backward move
Rollover soft surface
Stop
Turn while moving
Turn in place

9. Results

The data obtained from a basic WST-wheelchair skill test are represented graphically in (Figure 10). The presented system was tested with 15 test subjects, in an age range of 20–30 years. All the testing subjects were healthy and were asked to drive the wheelchair via eye and voice control modes separately through a fixed distance of 25 m. All participants operated the wheelchair up to the targeted distance separately for each movement (forward, backward, rollover soft surface, turn left and right). The main goal was to note the system's response time for both optical and vocal commands, separately.

Moreover, the system's other attributes, including mobility, rear and caster wheel motion, and compatibility with the user, were also analyzed. Wheelchair processing times were noted with a stopwatch and later compared graphically. The mean response times for each skill were computed.

All the test subjects successfully covered the 25-m distance without any intervention of spotter/researcher. The system was positively responsive, with 99% of participants without any false-positive results.

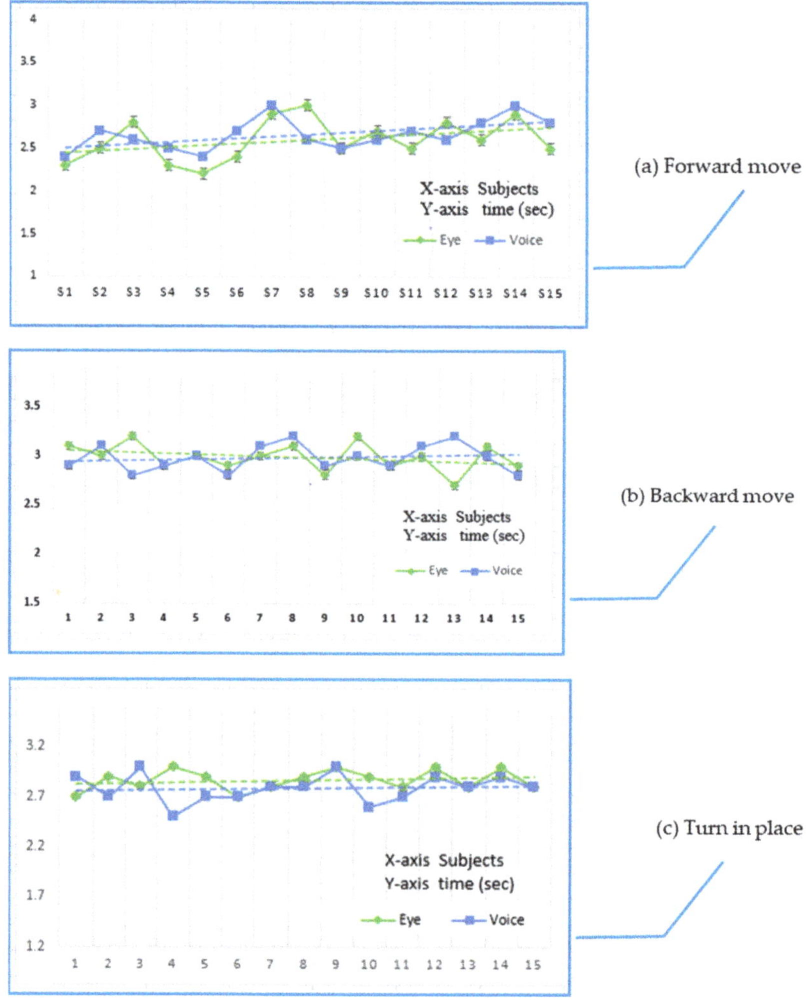

Figure 10. *Cont.*

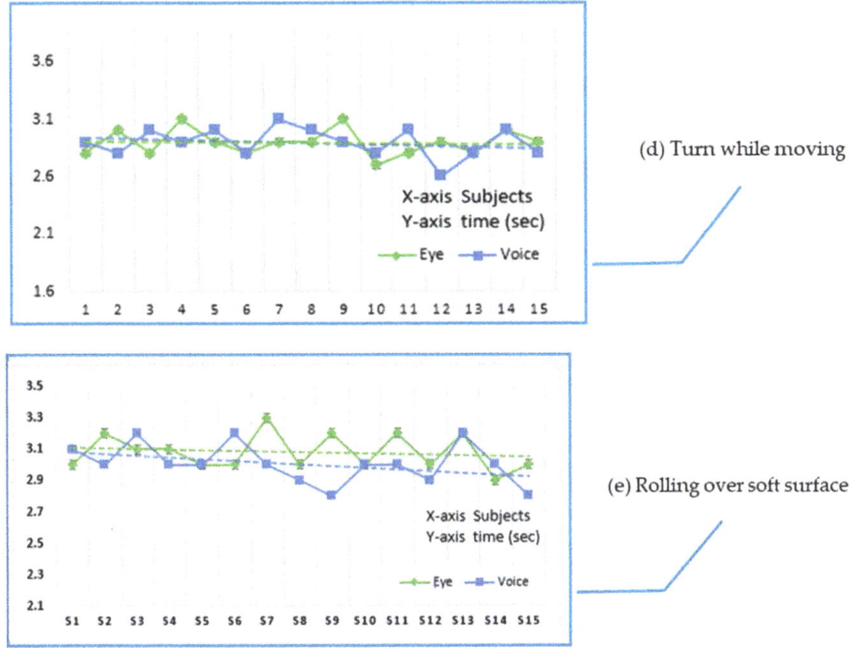

Figure 10. Graphical representation of a basic wheelchair skill test (WST). (**a**) Forward move on a plain surface. (**b**) Backward move over the plain surface. (**c**) Turn in place. (**d**) Turn while moving (**e**) Rolling over a soft surface.

Table 1 shows the system's mean response time for each basic skill from 15 participants. It is important to note that the participants had an age range of 20–30 years, disparate eye colors, weights, and vocal pitches. The system was tested separately for both (eye and voice) operating modes, and there was no significant difference between their operating proficiencies—the mean response time was calculated for each movement for each mode (Table 2).

Table 2. System response for WST.

Attribute	Mean Response Time
Forward move on a plain surface	Eye = 2.53 s, voice =2.8 s
Reverse move over a plain surface	Eye = 3.4 s, voice = 3.0 s
Turn in place	Eye = 2.86 s, voice = 2.9 s
Turn while moving	Eye = 2.78 s, voice = 2.89 s
Rolling over a soft surface (grass)	Eye = 3.46 s, voice = 3.5 s

10. Discussions

The aim was to develop a wireless system to assist patients with significant disabilities (stroke & quadriplegia). Thus, an eye and voice-controlled wheelchair system, which overcomes almost all the issues encountered in previous HMIs, is presented here. The present system removes the need to carry a personal computer or any wearable electrode band like EMG, hand and head gesture-controlled systems [4–6,17]. Furthermore, there are no harmful side effects on human eyes, as experience due to infrared accumulation system, which causes gradual but irreparable damage to the eyes [13]. Thus, the presented system that is safer, quickly responsive, compact, convenient, and user-friendly as compared to other existing wheelchair setups.

Moreover, a small basic WST-wheelchair skill test was also carried out to demonstrate the presented system's working capability and quantify its response times. The presented system successfully responds between 2.5–3.0 s on the plain surface; however, while moving on soft surface response time was slightly higher as it is relatively difficult to propel the wheelchair on such surfaces due to their high resistance and bumpy appearance (grass, dirt, carpet and dirt). Weight-relief is an important property that is necessary to be considered while designing any rehabilitation device. If such factors are ignored, it may take any body part under pressure causing pressure sores. In the present system, this aspect was carefully considered while integrating the components. The camera was adjusted such that the user does not need to put any extra effort into looking into the camera. Users can easily drive the wheelchair via eye or voice command while remaining in a comfortable position, thus avoiding potential tiredness. Overall, the designed system is proficient, feasible, comfortable, and safe to use. However, the system is not without its limitations. Although the image processing technique used has a relative superiority in processing, these techniques sometimes malfunction in the dark due to variation in illumination. In the existing setup, a 12-V LED is incorporated to compensate for this problem to some extent; however, in the near future, FPGA-field programmable gate array systems may be used to improve the processing speed and make the system more synchronized with environmental variations and user needs.

11. Conclusions

An eye and voice-controlled interface for a wheelchair to assist the mobility of physically impaired people has been designed so that they may be able to perform their daily life activities without additional support from a caregiver or healthcare professional.

Author Contributions: Conceptualization, methodology and drafting, S.A., A.W., I.K.N., M.S. (Muhammad Shafiq), A.N.P.; Supervision, S.I.B.; Project administration, A.W.; Testing, M.H.Z., M.S. (Moaz Sarwar), H.S., Writing, Review and editing A.W., S.A., I.K.N., A.N.P. All authors have read and agreed to the published version of the manuscript.

Funding: This research received no external funding. Passion of contribution in the continuously evoluting society for serving the physically handicaped people inspired the authors to pick this project and work together as a team.

Conflicts of Interest: There has been no conflict of interest.

References

1. Champaty, B.; Jose, J.; Pal, K.; Thirugnanam, A. Development of EOG based human machine interface control system for motorized wheelchair. In Proceedings of the 2014 Annual International Conference on Emerging Research Areas: Magnetics, Machines and Drives (AICERA/iCMMD), Kottayam, India, 24–26 July 2014; pp. 1–7.
2. Simpson, R.C. Smart wheelchairs: A literaturereview. *J. Rehabil. Res. Dev.* **2005**, *42*, 423. [CrossRef] [PubMed]
3. Parikh, S.P.; Grassi, V., Jr.; Kumar, V.; Okamoto, J., Jr. Integrating human inputs with autonomousbehaviors on an intelligent wheelchair platform. *IEEE Intell. Syst.* **2007**, *22*, 33–41. [CrossRef]
4. Valbuena, D.; Cyriacks, M.; Friman, O.; Volosyak, I.; Graser, A. Brain-computer interface for high-level control of rehabilitation robotic systems. In Proceedings of the 2007 IEEE 10th International Conference on Rehabilitation Robotics, Noordwijk, The Netherlands, 13–15 June 2007; pp. 619–625.
5. Choi, K.; Sato, M.; Koike, Y. A new, human-centered wheelchair system controlled by the EMG signal. In Proceedings of the 2006 IEEE International Joint Conference on Neural Network Proceedings, Vancouver, BC, Canada, 16–21 July 2006; pp. 4664–4671.
6. Viswanathan, P.; Zambalde, E.P.; Foley, G.; Graham, J.L.; Wang, R.H.; Adhikari, B.; Mihailidis, A.; Miller, W.C.; Mitchell, I.M. Intelligent wheelchair control strategies for older adults with cognitive impairment: User attitudes, needs, and preferences. *Auton. Robots* **2017**, *41*, 539–554. [CrossRef]
7. Hou, T.K. Arduino based voice controlled wheelchair. *J. Phys. Conf. Ser.* **2020**, *1432*, 012064.
8. Khairosfaizal, W.W.M.; Nor'aini, A.J. Eyes detection in facial images using circular hough transform. In Proceedings of the 2009 5th International Colloquium on Signal Processing & Its Applications, Kuala Lumpur, Malaysia, 6–8 March 2009; pp. 238–242.

9. Zhao, S.; Grigat, R.R. Robust eye detection under active infrared illumination. In Proceedings of the 18th International Conference on Pattern Recognition (ICPR'06), Hong Kong, China, 20–24 August 2006; Volume 4, pp. 481–484.
10. Wilson, P.I.; Fernandez, J. Facial feature detection using Haar classifiers. *J. Comput. Sci. Coll.* **2006**, *21*, 127–133.
11. Rad, A.A.; Faez, K.; Qaragozlou, N. Fast Circle Detection Using Gradient Pair Vectors. In Proceedings of the Seventh International Conference on Digital Image Computing: Techniques and Applications, Sydney, Australia, 10–12 December 2003; pp. 879–887.
12. Chen, B.C.; Wu, P.C.; Chien, S.Y. Real-time eye localization, blink detection, and gaze estimation system without infrared illumination. In Proceedings of the 2015 IEEE International Conference on Image Processing (ICIP), Quebec City, QC, Canada, 27–30 September 2015; pp. 715–719.
13. Helin, P. Physiotherapy and electromyography in muscle cramp. *Br. J. Sports Med.* **1985**, *19*, 230–231. [CrossRef] [PubMed]
14. Bhat, S.; Acharya, U.R.; Dadmehr, N.; Adeli, H. Clinical neurophysiological and automated EEG-based diagnosis of the alzheimer's disease. *Eur. Neurol.* **2015**, *74*, 202–210. [CrossRef] [PubMed]
15. Mirza, I.A.; Tripathy, A.; Chopra, S.; D'Sa, M.; Rajagopalan, K.; D'Souza, A.; Sharma, N. Mind-controlled wheelchair using an EEG headset and arduino microcontroller. In Proceedings of the 2015 International Conference on Technologies for Sustainable Development (ICTSD), Mumbai, India, 4–6 February 2015; pp. 1–5.
16. Tameemsultana, S.; Saranya, N.K. Implementation of head and finger movement based automatic wheel chair. *Bonfring Int. J. Power Syst. Integr. Circuits* **2011**, *1*, 48–51.
17. De Luca, C.J.; Gilmore, L.D.; Kuznetsov, M.; Roy, S.H. Filtering the surface EMG signal: Movement artifact and baseline noise contamination. *J. Biomech.* **2010**, *43*, 1573–1579. [CrossRef] [PubMed]
18. Goyal, D.; Saini, S.P.S. Accelerometer based hand gesture controlled wheelchair. *Int. J. Emerg. Technol.* **2013**, *4*, 15–20.
19. Barea, R.; Boquete, L.; Mazo, M.; López, E. Wheelchair guidance strategies using EOG. *J. Intell. Robot. Syst.* **2002**, *34*, 279–299. [CrossRef]
20. Arora, P.; Sharma, A.; Soni, A.S.; Garg, A. Control of wheelchair dummy for differently abled patients via iris movement using image processing in MATLAB. In Proceedings of the 2015 Annual IEEE India Conference (INDICON), New Delhi, India, 17–20 December 2015; pp. 1–4.
21. Kaehler, A.; Bradski, G. *Learning OpenCV 3: Computer Vision in C++ with the OpenCV Library*; O'Reilly Media, Inc.: Sevastopol, CA, USA, 2016.
22. Juhong, A.; Treebupachatsakul, T.; Pintavirooj, C. Smart eye-tracking system. In Proceedings of the 2018 International Workshop on Advanced Image Technology (IWAIT), Chiang Mai, Thailand, 7–9 January 2018; pp. 1–4.
23. Timm, F.; Barth, E. Accurate eye centre localisation by means of gradients. *Visapp* **2011**, *11*, 125–130.
24. Org, R.P. Raspberry Pi 3 Model B. 2016. Available online: https://www.raspberrypi.org/products/raspberry-pi-3-model-b/ (accessed on 2 June 2017).
25. Sprigle, S.; Huang, M. Impact of mass and weight distribution on manual wheelchair propulsion torque. *Assist. Technol.* **2015**, *27*, 226–235. [CrossRef] [PubMed]

Article

Expert Hypertension Detection System Featuring Pulse Plethysmograph Signals and Hybrid Feature Selection and Reduction Scheme

Muhammad Umar Khan [1,†], Sumair Aziz [1,*,†], Tallha Akram [2], Fatima Amjad [1], Khushbakht Iqtidar [3], Yunyoung Nam [4,*] and Muhammad Attique Khan [5]

1 Department of Electronics Engineering, University of Engineering and Technology Taxila, Taxila 47050, Pakistan; sa.umarkhan@gmail.com (M.U.K.); fatimaamjad1997@gmail.com (F.A.)
2 Department of Electrical and Computer Engineering, COMSATS University Islamabad, Wah Campus, Wah Cantonment, Islamabad 45550, Pakistan; tallha@ciitwah.edu.pk
3 Department of Computer and Software Engineering, College of Electrical and Mechanical Engineering, National University of Sciences and Technology, Islamabad 44000, Pakistan; khushbakht.iqtidar18@ce.ceme.edu.pk
4 Department of Computer Science and Engineering, Soonchunhyang University, Asan 31538, Korea
5 Department of Computer Science, HITEC University, Taxila 47080, Pakistan; attique.khan@hitecuni.edu.pk
* Correspondence: sumair.aziz@uettaxila.edu.pk (S.A.); ynam@sch.ac.kr (Y.N.)
† These authors contributed equally to this work.

Abstract: Hypertension is an antecedent to cardiac disorders. According to the World Health Organization (WHO), the number of people affected with hypertension will reach around 1.56 billion by 2025. Early detection of hypertension is imperative to prevent the complications caused by cardiac abnormalities. Hypertension usually possesses no apparent detectable symptoms; hence, the control rate is significantly low. Computer-aided diagnosis based on machine learning and signal analysis has recently been applied to identify biomarkers for the accurate prediction of hypertension. This research proposes a new expert hypertension detection system (EHDS) from pulse plethysmograph (PuPG) signals for the categorization of normal and hypertension. The PuPG signal data set, including rich information of cardiac activity, was acquired from healthy and hypertensive subjects. The raw PuPG signals were preprocessed through empirical mode decomposition (EMD) by decomposing a signal into its constituent components. A combination of multi-domain features was extracted from the preprocessed PuPG signal. The features exhibiting high discriminative characteristics were selected and reduced through a proposed hybrid feature selection and reduction (HFSR) scheme. Selected features were subjected to various classification methods in a comparative fashion in which the best performance of 99.4% accuracy, 99.6% sensitivity, and 99.2% specificity was achieved through weighted *k*-nearest neighbor (KNN-W). The performance of the proposed EHDS was thoroughly assessed by tenfold cross-validation. The proposed EHDS achieved better detection performance in comparison to other electrocardiogram (ECG) and photoplethysmograph (PPG)-based methods.

Keywords: pulse plethysmograph; biomedical signal processing; feature extraction; machine learning; feature selection and reduction; empirical mode decomposition; discrete wavelet transform; hypertension

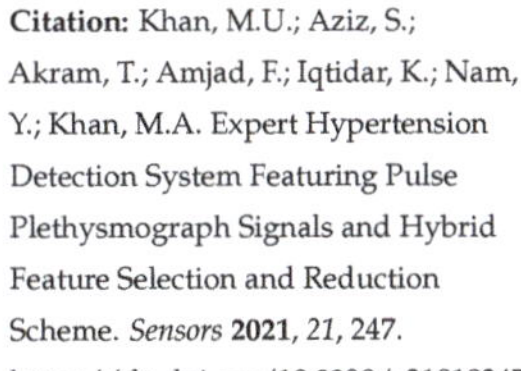

Citation: Khan, M.U.; Aziz, S.; Akram, T.; Amjad, F.; Iqtidar, K.; Nam, Y.; Khan, M.A. Expert Hypertension Detection System Featuring Pulse Plethysmograph Signals and Hybrid Feature Selection and Reduction Scheme. *Sensors* **2021**, *21*, 247. https://dx.doi.org/10.3390/s21010247

Received: 1 December 2020
Accepted: 24 December 2020
Published: 2 January 2021

1. Introduction

Hypertension, also known as high blood pressure, is one of the most common risk factor for cardiovascular disease (CVD) [1]. It is a very common condition in which a large amount of force from the blood pushes on the walls of the arteries leading towards heart diseases [2]. The main risk factors for hypertension include age, genetics, gender, lack of physical activity, bad diet practices, high cholesterol, excessive salt consumption, less intake of vegetables and fruit, smoking, obesity, family history, and other diseases such as kidney disease or diabetes [3]. According to the World Health Organization (WHO) statistics, 1.13 million of the world population suffers from hypertension, and more men are affected than women. One out of every four men suffers from high blood pressure issues [3]. It is a silent killer that affects the most significant tissues of the human body [4]. Indeed, many people are not aware they have hypertension [5]. In the US, an estimated 13 million people are unaware of their condition [6], while in China, 59% of people with hypertension are unaware of their condition [5]. In Pakistan, 18% of the adults are affected by hypertension, and 33% of the adults above the age of 45 were affected according to the National Health Survey Pakistan [7]. Prevalence rates of hypertension based on genetic and ethnic variations ranges from about 29% for Asians, 45% for black men, and around 46.3% for women [8].

Table 1 describes a blood pressure ranges of normal and hypertension in terms of systolic and diastolic pressures. Some of the common symptoms of hypertension include headaches, dizziness, migraine, lightheadedness, changes in vision, or fainting episodes [9]. Hypertension serves as the first step towards CVDs, but the most chronic effect of unchecked hypertension is stroke, which can lead to permanent paralysis of certain body parts. Prolonged and undetected hypertension can be fatal; therefore, its detection in the preliminary stages is crucial.

Table 1. Categorization of blood pressure.

Class	Systolic (mmHg)	Diastolic (mmHg)
Optimal	Less than 120	Less than 80
Normal	120 to 129	80 to 84
High Normal	130 to 139	85 to 89
Hypertension	More than or equal to 140	More than or equal to 90

Moreover, the world is currently suffering from the outbreak of a pandemic COVID-19 caused by the coronavirus SARS-CoV-2. It was reported that there are some specific comorbidities associated with a high risk of infection and increased severity of lung injury. Most of the common comorbidities in COVID-19 patients are hypertension (30%), cardiovascular disease (8%), and diabetes (19%) [10]. Therefore, it is not entirely surprising that the COVID-19 patients experiencing worst complications are hypertensive since hypertension is most frequent in older people and these elderlies are particularly at risk of being infected by a coronavirus [11]. Given the above information and statistics, it is clear that we need a technique for recognizing hypertension as early as possible to avoid significant damage to one's body.

Various techniques including physiological signals such as electrocardiogram (ECG) and photoplethymograph (PPG) are currently being used to detect hypertension. Detection of hypertension from PPG signals (MIMIC database) using continuous wavelet transform (CWT) and the GoogLeNet deep learning model [12] achieved an F1 score of 92.55%. This work relies on a deep learning model so it requires high processing power, large scale data sets, and more training time. The authors of [13] proposed a method based on pulse arrival time (PAT) features extracted from PPG and ECG signals. The *k*-nearest neighbor (KNN) classification method was employed to predict hypertension with an F1 score of 94.84%. The research achieved acceptable results but missing consideration of preprocessing the PPG signals as PPG suffers from motion artifacts and variation in light intensity. Identification of hypertension [14] from heart rate variability (HRV) signals yielded an accuracy of 85.47% using standard deviation of all NN intervals and multiple instance learning (MIL). However, HRV feature extraction for long-term data requires significant processing resources. A model [15] to detect hypertension obtained 93.33% accuracy using Savitzky–Golay filtering (SGF), entropy features extracted from ECG, and a support vector machine (SVM) classifier. The method achieved a considerable performance on a comparatively small data set consisting of 48 participants.

In [16], the authors proposed a hypertension detection framework based on five principal components extracted from HRV signals to achieve the highest accuracy of 85.5% with quadratic discriminant analysis (QDA). Rajput et al. [17] proposed a scheme to identify the low and high risk of hypertension. The scheme yielded classification accuracy 100% using optimal orthogonal wavelet filter back (OWFB), log, and fractal dimension features extracted from ECG. Despite promising results, this work suffers from a data imbalance problem. The authors in [18] proposed a method to detect ECG hypertensive signals using empirical mode decomposition (EMD) for preprocessing of the signals, yielding an accuracy of 97.7% through the KNN classifier. The extracted features were selected physically, making this process laborious. The method was only trained on a small data set. A system to detect hypertension using morphological descriptors derived from PPG with 92.31% accuracy is discussed in [19]. Identification of hypertension patients from ballistocardiograms (BCG) is presented in [20]. The system achieved a mean accuracy of 84.4% using class association rules (CAR) classifier and morphological features. The BCG signals were collected from patients lying on a smart mattress which has a limited availability.

Medical devices in hospitals can easily get affected by electromagnetic interference (EMI) in a complex electromagnetic environment [21,22]. ECG signals are usually affected by the EMI and preconditioning circuits. Changes in temperature and ambient lighting conditions impact the PPG signal acquisition. PPG signal acquisition is exposed to motion artifacts as well [23]. The frequency of the PPG signal is about 1–3 Hz [24], so it also requires a high order filter for signal denoising. The motivation behind this research was to investigate the feasibility of a new signal modality, i.e., pulse plethysmograph (PuPG). In contrast to PPG that uses light to detect the volume of blood flow in the finger, the PuPG senses the pressure changes in blood flow.

1.1. Main Contributions

In this research, novel PuPG signals were used to design the hypertension detection system. The PuPG signal includes considerable cardiac health characteristic information [25–27]. The PuPG signals are recently being used for emotions classification [28] and biometric systems [29] as well. The main contributions of this work are listed as follows:

- This is the first study that used PuPG-based signals for the detection of hypertension.
- To accurately detect the hypertension pattern, we extract a large number of multi-domain features from preprocessed PuPG signals through discrete wavelet transform (DWT) and EMD.
- To reduce the feature dimensions and redundancy while improving the discriminative power of features, we proposed a hybrid feature selection and reduction (HFSR) scheme.
- The proposed expert hypertension detection system (EHDS) comprises preprocessing through EMD, followed by the feature extraction, kernel principal component analysis (KPCA), and weighted *k*-nearest neighbor (KNN-W) classifier, achieved an accuracy of 99.4%, sensitivity of 99.6%, and specificity of 99.4%.

The rest of this paper is structured as follows: Section 2 gives details about the materials used in this study. Section 3 describes the details about the methods. Next, we present the results in Sections 4 and 5 discuss the proposed method and its comparative analysis. Section 6 concludes this research paper.

2. Materials

2.1. Data Acquisition

In this study, a portable pulse plethysmograph (PuPG) sensor PTN-104 (NISensors, iWorx Systems Inc., Dover, UK) in combination with NI myDAQ (National Instruments Corporation, Austin, TX, USA) was used for PuPG data acquisition. PTN-104 sensor is attached to the index finger of the subject to convert pulse pressure into an electrical voltage signal. The real-time integral of its output signal produces the same volume pulse signal as the expensive infrared PPG sensor. The PTN-104 is a rugged non-magnetic accelerometer, which is made up of piezoelectric material. IX-myDAQ (National Instruments Corporation, Austin, TX, USA) is a breakout board used for connecting the PTN-104 sensor and myDAQ (National Instruments Corporation, Austin, TX, USA) for data acquisition via mini DIN7 port. NI myDAQ is a low-cost data acquisition tool that converts analog signals to digital format and allows the users to analyze real-time data in NI LabVIEW software (National Instruments Corporation, Austin, TX, USA) on PC. The sampling frequency was set to be 1 kHz for PuPG data acquisition.

It is very essential to highlight the difference between PPG and PuPG signals and sensors. Both of them operate on completely different principles with different input parameters. Table 2 presents a comparison between various properties of both sensors such as input parameters, working principles, and the impact of noise on a signal acquisition. Figure 1 illustrates the output signals acquired from both sensors. It was observed that the PuPG signal carries more information as compared to the PPG signal since multiple frequencies contribute towards the dicrotic notch for PuPG.

Table 2. Difference between PPG and PuPG data acquisition.

Type	Photoplethysmograph (PPG) Sensor	Pulse Plethysmograph (PuPG) Sensor
Input signal	Optical signal	Pressure changes
Phenomenon	Blood volumetric changes are detected by measuring the amount of light transmitted or reflected by the sensor.	Blood volumetric changes are detected by the piezoelectric material of the sensor as pressure changes when the blood volume changes.
Noise Impact	Light signal can be easily impacted by any external light changes. Dirty hand can distort the light intensities.	Piezoelectric material based sensors are normaly temperature sensitive. Dirty hands or foreign material on hand or fingers does not have significant impact.

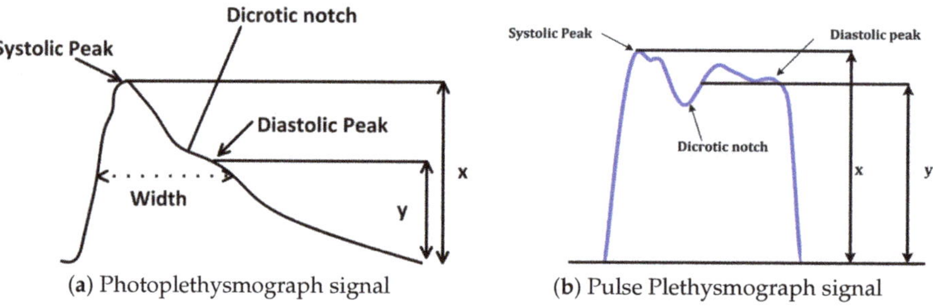

(a) Photoplethysmograph signal (b) Pulse Plethysmograph signal

Figure 1. Visual comparison of Photoplethysmograph (PPG) and Pulse Plethysmograph (PuPG) signals.

2.2. Data Set Description

Raw PuPG signals were acquired from the subjects. The data acquisition was carried out for developing a two-class data bank; one was hypertension and the other normal. A total of 700 signals were collected from hypertension subjects and 709 signals from normal subjects, with a timestamp of 10 s per signal and a sampling frequency of 1000 Hz. Subjects were advised to keep calm and remain static during data acquisition activity. Informed consent was obtained from all participants included in the research. Recording activity was performed between breakfast and lunch time. None of the involved subjects were smokers or diabetic. Table 3 shows the details of the subjects and the acquired data for this study. Figure 2 shows a comparison of raw PuPG signals collected from a normal subject and a subject suffering from hypertension. Sometimes acquired signals (both normal and hypertension) were affected by the circuit noise. The noise/power line distortion incurred due the embedded electronics of data acquisition setup can be seen as a sinusoidal oscillatory component (50 Hz) in the normal PuPG Hat of Figure 2.

Table 3. Summary of the self-collected PuPG data set.

Data Class	Subjects	Age Group	Samples
Hypertension	Male: 29 Female: 27	Male: 40–76 Female: 39–59	700
Normal	Male: 35 Female: 30	Male: 21–63 Female: 20–59	709
Overall	121	20–76	1409

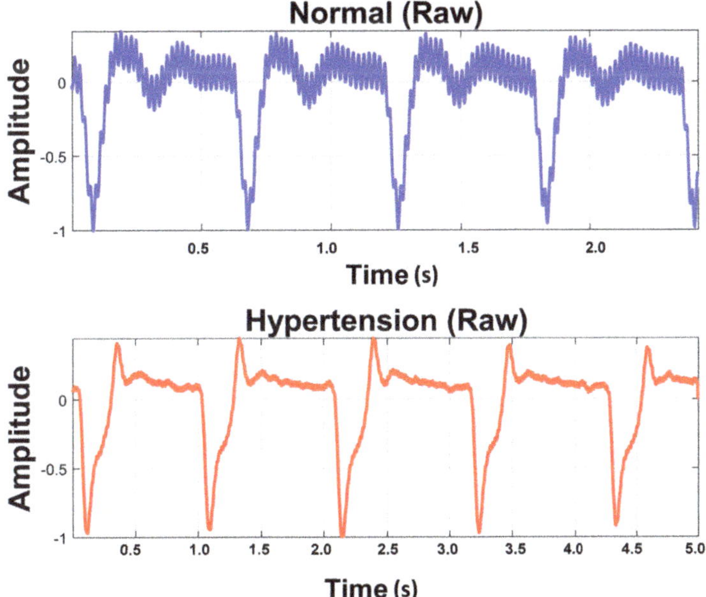

Figure 2. Raw PuPG signals of Normal and Hypertension classes.

3. Methods

3.1. Design of the Study

The proposed methodology in this research adopts the machine learning paradigm shown in Figure 3. It consists of four main stages, namely (i) preprocessing; (ii) feature extraction; (iii) hybrid feature selection and reduction, and (iv) classification. These stages are separated through a dotted line in Figure 3. Each step is elaborated in detail in forthcoming sections. This research adopts a comparative approach between two pattern analysis frameworks, i.e., method I and method II. Method I is comprised of discrete wavelet transform (DWT)-based preprocessing while method II adopts empirical mode decomposition (EMD) for signal denoising. The rest of the framework for both methods is the same. The feature values extracted, reduced feature vectors, and the performance of the classifiers vary for both methods due to the difference in preprocessing methods. All experiments were performed on MATLAB 2018a (The MathWorks, Inc., Natick, MA, USA) running on a personal computer with Core i7 (Intel Corporation, Santa Clara, CA, USA) processors and 32 GB RAM.

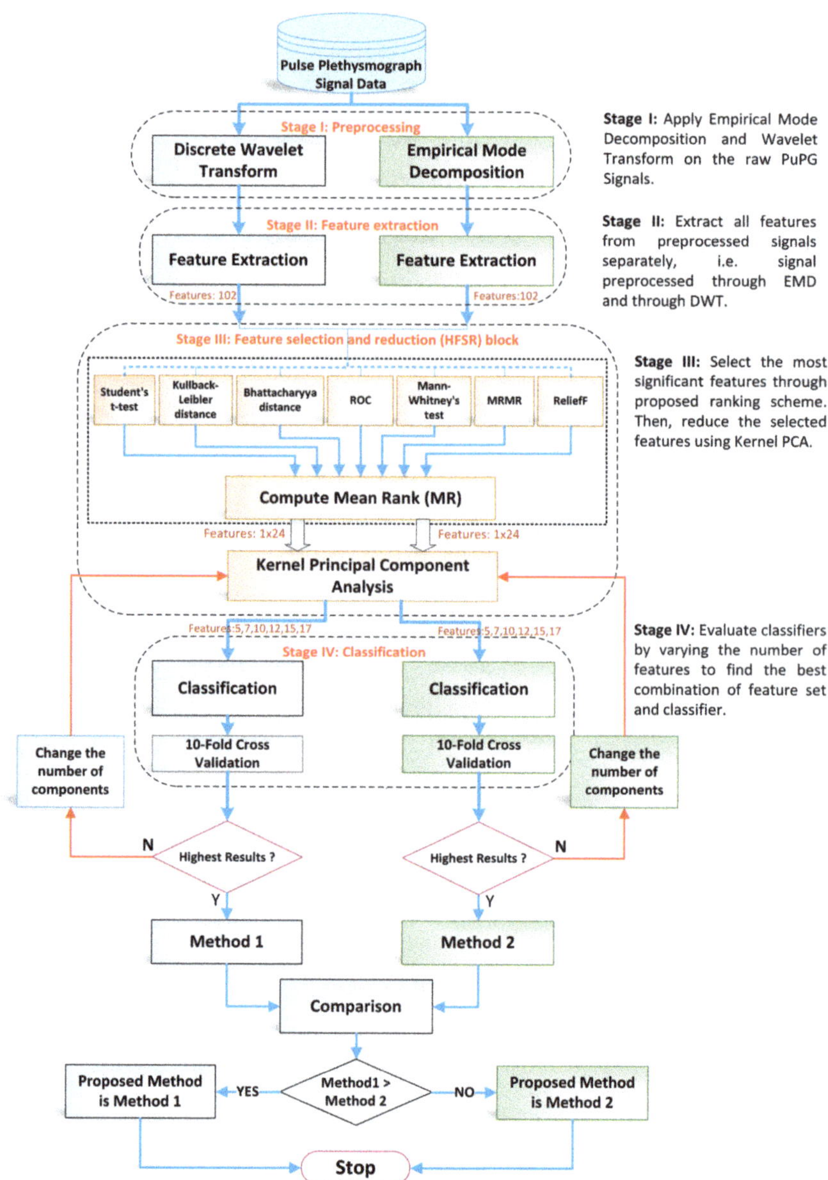

Figure 3. Overall flow chart of the proposed design methodology for detection of hypertension through pulse plethysmograph signals.

1. Preprocessing: It removes the irrelevant information and artifacts from the acquired PuPG signal data of normal and hypertension classes. Method I employs discrete wavelet transform (DWT) for signal denoising through frequency and mean relative energy-based criteria. Method II adopts empirical mode decomposition (EMD) for noise elimination through analysis of mean frequencies and energies of individual signal components extracted from normal and hypertension classes.
2. Feature extraction: It extracts a combination of 102 features from preprocessed PuPG through DWT and EMD separately. These include time, frequency, spectral, texture,

and cepstral features. The difference between signal classes is best captured through the extraction of a wide range of informative features.

3. Feature selection and reduction: This step eliminates features with redundant information through a hybrid feature selection and reduction (HFSR) method that is a combination of multiple feature ranking and transformation schemes. A high-dimensional feature vector is reduced through a new strategy of the averaging outcome of seven feature ranking methods, thus providing more reliable results. Next, we employed kernel principal component analysis (KPCA) to further decrease the feature dimension and represent significant information in fewer parameters. Extracted features in both method I and II are fed to the HFSR scheme to reduce the dimension of the resultant feature vector.
4. Classification:The final feature vectors extracted in both methods I and II of hypertension and normal classes are fed to a range of different classifiers, i.e, support vector machines (SVM), *k*-nearest neighbors (KNN), ensemble methods, decision trees (DT), and logistic regression (LR). Classification performance of both methods is evaluated through a baseline tenfold cross-validation strategy and compared with 5-, 15-, 20-, and 25-fold cross-validation.

3.2. Preprocessing

The acquired PuPG data were contaminated with noise and artifacts and include redundant information (Figure 2). These noise components needs to be eliminated for a robust performance of the proposed system. Therefore, we employed DWT and EMD-based preprocessing for signal denoising. Later on, we compared the preprocessing performance of both methods.

3.2.1. Discrete Wavelet Transform

The discrete wavelet transform (DWT) is a widely applied approach in biomedical signal processing applications [30–32]. DWT decomposes a signal into different resolutions by using a combination of high-pass and low-pass filters. Figure 4 illustrates the complete process of wavelet-based denoising [33] adopted in this research. Numerous filter coefficients have been developed for diverse types of signal analysis applications—for instance, Daubechies, Symlets, and Coiflets coefficients, etc.

In this study, we employed the Symlet wavelet due to its similarity with the shape of the PuPG signal under consideration [34,35]. Symlet wavelet yields the best results as compared to others due to its resemblances with the morphological characteristics of the PuPG signal.

Table 4 exhibits information about decomposition levels, frequency ranges, and mean relative energies of normal and hypertension data classes of PuPG signals. It can be observed that D_1, D_2, D_3, and D_4 signal components have high frequency range and include low mean relative energies; therefore, these components were eliminated while reconstructing a denoised signal. This is also endorsed by the fact that the PuPG signal has a very low frequency (normally less than 60 Hz). Figure 5 provides a graphical illustration of wavelet decomposition for normal and hypertension PuPG signals. Figure 6 presents the denoised signal generated as a result of applying DWT. High frequency noise visible in raw PuPG signal (Figure 2) is eliminated in the denoised version.

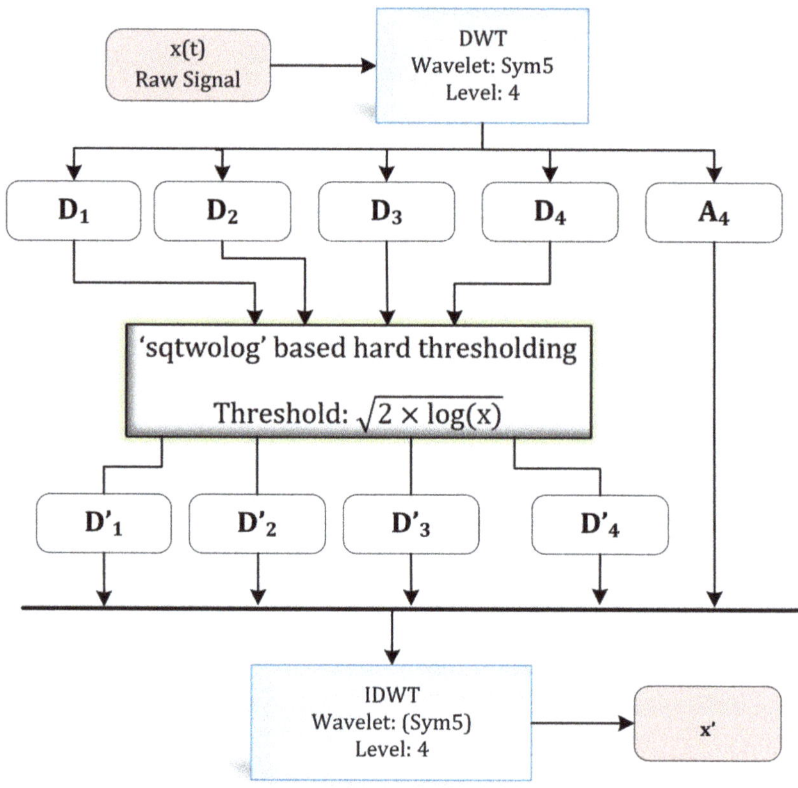

Figure 4. Wavelet-based denoising.

Table 4. Comparison of mean relative energies and frequency ranges of various decomposition levels for Normal and Hypertension classes.

Decomposition Levels	Frequency Range (Hz)	Mean Relative Energy (%)	
		Normal	Hypertension
D_1	250–500	0.07%	0.59%
D_2	122–256	0.09%	0.32%
D_3	61.1–128	0.19%	0.35%
D_4	30.6–63.9	0.46%	0.49%
D_5	15.3–31.9	1.93%	3.10%
D_6	7.65–16	14.77%	13.31%
D_7	3.84–7.97	31.03%	21.49%
D_8	1.94–3.99	29.11%	19.05%
D_9	1.03–1.99	21.11%	26.77%
D_{10}	0.594–0.958	0.16%	7.65%
A_{10}	0–0.431	1.08%	6.88%
A_4	0–31.2	99.19%	98.26%

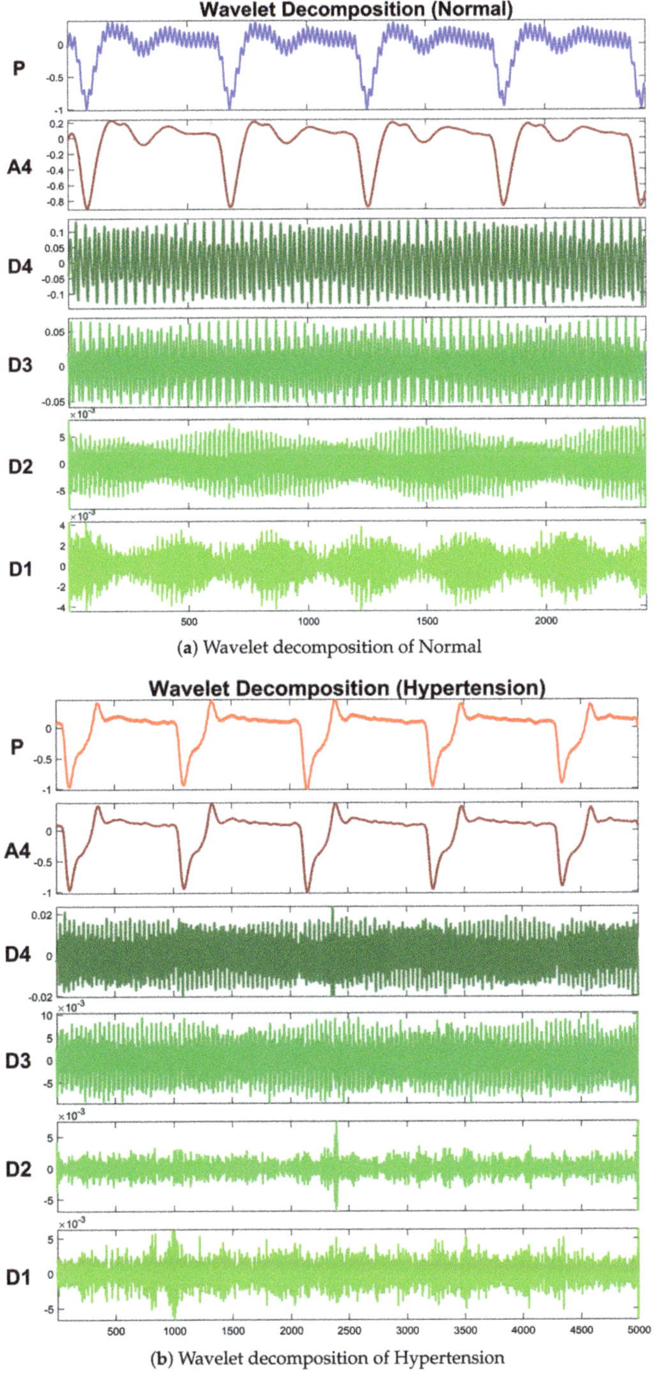

Figure 5. Wavelet decomposition of raw PuPG signals.

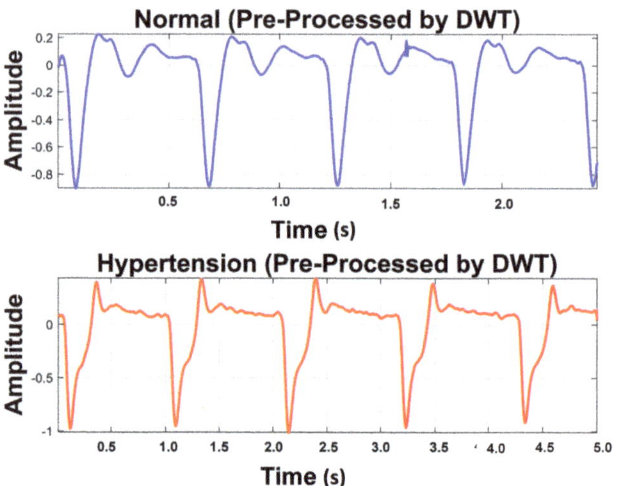

Figure 6. Denoised version of PuPG signal for Normal and Hypertension through DWT.

3.2.2. Empirical Mode Decomposition

EMD is an adaptive method that derives fundamental functions directly from the data [36]. EMD does not require any previously known value of the signal for its computation. The principal task for computing EMD of a given signal is to empirically determine the intrinsic oscillatory components through their particular time scales in a signal and subsequently disintegrate the signal into intrinsic mode functions (IMFs) [37]. Therefore, EMD provides remarkably better results for nonlinear and non-stationary biomedical signals.

Selection criteria of IMF have to satisfy two conditions;

- In the entire signal, the total number of local extrema and zero crossings must be equal to each other or differ by a maximum one.
- The average of the envelopes computed through local minima and local maxima must be zero.

The systematic approach to disintegrate the signal into its IMFs is known as the "sifting" process, explained in Figure 7.

The basic objective of applying EMD for preprocessing the PuPG signal was to decompose the distorted signal into its constituent IMFs as depicted in Figure 8. Considering the fact that some IMFs carry discriminative and characteristic information about various data classes while others include redundant and noisy content, the determination of the proper number of IMFs is a crucial step towards creating an effective signal denoising strategy.

It is perceived from Figure 8 and Table 5 that the first IMF includes mainly high-frequency content. Table 5 provides mean frequency and energy information of each IMF for normal and hypertension data classes of the PuPG signal. The first IMF also holds very little mean relative energy components for both classes, i.e., 0.00% and 1.02% for normal and hypertension classes, respectively. Therefore, it was discarded while reconstructing the denoised signal. All other IMFs and residual signals were added to form a denoised version of the PuPG signal. Figure 9 illustrates the PuPG signal denoised through the EMD process for normal and hypertension data. It is clear that high frequency noise that was visible in raw PuPG signal (Figure 2) is eliminated now.

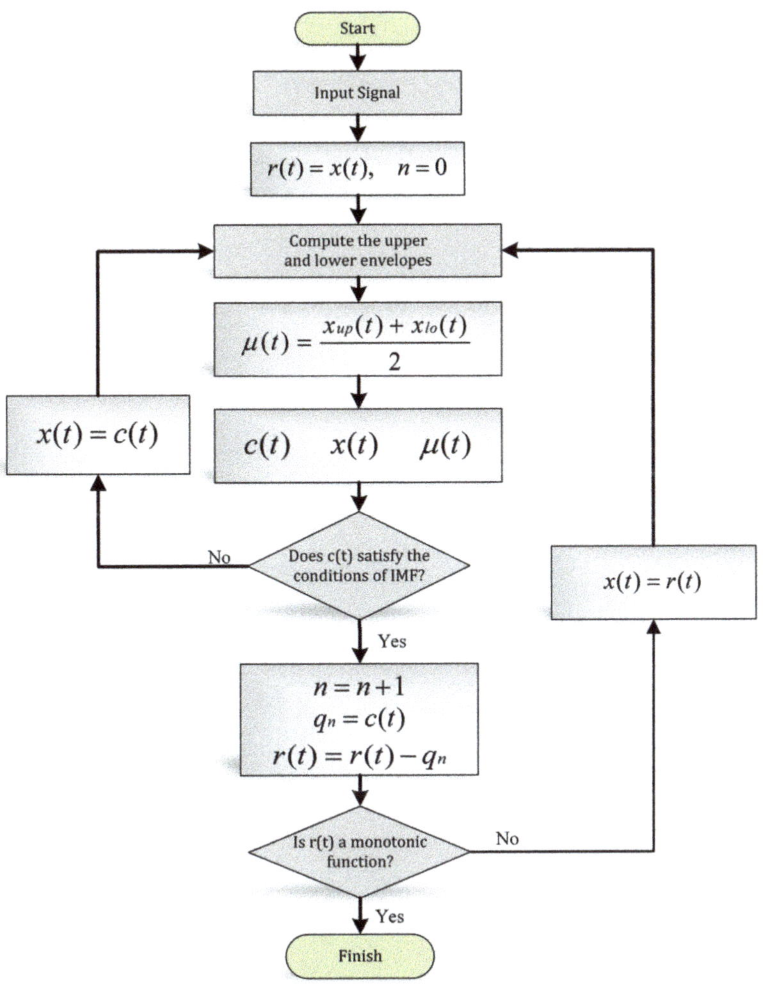

Figure 7. EMD algorithm (flow chart).

Table 5. Comparison of mean relative energies and frequency ranges of various intrinsic mode functions (EMD) for Normal and Hypertension classes. Bold font indicates the selected components.

Components	Normal		Hypertension	
	Mean Frequency Range (Hz)	Mean Relative Energy (%)	Mean Frequency Range (Hz)	Mean Relative Energy (%)
IMF_1	103–483	0.00	86.5–484	1.02
IMF_2	**11.3–60.2**	**0.14**	**40.7–219**	**0.35**
IMF_3	**3.09–14**	**30.34**	**3.3–61**	**2.04**
IMF_4	**2.99–12.2**	**4.97**	**3.34–23.3**	**6.65**
IMF_5	**1.28–10**	**22.76**	**2.98–11.4**	**14.95**
Residual	**0.129–5.55**	**41.78**	**0.0197–4.33**	**74.99**

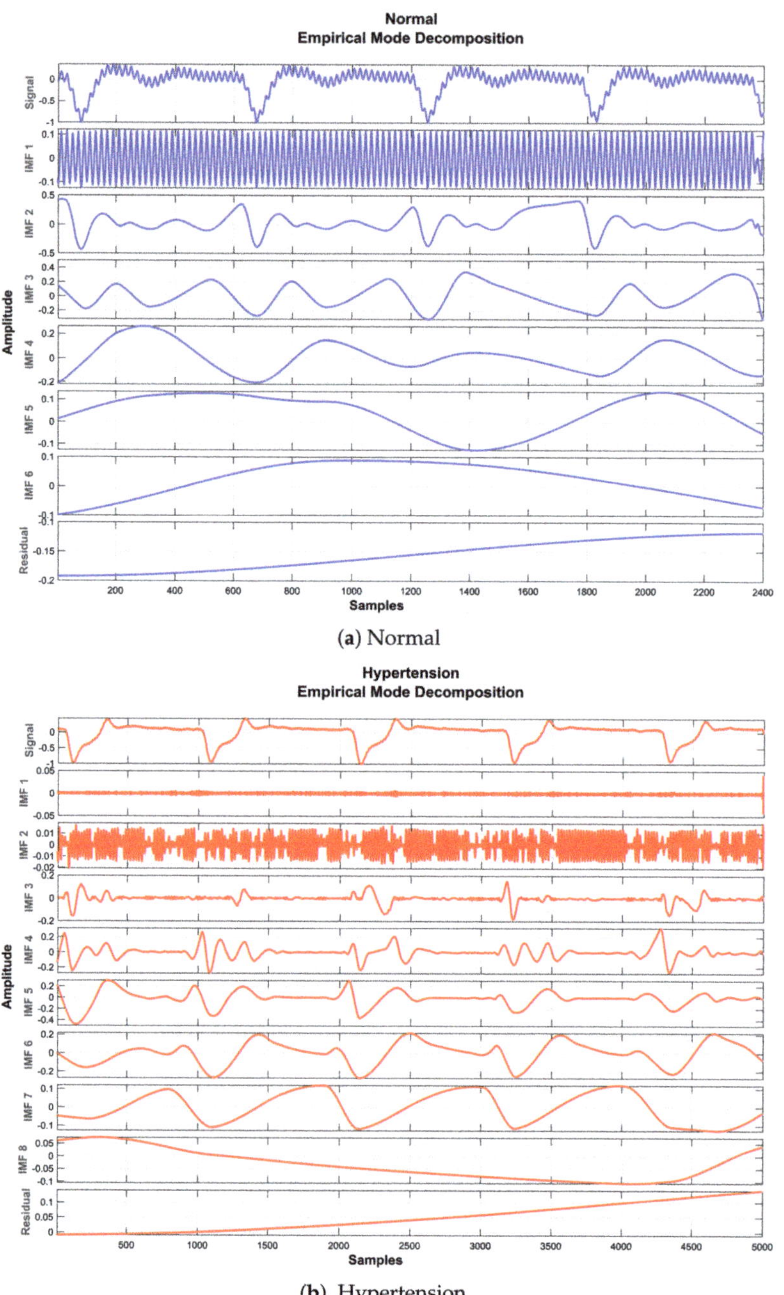

Figure 8. EMD decomposition of raw PuPG signals.

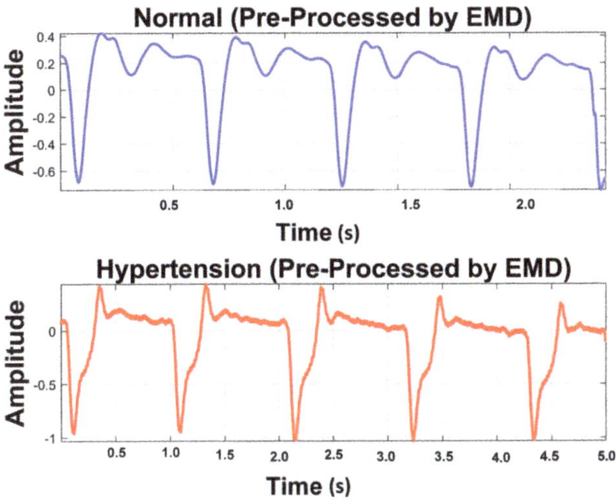

Figure 9. Preprocessed signal using EMD.

3.3. Feature Extraction

The objective of the feature extraction stage is to extract significant features from the biomedical signals of various classes that contributes towards an effective classification performance. In this study, a total of 102 features were separately computed from the PuPG signal denoised through DWT and EMD. Table 6 lists all the extracted features along with their statistical measures of mean and standard deviation (STD) for method I (DWT) and method II (EMD). We extracted time domain [38–45], spectral [46,47], fractal and chaos [48,49], chroma [50,51], cepstral [52], and texture features [53] and analyzed them statistically.

Table 6. Statistical data of all extracted features for both methods.

Feature	Method I				Method II			
	Normal		Hypertension		Normal		Hypertension	
	Mean	STD	Mean	STD	Mean	STD	Mean	STD
Mean	0.008	0.027	0.017	0.031	0.001	0.049	0.013	0.053
Standard Deviation	0.253	0.032	0.250	0.046	0.254	0.032	0.248	0.045
Skewness	−1.959	0.522	−2.144	0.651	−1.997	0.576	−2.220	0.641
Kurtosis	6.993	1.499	8.083	3.864	7.139	1.688	8.297	3.944
Peak to Peak Value	1.380	0.199	1.398	0.140	1.377	0.178	1.375	0.144
Root Mean Square	0.255	0.033	0.252	0.047	0.258	0.038	0.254	0.046
Crest Factor	1.659	0.939	1.635	0.381	1.587	0.991	1.523	0.536
Shape Factor	1.484	0.134	1.458	0.159	1.522	0.163	1.465	0.167
Impulse Factor	2.435	1.297	2.382	0.593	2.371	1.376	2.171	0.632
Margin Factor	15.22	14.64	15.12	7.27	15.16	15.56	13.39	6.45
Energy	389.6	209.7	437.4	207.6	393.9	207.4	448.5	222.5
Peak to RMS Value	3.894	0.623	4.094	0.991	3.921	0.663	4.108	1.102
Root Sum of Squares	18.933	5.600	20.292	5.089	19.069	5.525	20.480	5.412
Shannon Energy	549.7	312.3	618.2	279.7	526.6	295.8	686.5	414.4
Log Energy	−27,888	15,515	−34,569	18,644	−28,509	16,289	−32,613	16,387
Mean Absolute Deviation	0.169	0.026	0.169	0.043	0.169	0.028	0.168	0.043
Median Absolute Deviation	0.074	0.026	0.071	0.030	0.071	0.033	0.059	0.024

Table 6. *Cont.*

Feature	Method I				Method II			
	Normal		Hypertension		Normal		Hypertension	
	Mean	STD	Mean	STD	Mean	STD	Mean	STD
Average Frequency	0.002	0.001	0.001	0.001	0.002	0.003	0.002	0.002
Jitter	137.1	180.9	85.5	150.4	159.0	248.8	52.0	95.8
Spectral Mean	3.144	5.928	0.771	1.960	8.623	15.700	2.764	6.372
Spectral Standard Deviation	3.401	4.833	1.549	3.463	7.571	11.498	3.031	6.016
Spectral Skewness	2.361	1.496	3.763	1.295	0.996	1.360	1.797	1.908
Specral Kurtosis	11.331	8.235	20.190	9.589	5.588	4.947	10.564	8.751
Spectral Centroid	9.442	0.202	9.915	1.481	9.771	1.576	10.768	2.216
Spectral Flux	0.008	0.002	0.008	0.002	0.008	0.002	0.008	0.002
Spectral Roll-off	91.699	1.010	96.587	8.383	97.036	10.371	136.257	44.372
Spectral Flatness	0.025	0.014	0.051	0.035	0.063	0.065	0.171	0.138
Spectral Crest	0.642	0.012	0.621	0.040	0.631	0.030	0.593	0.059
Spectral Decrease	−4.333	0.228	−4.013	0.658	−4.169	0.538	−3.656	0.830
Spectral Slope	−0.023	0.003	−0.023	0.003	−0.024	0.003	−0.023	0.005
Spectral Spread	18.505	0.199	19.029	1.178	19.328	1.617	23.521	5.818
Mean Frequency	4.347	0.949	4.999	3.451	4.989	3.063	6.446	4.440
Median Frequency	3.574	0.803	2.972	0.880	3.811	3.573	2.846	0.753
Spurious-free Dynamic Range	3.073	6.019	2.196	2.107	3.163	6.451	2.056	1.812
Signal to Noise Distortion	−0.885	6.041	−2.036	3.101	−0.755	6.819	−2.097	3.030
Total Harmonic Distortions	−2.376	5.452	−0.938	4.964	−3.144	6.562	−0.396	4.110
1st Coeffient of MFCC	−44.99	0.37	−44.716	0.512	−44.84	0.60	−44.45	0.80
2nd Coeffient of MFCC	6.268	0.523	6.661	0.714	6.480	0.846	7.028	1.122
3rd Coeffient of MFCC	5.976	0.499	6.350	0.683	6.169	0.802	6.690	1.066
4th Coeffient of MFCC	5.508	0.462	5.851	0.634	5.671	0.733	6.148	0.976
1st Coeffient of GFCC	−7.183	0.430	−6.762	0.610	−7.027	0.793	−6.220	1.266
2nd Coeffient of GFCC	1.844	0.063	1.869	0.071	1.522	0.419	1.119	0.574
3rd Coeffient of GFCC	0.553	0.138	0.367	0.269	0.643	0.105	0.492	0.206
4th Coeffient of GFCC	0.301	0.024	0.266	0.033	0.392	0.109	0.408	0.109
1st Coefficient of Chroma Vector	0.383	0.235	0.750	0.501	0.653	0.532	2.126	1.930
2nd Coefficient of Chroma Vector	0.416	0.258	0.773	0.518	0.663	0.546	2.130	1.948
3rd Coefficient of Chroma Vector	0.433	0.269	0.842	0.575	0.742	0.672	2.129	2.144
4th Coefficient of Chroma Vector	0.623	0.378	1.297	0.942	0.700	0.568	2.011	2.046
5th Coefficient of Chroma Vector	0.564	0.337	1.212	0.872	0.691	0.534	2.044	1.935
6th Coefficient of Chroma Vector	0.527	0.320	1.230	0.987	0.748	0.563	2.227	2.267
7th Coefficient of Chroma Vector	0.483	0.296	1.069	0.760	0.705	0.524	2.107	1.893
8th Coefficient of Chroma Vector	0.451	0.279	0.982	0.696	0.686	0.528	2.071	1.863
9th Coefficient of Chroma Vector	0.429	0.268	0.908	0.638	0.679	0.529	2.099	1.929
10th Coefficient of Chroma Vector	0.400	0.251	0.878	0.609	0.651	0.537	2.087	1.848
11th Coefficient of Chroma Vector	0.373	0.232	0.776	0.537	0.668	0.522	2.106	1.954
12th Coefficient of Chroma Vector	0.348	0.225	0.705	0.474	0.622	0.522	2.078	1.890
Enhanced Mean Absolute Value	0.297	0.039	0.302	0.055	0.294	0.052	0.301	0.049
Enhanced Wavelength	236.4	133.5	413.7	319.8	284.3	198.6	665.7	515.0
Wavelength	36.83	21.94	85.59	96.94	54.42	46.94	193.47	186.80
Slope Sign Change	45.1	86.9	508.4	549.0	1039.5	1657.6	3463.1	2792.1
Average Amplitude Change	0.006	0.003	0.010	0.009	0.009	0.009	0.021	0.018
Difference Absolute Std. Dev.	0.009	0.003	0.013	0.010	0.014	0.010	0.027	0.021
Log Detector	0.108	0.030	0.118	0.037	0.107	0.043	0.117	0.036
Modified Mean Absolute Value	0.130	0.025	0.133	0.034	0.130	0.033	0.132	0.030
Modified Mean Absolute Value 2	0.083	0.022	0.089	0.026	0.084	0.027	0.087	0.020
Pulse Percentage Rate	0.939	0.029	0.953	0.027	0.937	0.045	0.957	0.035

Table 6. *Cont.*

Feature	Method I				Method II			
	Normal		Hypertension		Normal		Hypertension	
	Mean	STD	Mean	STD	Mean	STD	Mean	STD
Simple Square Integral	389.6	209.7	437.4	207.6	393.9	207.4	448.5	222.5
Willison Amplitude	1153.6	765.6	2670.8	2603.9	1836.0	1707.6	4551.3	3502.4
Maximum Fractal Length	−0.463	0.384	−0.136	0.740	−0.174	0.594	0.428	1.121
Root Squared Zero Order Moment	2.592	0.032	2.600	0.026	2.593	0.031	2.601	0.028
Root Squared 2nd Order Moment	2.068	0.066	2.036	0.062	1.984	0.115	1.913	0.130
Root Squared 4th Order Moment	2.045	0.077	2.001	0.078	1.891	0.159	1.794	0.205
Sparseness	0.535	0.064	0.582	0.086	0.655	0.137	0.747	0.188
Irregularity Factor	−0.464	0.037	−0.445	0.047	−0.446	0.058	−0.406	0.061
Waveform Length Ratio	−0.065	0.703	−0.354	0.230	−0.648	0.604	−0.721	0.727
Complexity	0.502	0.222	0.706	0.253	0.897	0.524	1.314	0.515
Mobility	0.038	0.011	0.057	0.038	0.055	0.035	0.115	0.079
Higuchi's Fractal Dimension	1.054	0.052	1.149	0.119	1.183	0.240	1.490	0.401
Katz Fractal Dimension	1.000	0.000	1.000	0.000	1.000	0.000	1.000	0.000
Lyapunov Exponent	437.4	49.5	394.8	49.3	362.9	78.8	239.1	92.2
Approximate Entropy	0.104	0.049	0.155	0.134	0.127	0.080	0.316	0.312
Correlation Dimension	1.687	0.168	1.733	0.191	1.676	0.202	1.731	0.300
1st Coefficient of LTP	259.6	128.8	291.5	122.1	259.6	128.8	291.5	122.1
2nd Coefficient of LTP	40.790	30.943	56.809	35.998	40.790	30.943	56.809	35.998
3rd Coefficient of LTP	24.574	23.717	41.723	32.083	24.574	23.717	41.723	32.083
4th Coefficient of LTP	16.381	15.761	30.738	22.667	16.381	15.761	30.738	22.667
5th Coefficient of LTP	18.472	15.755	33.411	24.881	18.472	15.755	33.411	24.881
6th Coefficient of LTP	25.205	21.607	45.518	33.715	25.205	21.607	45.518	33.715
7th Coefficient of LTP	17.699	16.352	29.035	22.609	17.699	16.352	29.035	22.609
8th Coefficient of LTP	26.261	23.075	39.645	28.386	26.261	23.075	39.645	28.386
9th Coefficient of LTP	47.483	32.844	58.163	35.132	47.483	32.844	58.163	35.132
10th Coefficient of LTP	207.278	94.769	201.809	72.537	207.278	94.769	201.809	72.537
11th Coefficient of LTP	269.5	137.7	302.1	121.7	269.5	137.70	302.1	121.7
12th Coefficient of LTP	41.733	31.354	59.773	38.092	41.733	31.354	59.773	38.092
13th Coefficient of LTP	24.006	23.096	41.071	31.294	24.006	23.096	41.071	31.294
14th Coefficient of LTP	16.784	15.900	30.199	23.471	16.784	15.900	30.199	23.471
15th Coefficient of LTP	18.506	15.140	33.390	25.583	18.506	15.140	33.390	25.583
16th Coefficient of LTP	26.148	22.755	44.177	31.682	26.148	22.755	44.177	31.682
17th Coefficient of LTP	16.898	16.559	30.149	21.910	16.898	16.559	30.149	21.910
18th Coefficient of LTP	25.790	22.892	39.312	26.784	25.790	22.892	39.312	26.784
19th Coefficient of LTP	46.534	33.045	56.298	33.647	46.534	33.045	56.298	33.647
20th Coefficient of LTP	197.773	85.090	191.858	74.487	197.773	85.090	191.858	74.487

These features were subjected to the feature selection step (HFSR) to recognize the features with maximum discriminative content among normal and hypertension classes.

3.4. Hybrid Feature Selection and Reduction

Feature selection is one of the key steps in the modern pattern recognition and machine learning paradigms. The extracted features may include redundant information and irrelevant and noisy parameters. A two-stage hybrid feature selection and reduction (HFSR) strategy was designed to select and transform the best distinctive features as shown in Figure 10. The first stage ranks the input features through seven different methods and the second stage transforms the selected ranked features to further reduce dimensionality.

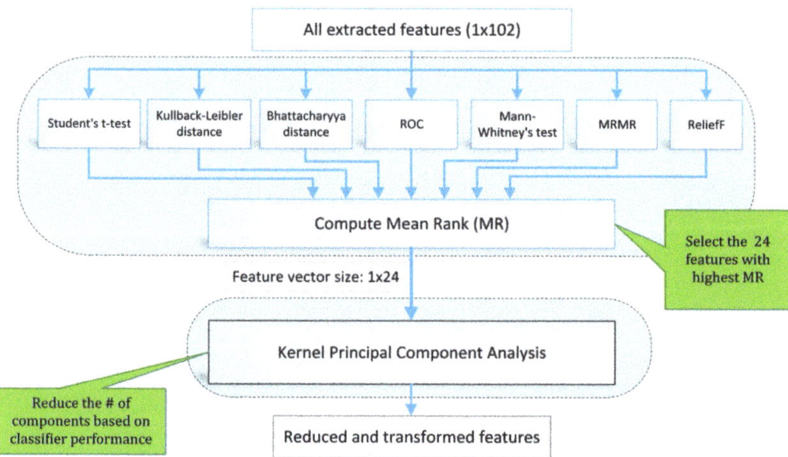

Figure 10. Block diagram (feature selection and reduction method).

3.4.1. Feature Selection Scheme

Feature selection routines serve to enhance the performance of classifiers by reducing the feature dimensions as well as decreasing the computational time [54,55]. Feature selection methods are categorized as filter methods and wrapper methods. Filter type feature selection methods employ feature ranking techniques based on the applied statistical measure for selecting a suitable feature. In wrapper type feature selection techniques, a feature subset is selected recursively based on the overall model performance. The selection criterion computes the variation in model performance that decides the addition or removal of a feature from the subset.

To address the limitations of individual feature selection approaches, we employed a hybrid scheme of feature selection by combining seven feature ranking methods through a voting strategy. Figure 10 illustrates the hybrid scheme of feature selection and reduction. In this scheme, seven state-of-the-art feature ranking techniques, namely student *t*-test (TT), Kullback–Leibler distance (KLD) [56], Bhattacharya distance (BD) [57], Mann–Whitney's test (MWT), ReliefF (RRF) [58], minimum redundancy maximum relevance (MRMR) [59,60], and receiver operating characteristic curve (ROC) were employed to rank the feature individually. Ranking assigned to each feature by all feature ranking methods is combined to calculate the mean rank (MR) value. A threshold is applied to MR value for feature selection.

Table 7 provides the sorted lists of the best forty features with the highest MR values for method I (features extracted from signal preprocessed through DWT). Rank assigned to individual features by each ranking method is also computed. The top 24 features are highlighted in Table 6 were forwarded to the next stage. It was perceived that if a feature ranking method assigns a high rank to a particular feature that failed to get high scores from other methods, it gets rejected due to the hybrid scheme of feature selection. For instance, consider the Root Sum of Squares feature that received the rank value of 99 from the ROC method, but gets scores of 53, 49, 52, 18, 58, and 72 from TT, KLD, BD, MWT, MRMR, and RRF, respectively. It achieved an MR value of 57.29 that is below the selection criterion, so it was rejected from the final feature vector of 1×24 dimensions. Table 8 enlists the top forty features with the highest MR values for method II, i.e., features extracted from the signal preprocessed through EMD. The rank value assigned by an individual feature ranking method to a specific feature can be examined. One to one comparison of the top ten MR values of method I in Table 7 and method II in Table 8 reveals that the magnitude of MR values of method II (81–70) is higher than that for method I (73–67).

Table 7. List of features extracted for method I and sorted with respect to mean rank (MR) value. Bold font indicates the top 24 ranked features.

Feature	TT	KLD	BD	ROC	MWT	MRMR	RRF	MR
3rd Coefficient of LTP	12	93	93	83	32	102	98	73.29
6th Coefficient of Chroma Vector	38	98	76	79	78	94	49	73.14
Lyapunov Exponent	93	92	92	70	81	10	70	72.57
Sparseness	70	90	90	30	62	79	84	72.14
Jitter	78	58	58	74	77	85	74	72.00
9th Coefficient of LTP	57	83	83	75	82	92	24	70.86
Spectral Decrease	62	28	61	85	75	93	83	69.57
4th Coefficient of MFCC	96	41	41	95	40	72	99	69.14
Irregularity	35	94	100	57	88	90	13	68.14
1st Coeffient of MFCC	99	23	23	100	37	96	97	67.86
Waveform Length Ratio	13	100	94	91	9	100	66	67.57
3rd Coeffient of MFCC	79	97	97	84	39	55	17	66.86
1st Coefficient of Chroma Vector	71	86	86	98	26	48	53	66.86
Spectral Roll-off	81	43	79	66	56	59	81	66.43
Spectral Crest	61	80	60	26	92	71	75	66.43
6th Coefficient of Chroma Vector	100	76	99	31	61	45	52	66.29
7th Coefficient of Chroma Vector	37	99	98	1	96	42	86	65.57
Median Frequency	88	35	78	71	102	78	2	64.86
2nd Coefficient of Chroma Vector	89	7	96	89	41	81	51	64.86
Spectral Centroid	76	79	43	39	86	69	58	64.29
Difference Absolute Std. Dev. Value	84	38	37	81	79	63	64	63.71
Shape Factor	59	53	53	54	72	80	73	63.43
Spectral Mean	43	77	77	96	46	40	61	62.86
Simple Square Integral	4	72	72	22	89	89	92	62.86
3rd Coefficient of GFCC	95	29	88	94	34	53	45	62.57
4th Coefficient of Chroma Vector	40	96	30	87	85	52	47	62.43
Root Mean Square	45	56	56	77	43	87	69	61.86
Signal to Noise Distortion	97	69	69	47	74	31	42	61.29
9th Coefficient of Chroma Vector	72	12	95	88	27	76	57	61.00
Mean Absolute Deviation	58	48	48	49	76	82	62	60.43
Root Squared 2nd Order Moment	91	82	71	8	59	20	91	60.29
Root Squared 4th Order Moment	101	71	82	82	4	66	14	60.00
10th Coefficient of Chroma Vector	36	95	85	28	73	41	56	59.14
12th Coefficient of Chroma Vector	90	89	89	59	10	37	38	58.86
1st Coefficient of GFCC	69	87	87	61	66	38	3	58.71
Mean	80	26	26	80	80	68	44	57.71
Enhanced Mean Absolute Value	21	73	73	62	58	54	63	57.71
Root Sum of Squares	53	49	52	99	18	58	72	57.29
2nd Coefficient of LTP	82	70	70	35	36	12	95	57.14
Katz Fractal Dimension	67	3	3	67	90	75	90	56.43

Table 8. List of features extracted for method II and sorted with respect to MR value. Bold font indicates the top 24 ranked features.

Feature	TT	KLD	BD	ROC	MWT	MRMR	RRF	MR
7th Coefficient of Chroma Vector	100	98	98	26	96	97	55	81.43
4th Coefficient of Chroma Vector	89	96	76	73	85	52	96	81.00
Mobility	93	64	64	67	68	91	95	77.43
Spectral Centroid	68	78	78	77	45	98	88	76.00
Enhanced Mean Absolute Value	71	95	95	28	98	56	87	75.71
9th Coefficient of LTP	64	101	101	64	93	79	19	74.43
1st Coefficient of GFCC	95	35	97	98	38	99	54	73.71
7th Coefficient of LTP	57	93	93	57	101	24	89	73.43
Slope Sign Change	74	73	89	79	89	89	13	72.29
Maximum Fractal Length	3	72	74	91	90	90	70	70.00
3rd Coefficient of MFCC	38	97	22	95	102	80	53	69.57
6th Coefficient of Chroma Vector	69	80	96	90	27	68	56	69.43
8th Coefficient of Chroma Vector	61	99	99	69	66	49	40	69.00
Enhanced Wavelength	81	85	85	68	94	16	50	68.43
Pulse Percentage Rate	4	94	100	42	73	76	77	66.57
Root Squared Zero Order Moment	63	84	84	30	69	72	46	64.00
Crest Factor	51	68	68	80	72	32	72	63.29
Modified Mean Absolute Value 2	42	90	90	7	88	64	62	63.29
Spectral Crest	87	58	45	40	47	85	75	62.43
2nd Coeffient of MFCC	37	69	69	84	77	48	51	62.14
1st Coeffient of MFCC	77	9	9	100	99	38	102	62.00
Average Frequency	80	29	54	47	54	86	82	61.71
4th Coefficient of LTP	34	65	65	65	42	74	84	61.29
Willison Amplitude	91	74	72	15	16	77	83	61.14
Spectral Spread	99	47	58	85	60	9	68	60.86
3rd Coefficient of LTP	65	91	91	5	65	2	101	60.00
Root Squared 4th Order Moment	70	66	71	14	97	62	36	59.43
Lyapunov Exponent	32	63	63	8	62	81	100	58.43
2nd Coeffient of GFCC	40	87	87	89	40	50	14	58.14
3rd Coeffient of GFCC	73	22	10	66	39	96	99	57.86
Correlation Dimension	14	70	70	101	57	1	91	57.71
Root Squared 2nd Order Moment	101	5	66	36	31	100	61	57.14
5th Coefficient of Chroma Vector	20	76	80	97	4	36	86	57.00
2nd Coefficient of Chroma Vector	75	41	41	88	78	73	2	56.86
11th Coefficient of Chroma Vector	66	39	39	74	95	47	37	56.71
Log Energy	45	75	52	53	19	88	60	56.00
10th Coefficient of Chroma Vector	90	38	38	19	100	69	38	56.00
5th Coefficient of LTP	8	82	82	83	5	101	30	55.86
1st Coefficient of LTP	11	92	92	24	83	57	31	55.71

3.4.2. Feature Reduction Using Kernel PCA

PCA applies orthogonal transformation to transform a group of likely correlated features into a set of linearly independent features known as principal components. These principal components represent the normalized linear combinations of the original features. It includes information about the most powerful variations present in the data set. The first principal component holds maximum variance information of the data set.

Kernel PCA (KPCA) [61,62] enhances the original PCA to non-linear data distribution problems through a kernel function. A kernel function projects low-dimensional feature data to a higher-dimensional feature space, where it becomes linearly separable [63].

The previous stage of hybrid feature selection reduced the feature dimensions to 1×24 which are fed to KPCA to further decrease dimensions for both methods I and II. Components of KPCA were selected recursively based on the classification performance through tenfold cross-validation. Separate sets of 5, 7, 10, 12, 15, and 17 components were picked for methods I and II to investigate the classification performance for differentiating normal and hypertension signal classes of PuPG signals.

3.5. Classification

To perform the classification of normal and hypertension classes of PuPG signal data set, this study employed a range of classification methods through tenfold cross-validation schemes. The classification methods opted in this study were SVM-Linear (SVM-L), SVM-Quadratic (SVM-Q), SVM-Cubic (SVM-C), SVM-Fine Gaussian (SVM-FG), SVM-Medium Gaussian (SVM-MG), KNN-Fine (KNN-F), KNN-Medium (KNN-M), KNN-Cosine distance (KNN-Cos), KNN-Cubic (KNN-C), KNN-Weighted (KNN-W), Decision Trees (DT), Linear Discriminant (LD), Logistic Regression (LR), Gaussian Naive Baise (NBG), Kernel Naive Baise (NBK), Ensemble Boosted Trees (Eboost), Ensemble Bagged Trees (EBT), Ensemble Subspace Discriminant (ESD), and Ensemble Subspace KNN (ESKNN). The tenfold cross-validation was also compared with 5-, 15-, and 20-fold cross-validation and 80–20% and 75–25% train-test experiments. All experiments were implemented on MATLAB 2018a on a personal computer with Core i7 with 32 GB RAM.

4. Results

In this study, the PuPG signal data set comprising two classes (Normal and Hypertension) was first preprocessed through DWT and EMD to develop methods I and II respectively. We obtained 102 features for each method, i.e., DWT and EMD. These features were subjected to the HFSR framework to reduce the computational complexity and feature vector dimensions. Standard statistical parameters of Accuracy (Acc), Sensitivity (Sen), Specificity (Sp), and Error rate (Err) were used to measure the classification performance.

4.1. Method I

In this research, a comparative analysis was performed via preprocessing the PuPG signal through DWT and EMD. This section presents the results yielded by preprocessing through DWT and succeeding processes of feature extraction, selection, and classification. Various feature sets, namely S_1, S_2, S_3, S_4, S_5, and S_6 were formed by randomly choosing 5, 7, 10, 12, 15, and 17 transformed features. These feature components were fed to several classification methods to examine the diagnostic performance through tenfold cross-validation. Table 9 presents consolidated result analysis of various classification methods for features sets S_1 (5 components), S_2 (7 components), and S_3 (10 components). Table 10 illustrates comprehensive analysis of classification performance over different classifiers for feature sets S_4 (12 components), S_5 (15 components), and S_6 (17 components). As expressed in Table 10, Ensemble Subspace KNN classifier scores highest average accuracy of 98.4%, for 12 feature components, i.e., S_4 feature set.

Figure 11 shows the performance in terms of accuracy for different feature sets in various classifiers for distinguishing normal and hypertension classes using PuPG signals. Figure 12 demonstrates the specificity performance of several classifiers for various features sets from DWT based preprocessing method. Figure 13 presents a graphical comparison of the sensitivity performance of several classifiers for different feature combinations.

NBG classifier achieves highest specificity performance of 100% for feature sets S_3, S_4, S_5, and S_6 (Figure 12), but it reaches maximum sensitivities of 26%, 26%, 32%, and 34% for the same feature sets (Figure 13); therefore, it results in significant reduction of overall classifier accuracy of NBG. The sensitivity performance is 100% for several classifiers (LD, LR, NBG, SVM-FG, SVM-MG, EBT) for feature set S_1 (Figure 13), but the specificity performance is comparatively low.

Table 9. Consolidated result analysis of feature sets (S_1, S_2, S_3) for method I with various classifiers.

Classifier	S_1 (5 Components)				S_2 (7 Components)				S_3 (10 Components)			
	Acc	Sp	Sen	Err	Acc	Sp	Sen	Err	Acc	Sp	Sen	Err
DT	0.874	0.89	0.89	0.126	0.924	0.92	0.93	0.076	0.934	0.94	0.93	0.066
LD	0.688	0.3	1	0.312	0.688	0.3	1	0.312	0.669	0.3	0.97	0.331
LR	0.688	0.3	1	0.312	0.688	0.3	1	0.312	0.688	0.3	1	0.312
NBG	0.688	0.3	1	0.312	0.688	0.3	1	0.312	0.59	1	0.26	0.41
NBK	0.804	0.91	0.72	0.196	0.83	0.92	0.76	0.17	0.893	0.91	0.88	0.107
SVM-L	0.479	0.56	0.41	0.521	0.587	0.11	0.97	0.413	0.498	0.37	0.6	0.502
SVM-Q	0.527	0.65	0.43	0.473	0.527	0.08	0.89	0.473	0.546	0.41	0.65	0.454
SVM-C	0.47	0.4	0.53	0.53	0.555	0.03	0.98	0.445	0.524	0	0.94	0.476
SVM-FG	0.688	0.3	1	0.312	0.688	0.30	1	0.312	0.688	0.3	1	0.312
SVM-MG	0.688	0.3	1	0.312	0.688	0.30	1	0.312	0.688	0.3	1	0.312
KNN-F	0.937	0.9	0.97	0.063	0.972	0.96	0.98	0.028	0.984	0.97	0.99	0.016
KNN-M	0.792	0.68	0.88	0.208	0.864	0.81	0.91	0.136	0.905	0.86	0.94	0.095
KNN-Cos	0.685	0.3	0.99	0.315	0.681	0.3	0.99	0.319	0.685	0.3	0.99	0.315
KNN-C	0.672	0.68	0.66	0.328	0.871	0.83	0.9	0.129	0.896	0.84	0.94	0.104
KNN-W	0.921	0.88	0.95	0.079	0.965	0.96	0.97	0.035	0.978	0.97	0.98	0.022
Eboost	0.918	0.89	0.94	0.082	0.864	0.74	0.96	0.136	0.555	0	1	0.445
EBT	0.688	0.3	1	0.312	0.972	0.95	0.99	0.028	0.943	0.93	0.95	0.057
ESD	0.94	0.92	0.95	0.06	0.688	0.3	1	0.312	0.681	0.3	0.99	0.319
ESKNN	0.915	0.91	0.91	0.085	0.984	0.98	0.99	0.016	0.981	0.97	0.99	0.019

Table 10. Consolidated result analysis of feature sets (S_4, S_5, S_6) for method I with various classifiers. Bold font indicates best results.

Classifier	S_4 (12 Components)				S_5 (15 Components)				S_6 (17 Components)			
	Acc	Sp	Sen	Err	Acc	Sp	Sen	Err	Acc	Sp	Sen	Err
DT	0.959	0.96	0.96	0.041	0.959	0.94	0.98	0.041	0.972	0.96	98	0.028
LD	0.581	0.3	0.99	0.419	0.662	0.3	0.95	0.338	0.691	0.32	0.99	0.309
LR	0.688	0.3	1	0.312	0.688	0.3	1	0.312	0.675	0.35	0.94	0.325
NBG	0.59	1	0.26	0.41	0.625	1	0.32	0.375	0.631	1	0.34	0.369
NBK	0.868	0.85	0.89	0.132	0.877	0.92	0.84	0.123	0.88	0.91	0.95	0.12
SVM-L	0.524	0.26	0.74	0.476	0.543	0.12	0.88	0.457	0.536	0.07	0.91	0.464
SVM-Q	0.552	0.22	0.82	0.448	0.536	0.33	0.7	0.464	0.546	0.02	0.97	0.454
SVM-C	0.524	0.8	0.88	0.476	0.517	0	0.93	0.483	0.514	0.05	0.89	0.486
SVM-FG	0.688	0.3	1	0.312	0.688	0.3	1	0.312	0.7	0.33	1	0.3
SVM-MG	0.665	0.3	1	0.335	0.688	0.3	1	0.312	0.694	0.31	1	0.306
KNN-F	0.981	0.97	0.99	0.019	0.975	0.97	0.98	0.025	0.915	0.88	0.94	0.085
KNN-M	0.918	0.85	0.97	0.082	0.912	0.85	0.96	0.088	0.659	0.7	0.63	0.341
KNN-Cos	0.688	0.3	1	0.312	0.688	0.3	1	0.312	0.685	0.3	0.99	0.315
KNN-C	0.905	0.85	0.95	0.095	0.909	0.85	0.95	0.091	0.909	0.89	0.93	0.091
KNN-W	0.981	0.98	0.98	0.019	0.975	0.97	0.98	0.025	0.978	0.97	0.98	0.022
Eboost	0.555	0	1	0.445	0.555	0	1	0.445	0.555	0	1	0.445
EBT	0.965	0.94	0.98	0.035	0.972	0.97	0.97	0.028	0.94	0.92	0.95	0.06
ESD	0.688	0.3	1	0.312	0.666	0.3	0.96	0.334	0.681	0.3	0.99	0.319
ESKNN	**0.984**	**0.97**	**0.99**	**0.016**	0.975	0.97	0.98	0.025	0.981	0.99	0.99	0.019

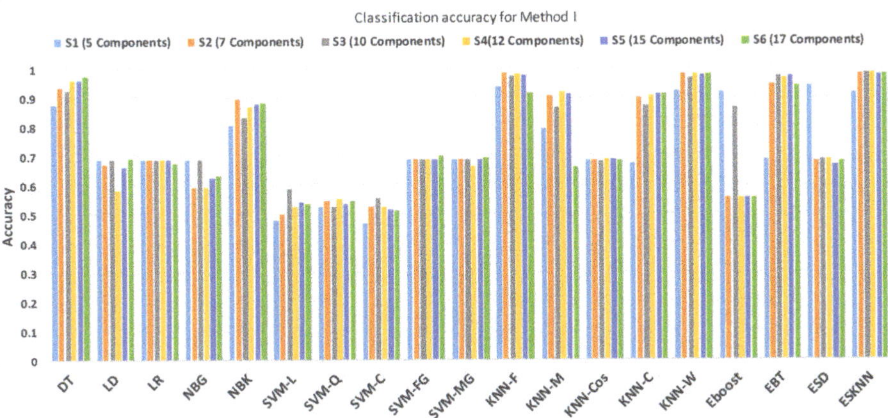

Figure 11. Performance of accuracy for different feature sets in various classifiers for PuPG signal through method I.

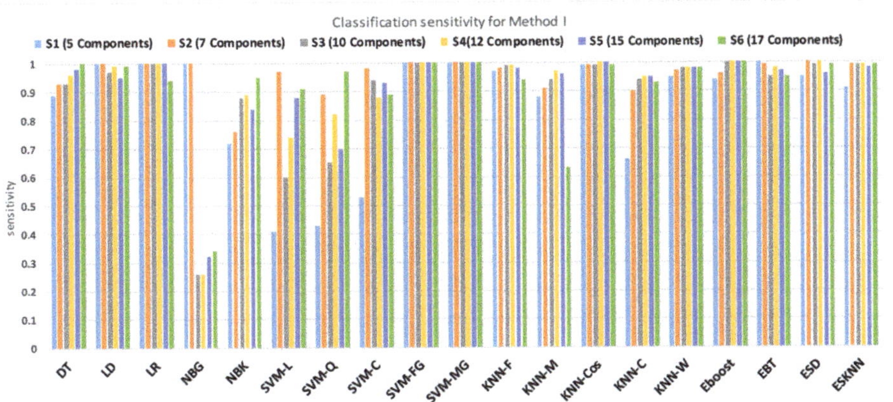

Figure 12. Performance of sensitivity for different feature sets in various classifiers for PuPG signal through method I.

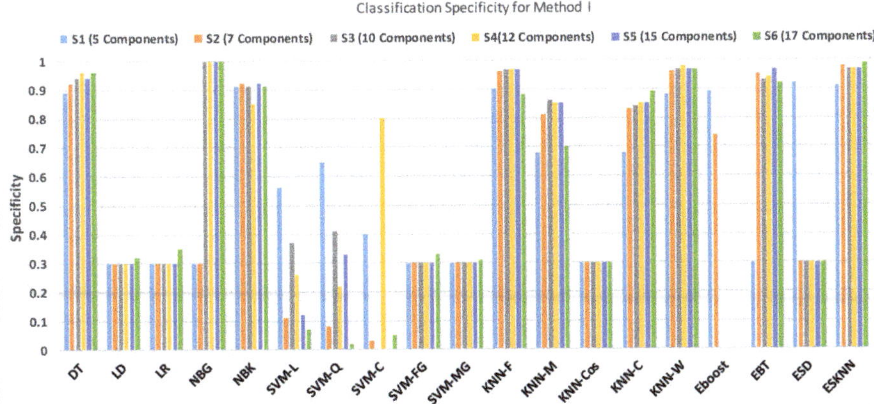

Figure 13. Performance of specificity for different feature sets in various classifiers for PuPG signal through method I.

Figure 14 shows the classification performance results in the form of a confusion matrix for best configurations such as ESKNN classifier with S_4 (12 feature components). The sensitivity of classification is 99%, which means that out of 700 PuPG signals of hypertension, 693 were correctly predicted as hypertension data class while testing, whereas only seven were misclassified as healthy class. The classifier achieved a 98% specificity performance. Out of 709 healthy PuPG signal samples, 695 were accurately predicted as healthy class, whereas the remaining 14 signals were misclassified.

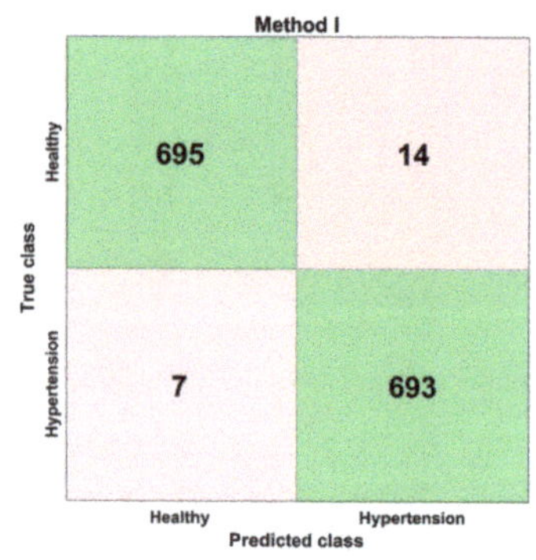

(**a**) Confusion matrix in terms of numbers

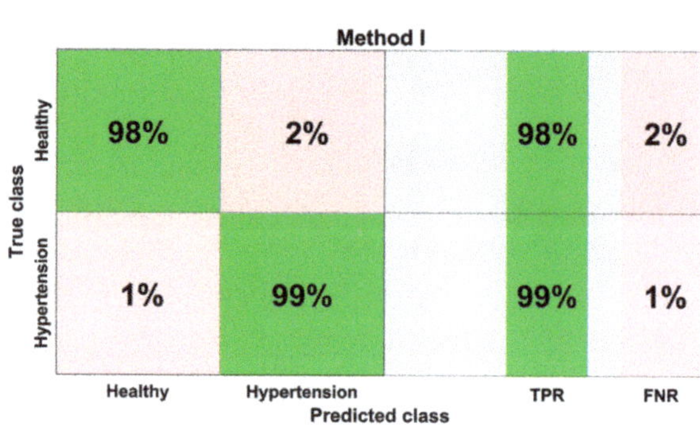

(**b**) Confusion matrix in terms of percentage

Figure 14. Confusion matrix for method I.

Table 11 includes the extensive experimentation results to avoid the classifier overfitting. The selected configuration was tested through 5-, 10-, 15-, and 20-fold cross-validation and 20% and 25% train-test holdout validations.

Table 11. Validation of the selected scheme of method I.

Evaluation	Classes	Accuracy	True Positive Rate	False Negative Rate
5-Fold Cross-Validation	Healthy Hypertension	0.983	0.98 0.99	0.02 0.01
10-Fold Cross-Validation	Healthy Hypertension	0.984	0.98 0.99	0.02 0.01
15-Fold Cross-Validation	Healthy Hypertension	0.984	0.98 0.99	0.02 0.01
20-Fold Cross-Validation	Healthy Hypertension	0.984	0.98 0.99	0.02 0.01
20% Hold Out Validation	Healthy Hypertension	0.978	1 0.94	0 0.06
25% Hold Out Validation	Healthy Hypertension	0.989	0.98 1	0.02 0

4.2. Method II

This section is primarily focused on the second method that is under discussion for this research. It encompasses the results of the classification of the features extracted after the preprocessing of the PuPG signal via EMD. A certain number of feature sets were chosen that were the result of the HFSR. The feature sets comprising of 5, 7, 10, 12, 15, and 17 transformed features were chosen and named S_1, S_2, S_3, S_4, S_5, and S_6, respectively. These feature components were fed to a various number of classifiers for classification and their performance was tested through tenfold cross-validation.

Table 12 depicts the results obtained after the classification of the feature set S_1 (5 components), S_2 (7 components), and S_3 (10 components) on using a selection of various classifiers. Table 13 shows the outcomes of various classification techniques applied on feature sets S_4 (12 components), S_5 (15 components), and S_6 (17 components). Analysis of both Tables 12 and 13 show that a maximum average accuracy using the least number of features is 99.4%. This accuracy is the result of the weighted KNN classification method applied on the feature set S_1.

Table 12. Feature analysis table (S_1, S_2, S_3) for method II. Bold font indicates the best results.

Classifier	S_1 (5 Components)				S_2 (7 Components)				S_3 (10 Components)			
	Acc	Sp	Sen	Err	Acc	Sp	Sen	Err	Acc	Sp	Sen	Err
DT	0.974	0.98	0.97	0.026	0.983	0.98	0.99	0.017	0.989	0.98	0.99	0.011
LD	0.619	0.24	1	0.381	0.619	0.24	1	0.381	0.568	0.24	0.9	0.432
LR	0.679	0.97	0.39	0.321	0.679	0.97	0.39	0.321	0.679	0.97	0.39	0.321
NBG	0.619	0.24	1	0.381	0.619	0.24	1	0.381	0.268	1	0.26	0.732
NBK	0.946	0.97	0.92	0.054	0.946	0.98	0.91	0.054	0.94	0.97	0.91	0.06
SVM-L	0.48	0.49	0.47	0.52	0.497	0.59	0.41	0.503	0.523	0.52	0.53	0.477
SVM-Q	0.51	0.53	0.5	0.49	0.511	0.24	0.78	0.489	0.497	0.46	0.53	0.503
SVM-C	0.49	0.3	0.69	0.51	0.491	0.22	0.77	0.509	0.491	0.23	0.76	0.509
SVM-FG	0.668	1	0.34	0.332	0.662	1	0.32	0.338	0.665	0.99	0.34	0.335
SVM-MG	0.619	0.24	1	0.381	0.614	0.51	0.72	0.386	0.597	0.73	0.46	0.403
KNN-F	0.99	0.99	0	0.01	0.893	0.98	0.984	0.107	0.991	0.99	0.99	0.009
KNN-M	0.957	0.93	0.98	0.043	0.969	0.95	0.99	0.031	0.972	0.95	0.99	0.028
KNN-Cos	0.631	0.27	0.99	0.369	0.639	0.3	0.98	0.361	0.636	0.3	0.98	0.364
KNN-C	0.957	0.93	0.98	0.043	0.966	0.94	0.99	0.034	0.969	0.94	0.99	0.031
KNN-W	**0.994**	**0.992**	**0.996**	**0.006**	0.986	0.94	0.99	0.014	0.992	0.99	0.99	0.008
Eboost	0.489	0.39	0.59	0.511	0.489	0.39	0.59	0.511	0.489	0.39	0.59	0.511
EBT	0.98	0.97	0.99	0.02	0.986	0.98	0.99	0.014	0.986	0.98	0.99	0.014
ESD	0.619	0.24	1	0.381	0.619	0.24	1	0.381	0.571	0.24	0.9	0.429
ESKNN	0.991	0.99	0.99	0.009	0.983	0.99	0.98	0.017	0.991	0.99	0.99	0.009

Table 13. Feature analysis table (S_4, S_5, S_6) for method II.

Classifier	S_4 (12 Components)				S_5 (15 Components)				S_6 (17 Components)			
	Acc	Sp	Sen	Err	Acc	Sp	Sen	Err	Acc	Sp	Sen	Err
DT	0.992	0.99	0.99	0.008	0.972	0.95	0.99	0.028	0.983	0.98	0.98	0.017
LD	0.548	0.32	0.78	0.452	0.565	0.34	0.8	0.435	0.665	1	0.33	0.335
LR	0.679	0.97	0.39	0.321	0.679	0.97	0.39	0.321	0.676	0.97	0.38	0.324
NBG	0.636	0.97	0.39	0.364	0.662	1	0.32	0.338	0.665	1	0.33	0.335
NBK	0.92	0.96	0.88	0.08	0.926	0.97	0.89	0.074	0.909	0.95	0.87	0.091
SVM-L	0.531	0.27	0.79	0.469	0.486	0.23	0.74	0.514	0.5	0.24	0.76	0.5
SVM-Q	0.503	0.19	0.82	0.497	0.469	0.15	0.79	0.531	0.514	0.15	0.88	0.486
SVM-C	0.472	0	0.94	0.528	0.472	0.1	0.85	0.528	0.486	0	0.97	0.514
SVM-F	0.662	1	0.32	0.338	0.662	1	0.32	0.338	0.696	1	0.39	0.304
SVM-MG	0.665	1	0.33	0.335	0.662	1	0.32	0.338	0.696	1	0.39	0.304
KNN-F	0.991	0.99	0.99	0.009	0.983	0.98	0.98	0.017	0.989	0.98	0.99	0.011
KNN-M	0.949	0.94	0.96	0.051	0.96	0.94	0.98	0.04	0.94	0.97	0.91	0.06
KNN-Cos	0.639	0.28	1	0.361	0.628	0.28	0.97	0.372	0.645	0.3	0.99	0.355
KNN-C	0.946	0.94	0.95	0.054	0.963	0.94	0.98	0.037	0.94	0.98	0.9	0.06
KNN-W	0.991	0.99	0.99	0.009	0.986	0.99	0.98	0.014	0.993	0.99	0.99	0.007
Eboost	0.489	0.39	0.59	0.511	0.534	0.48	0.59	0.466	0.489	0.39	0.59	0.511
EBT	0.989	0.98	0.99	0.011	0.966	0.97	0.96	0.034	0.986	0.99	0.98	0.014
ESD	0.577	0.26	0.89	0.423	0.563	0.39	0.73	0.437	0.665	1	0.33	0.335
ESKNN	0.991	0.99	0.99	0.009	0.983	0.98	0.98	0.017	0.991	0.99	0.99	0.009

Figure 15 shows a comparison of the performance of various classifiers based on the accuracy achieved as a result of distinguishing hypertension and normal PuPG signal. Figure 16 depicts the comparison result of various classifiers based on their specificities after using EMD as the preprocessing technique. Figure 17 represents the comparison of the sensitivities of various classification methods. NBG classifier achieves the highest specificity performance of 100% for feature sets S_3 (Figure 16), but it reaches maximum sensitivities of 26% for the same feature set (Figure 17). The sensitivity performance is 100% for several classifiers (LD, NBG, SVM-MG, ESD) for feature set S_1 (Figure 17), but the specificity performance is comparatively low.

Figure 18 illustrates the best classification performance in the form of a confusion matrix for selected features set (S_1) with KNN-W classifier. The sensitivity of classification is more than 99%, which means only one out of 700 PuPG signals was wrong predicted as hypertension data class, whereas the remaining 699 PuPG signals were correctly identified as hypertension. Out of 709 healthy PuPG signals, 702 were correctly predicted as healthy, achieving specificity of 99%. The overall average classification accuracy in the best configuration with the KNN-W classifier was 99.4%.

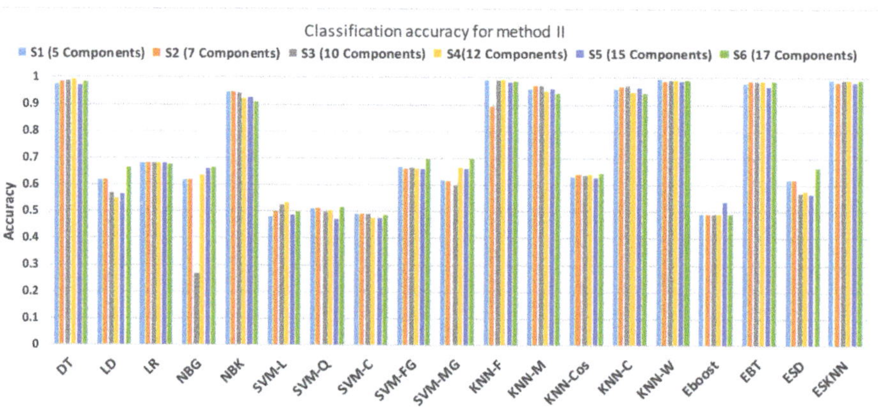

Figure 15. Performance of accuracy for different feature sets in various classifiers for PuPG signal through method II.

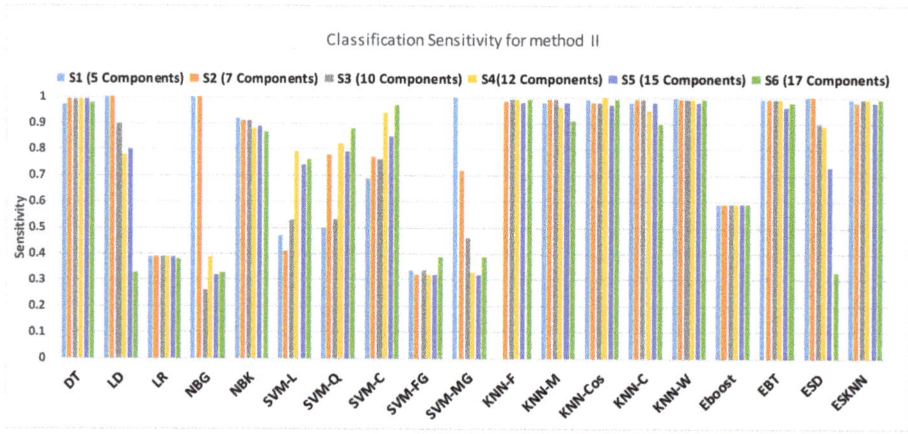

Figure 16. Performance of sensitivity for different feature sets in various classifiers for PuPG signal through method II.

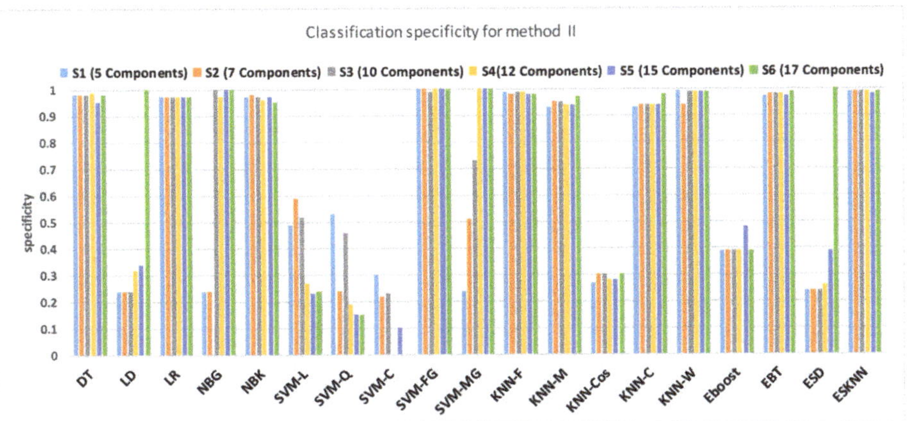

Figure 17. Performance of specificity for different feature sets in various classifiers for PuPG signal through method II.

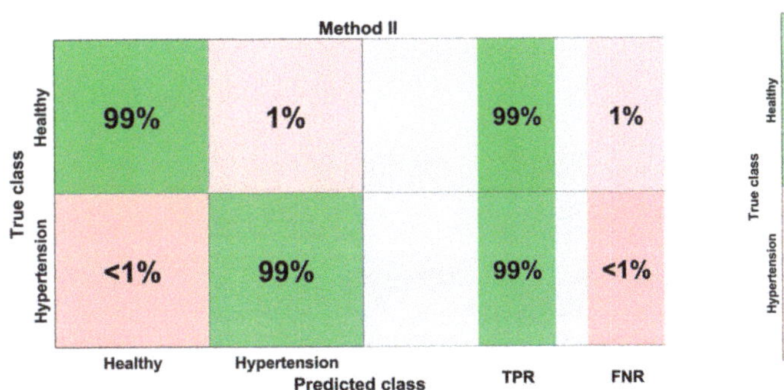

(**a**) Confusion matrix in terms of percentage

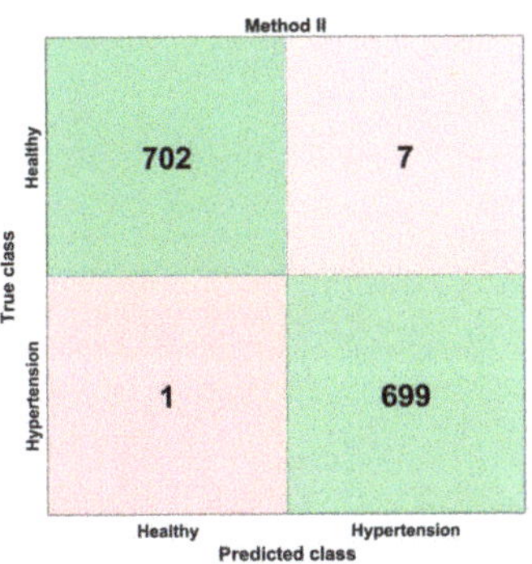

(**b**) Confusion matrix in terms of numbers

Figure 18. Confusion matrix for method II.

Table 14 includes the results of comprehensive experimentation which is performed to avoid the classifier overfitting. The selected framework was examined through 5-, 10-, 15-, and 20-fold cross-validation and 20% and 25% train-test holdout validations. For all experimental settings, the proposed scheme achieved more than 98% accuracy.

Table 14. Validation of the selected scheme of method II.

Evaluation	Classes	Accuracy	True Positive Rate	False Negative Rate
5 Fold Cross-Validation	Healthy Hypertension	0.986	0.99 0.98	0.01 0.02
10 Fold Cross-Validation	Healthy Hypertension	0.994	0.99 >0.99	0.01 <0.01
15 Fold Cross-Validation	Healthy Hypertension	0.994	0.99 >0.99	0.01 <0.01
20 Fold Cross-Validation	Healthy Hypertension	0.997	0.99 1	0.01 0
20% Hold Out Validation	Healthy Hypertension	0.986	1 0.97	0 0.03
25% Hold Out Validation	Healthy Hypertension	0.989	0.98 1	0.02 0

4.3. Method I versus Method II: A Comparative Analysis

This section aims to compare both methods I and II analytically. Based on this comparison, we figure out the best working solution for the detection of hypertension through PuPG signals. Method I comprises of preprocessing of PuPG signals through DWT, followed by feature extraction. Extracted features were subjected to the HFSR scheme and finally classified through Ensemble Subspace KNN. Method II consists of EMD-based signal preprocessing followed by feature extraction. Features were fed to KNN-W classifier for distinguishing normal and hypertension data classes after being reduced through the HFSR approach.

Table 15 shows the performance comparison of methods I and II in terms of average accuracy, sensitivity, specificity, error, and number of features. Method I achieves classification performance of 98.4% accuracy, 97% sensitivity, and 99% specificity using 12 transformed features. Method II obtains 99.4%, 99.2%, and 99.6% results of classification accuracy, sensitivity, and specificity respectively through only five reduced features.

Table 15. Performance comparison of methods I and II.

Performance	Method I	Method II
Accuracy	98.40%	99.40%
Sensitivity	97.00%	99.20%
Specificity	99.00%	99.60%
Error	0.02%	0.60%
# of features	12	5

Comparative analysis of both methods establishes that method II outperforms method I in terms of achieving better classification accuracy on a reduced number of features. This might be due to the fact that the accuracy achieved in the case of DWT highly depends on the proper wavelet basis selection [64]. The selection of an appropriate basis is challenging especially for non-stationary data [65]. On the other hand, EMD is a fully data-driven, adaptive, and basis-less transformation [66]. Moreover, the IMF selection process of EMD based on relative energy and mean frequency has assisted the selection of useful discriminative signal characteristics.

Figure 19 presents the finalized EHDS (expert hypertension detection system) based on PuPG signal analysis. EHDS first takes raw PuPG signal as input and performs preprocessing through EMD by rejecting the irrelevant IMFs. Next, only 24 significant features highlighted by the hybrid selection scheme are extracted and reduced through KPCA. The final transformed 1×5 feature vector is fed to KNN-W to distinguish the normal and hypertension data classes. Figure 20 illustrates the classification performance of the proposed EHDS as a function of the number of transformed features. It can be observed that the proposed EHDS achieves the optimum performance on only five transformed features. The classification performance shows no notable improvement with the increase in the number of features.

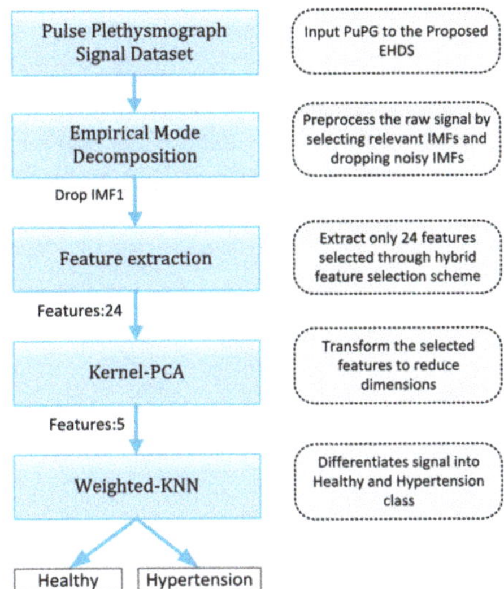

Figure 19. Proposed EHDS block diagram.

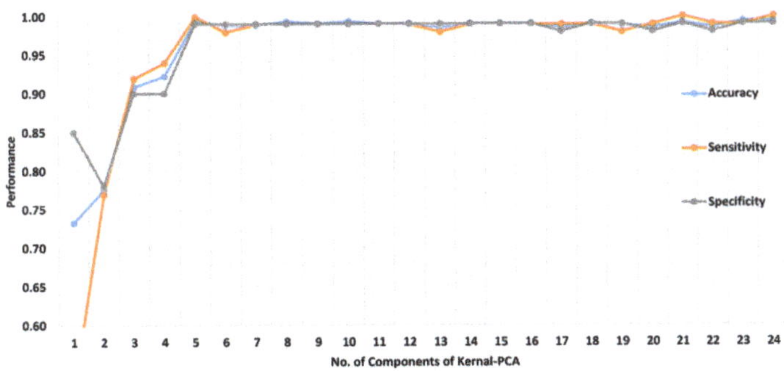

Figure 20. Performance of method II in terms of accuracy, sensitivity, and specificity for 1 to 24 transformed features.

5. Discussion

Human blood vessels and the microcirculation system experience transformations with the rise in blood pressure (BP); these changes are exceptionally obvious for patients with severe hypertension. PuPG signals carry a wealth of information about the cardiac health [25–27]. The PuPG signal reflects physical changes in blood volume pressure in blood vessels during the cardiac cycle. The features extracted in this study indicate the changes in Normal and Hypertension PuPG signals acquired from various subjects. The high classification performance of EHDS reflects the association of extracted transformed features with the physiological characteristics of the cardiac condition of the subject. Thus, the proposed expert system may provide a good approximation of the presence or absence of non-communicable diseases such as hypertension.

Table 16 presents a performance comparison of the recent studies. A diagnostic index for the classification of low and high-risk hypertension classes attaining accuracy of 100% was proposed by [17]. In contrast, our work is targeted towards the classification of Normal and Hypertension classes through PuPG signals. In another study, [18] developed a computational intelligence tool based on ECG signals for the classification of normal and hypertension. EMD was employed in the signal preprocessing stage, followed by nonlinear feature extraction from the decomposed IMFs. Extracted features were ranked through Student's *t*-test. The highest classification accuracy of 97.70% was obtained through the KNN classifier with tenfold cross-validation. A photoplethysmograph (PPG) based detection of hypertension was proposed by [19]. A total of 125 features of various types were extracted and reduced through MRMR. The authors reported the best classification performance with KNN-W, specifically to be 100%, 85.71%, and 92.31% for positive predictive value, sensitivity, and F1-score, respectively.

Table 16. Comparison with previous works.

Ref.	Modality	Preprocessing	Features	Feature Reduction	Classification	Data Set	Results
[12]	PPG	CWT	GoogLeNet	-	GoogLeNet	MIMIC	F1 score: 92.55%
[13]	PPG and ECG	-	PAT and morphological features	-	KNN	MIMIC	F1 score: 94.84%
[14]	HRV	-	Standard deviation of NN intervals	-	MIL	Self-collected data set Hypertension 24 and Normal: 19	Accuracy: 85.47%
[15]	ECG	SGF	Entropy features	-	SVM	Self-collected data set Hypertension: 61 and Normal: 67	Accuracy: 93.33%
[16]	HRV	-	Statistical, spectral, geometrical, wavelet, fractal, and non-linear features	PCA	QDA	Self-collected data set Hypertension: 41 Normal: 30	Accuracy: 85.5%
[17]	ECG	OWFB	Fractal dimension and energy features	Student's *t*-test	Diagnosis index	PhysioNet database High-risk Hypertension: 17 subjects Low-risk Hypertension: 122 subjects Total: 139 subjects	100% between low-risk and high-risk classes
[18]	ECG	EMD	Entropy features	Student's *t*-test	KNN classifier	MIT BIH Sinus rhythm database, SHAREE database: Normal: 18 signals Hypertension: 139 signals	Accuracy: 97.70% Sensitivity: 98.90% Specificity: 89.10%
[19]	PPG	Chebyshev II	Time and morphological features	MRMR	KNN-W	Hypertension: 35 Normal: 48 Total: 83	Positive Predictive Value: 100% Sensitivity: 85.71% F1-score: 92.31%
[20]	BCG		Morphological features	-	CAR	Self-collected data set Hypertension: 61 and Normal: 67	Accuracy: 84.4%
This study	PuPG	EMD	Time, frequency, cepstral, fractal, and chaotic features	HFSR	KNN-W	Self-collected data set Hypertension: 56 Normal: 65	Accuracy: 99.7% Sensitivity: 99.2% Specificity: 99.4%

The current research is focused on the classification between normal and hypertension data through PuPG signals. To the best of author's knowledge, this is the first study that uses the PuPG signals for discriminating among normal and hypertension with high precision. The current method achieves better performance than the existing ECG- [15,17,18], PPG- [12,19], HRV- [14,16], and BCG-based [20] approaches. Our method also outperforms the fusion-based method for detection of hypertension that utlized a combination of PPG and ECG [13].

The proposed expert system could play a vital role in the early detection of hypertension in low- and middle-income countries. It is important to mention that an estimated 1.04 billion population suffered from hypertension in 2010 [67]. A non-invasive technique based on PuPG signals analysis proposed in this research could be used for the detection of non-communicable diseases.

6. Conclusions

Early detection of hypertension or high blood pressure is extremely significant since it does not cause any obvious symptoms in many people; hence, it can harm the heart, the kidneys, and even the brain. In this study, we proposed an automated detection system for hypertension from PuPG signals for timely and precise screening of disease. First, PuPG signals were preprocessed through EMD, followed by feature extraction of various types. Highly discriminative features were selected through the proposed HFSR scheme that consisted of feature reduction and selection methods. The resultant reduced features of dimension 1×5 were subjected to various classification methods. The KNN-W classifier achieved the best performance in terms of accuracy, sensitivity, and specificity of 99.4%, 99.2%, and 99.6%, respectively. To compute the model performance and avoid overfitting, 5-, 10-, 15-, and 20-fold cross-validations were employed. The proposed method was also compared with the DWT based preprocessing scheme followed by the same feature extraction, selection (HFSR), and classification pipeline. The main advantages of this research are as follows:

- The proposed EHDS system is based on the non-invasive methodology of PuPG signals.
- The EHDS is reliable and less computational intensive with high accuracy.
- The EHDS avoids overfitting as it is validated through 5-, 10-, 15-, and 20-fold cross-validation.
- The proposed approach does not only rely on morphological characteristics of the acquired signal.
- The method can be completely automated, and it works with all qualities of PuPG signals.

Despite the enormous advantages of the proposed method, it has a few limitations.

- The data set used in this research is yet small, with each sample with a length of 10 s.
- The procedure of initial feature extraction and selection of proper IMFs in EMD made the overall process strenuous and time-consuming.

The proposed study conducted a comprehensive comparison of preprocessing schemes (DWT and EMD), feature analysis, selection, and classification as illustrated in Figure 3. The computational complexity of the proposed is significantly low due to the fact that it operates on trained classifier models, therefore eliminating the training computational cost (Figure 19). The proposed system has the potential to be deployed in clinical environments and intensive care units where it can contribute to lessen the workload of medical professionals through its accurate detection and timely diagnosis. In future works, our research group aims to increase the data set size and apply deep learning models to automate the feature extraction process. The proposed framework is intended to be implemented on portable embedded platforms.

Author Contributions: Conceptualization, M.U.K. and S.A.; methodology, M.U.K. and S.A.; software, K.I. and T.A.; validation, F.A. and K.I.; formal analysis, T.A. and K.I.; investigation, K.I. and F.A.; writing—original draft preparation, S.A. and M.U.K.; writing—review and editing, M.A.K. and Y.N.; funding acquisition, Y.N.; project administration, T.A.; data curation, F.A. and M.A.K. All authors have read and agreed to the published version of the manuscript.

Funding: This research was supported by Korea Institute for Advancement of Technology (KIAT) grant funded by the Korea Government (MOTIE) (P0012724, The Competency Development Program for Industry Specialist) and the Soonchunhyang University Research Fund.

Institutional Review Board Statement: Not applicable.

Informed Consent Statement: Informed consent was obtained from all subjects involved in the study.

Data Availability Statement: Data will be available on request.

Conflicts of Interest: The authors declare no conflict of interest.

References

1. Tayefi, M.; Esmaeili, H.; Karimian, M.S.; Zadeh, A.A.; Ebrahimi, M.; Safarian, M.; Nematy, M.; Parizadeh, S.M.R.; Ferns, G.A.; Ghayour-Mobarhan, M. The application of a decision tree to establish the parameters associated with hypertension. *Comput. Methods Programs Biomed.* **2017**, *139*, 83–91. [CrossRef] [PubMed]
2. WHO. Cardiovascular Diseases. 2017. Available online: https://www.who.int/news-room/fact-sheets/detail/cardiovascular-diseases-(cvds) (accessed on 23 October 2020).
3. WHO. Hypertension, 13 September 2019. Available online: https://www.who.int/news-room/fact-sheets/detail/hypertension (accessed on 23 October 2020).
4. Poddar, M.; Kumar, V.; Sharma, Y.P. Heart rate variability based classification of normal and hypertension cases by linear-nonlinear method. *Def. Sci. J.* **2014**, *64*, 542. [CrossRef]
5. Li, W.; Gu, H.; Teo, K.K.; Bo, J.; Wang, Y.; Yang, J.; Wang, X.; Zhang, H.; Sun, Y.; Jia, X.; et al. Hypertension prevalence, awareness, treatment, and control in 115 rural and urban communities involving 47,000 people from China. *J. Hypertens.* **2016**, *34*, 39–46. [CrossRef] [PubMed]
6. Wall, H.K.; Hannan, J.A.; Wright, J.S. Patients with undiagnosed hypertension: Hiding in plain sight. *JAMA* **2014**, *312*, 1973–1974. [CrossRef] [PubMed]
7. Saleem, F.; Hassali, A.A.; Shafie, A.A. Hypertension in Pakistan: Time to take some serious action. *Br. J. Gen. Pract.* **2010**, *60*, 449–450. [CrossRef] [PubMed]
8. Benjamin, E.J.; Blaha, M.J.; Chiuve, S.E.; Cushman, M.; Das, S.R.; Deo, R.; De Ferranti, S.D.; Floyd, J.; Fornage, M.; Gillespie, C.; et al. Heart disease and stroke statistics—2017 update. *Circulation* **2017**, *135*, e146–e603. [CrossRef]
9. Goodhart, A.K. Hypertension from the patient's perspective. *Br. J. Gen. Pract.* **2016**, *66*, 570. [CrossRef]
10. Huang, C.; Wang, Y.; Li, X.; Ren, L.; Zhao, J.; Hu, Y.; Zhang, L.; Fan, G.; Xu, J.; Gu, X.; et al. Clinical features of patients infected with 2019 novel coronavirus in Wuhan, China. *Lancet* **2020**, *395*, 497–506. [CrossRef]
11. Schiffrin, E.L.; Flack, J.M.; Ito, S.; Muntner, P.; Webb, R.C. Hypertension and COVID-19. *Am. J. Hypertens.* **2020**, *33*, 373–374. [CrossRef]
12. Liang, Y.; Chen, Z.; Ward, R.; Elgendi, M.J.B. Photoplethysmography and deep learning: Enhancing hypertension risk stratification. *Biosensors* **2018**, *8*, 101. [CrossRef]
13. Liang, Y.; Chen, Z.; Ward, R.; Elgendi, M. Hypertension assessment via ECG and PPG signals: An evaluation using MIMIC database. *Diagnostics* **2018**, *8*, 65. [CrossRef] [PubMed]
14. Lan, K.c.; Raknim, P.; Kao, W.F.; Huang, J.H. Toward hypertension prediction based on PPG-derived HRV signals: A feasibility study. *J. Med. Syst.* **2018**, *42*, 103. [CrossRef] [PubMed]
15. Ni, H.; Cho, S.; Mankoff, J.; Yang, J.J.; Computing, H. Automated recognition of hypertension through overnight continuous HRV monitoring. *J. Ambient. Intell. Humaniz. Comput.* **2018**, *9*, 2011–2023. [CrossRef]
16. Kublanov, V.S.; Dolganov, A.Y.; Belo, D.; Gamboa, H.J. Comparison of machine learning methods for the arterial hypertension diagnostics. *Appl. Bionics Biomech.* **2017**, *2017*, 5985479. [CrossRef] [PubMed]
17. Rajput, J.S.; Sharma, M.; Acharya, U.R. Hypertension Diagnosis Index for Discrimination of High-Risk Hypertension ECG Signals Using Optimal Orthogonal Wavelet Filter Bank. *Int. J. Environ. Res. Public Health* **2019**, *16*, 4068. [CrossRef]
18. Soh, D.C.K.; Ng, E.; Jahmunah, V.; Oh, S.L.; San, T.R.; Acharya, U.R. A computational intelligence tool for the detection of hypertension using empirical mode decomposition. *Medicine* **2020**, *118*, 103630. [CrossRef]
19. Liang, Y.; Chen, Z.; Ward, R.; Elgendi, M. Hypertension assessment using photoplethysmography: A risk stratification approach. *J. Clin. Med.* **2019**, *8*, 12. [CrossRef]
20. Liu, F.; Zhou, X.; Wang, Z.; Cao, J.; Wang, H.; Zhang, Y. Unobtrusive mattress-based identification of hypertension by integrating classification and association rule mining. *Sensors* **2019**, *19*, 1489. [CrossRef]
21. Baranchuk, A.; Kang, J.; Shaw, C.; Campbell, D.; Ribas, S.; Hopman, W.M.; Alanazi, H.; Redfearn, D.P.; Simpson, C.S. Electromagnetic interference of communication devices on ECG machines. *Clin. Cardiol. Int. Index. Peer Rev. J. Adv. Treat. Cardiovasc. Dis.* **2009**, *32*, 588–592. [CrossRef]
22. Klein, A.; Djaiani, G. Mobile phones in the hospital–past, present and future. *Anaesthesia* **2003**, *58*, 353–357. [CrossRef]
23. Kranjec, J.; Beguš, S.; Geršak, G.; Drnovšek, J. Non-contact heart rate and heart rate variability measurements: A review. *Biomed. Signal Process. Control.* **2014**, *13*, 102–112. [CrossRef]
24. Ave, A.; Fauzan, H.; Adhitya, S.R.; Zakaria, H. Early detection of cardiovascular disease with photoplethysmogram (PPG) sensor. In Proceedings of the 2015 International Conference on Electrical Engineering and Informatics (ICEEI), Denpasar, Indonesia, 10–11 August 2015; pp. 676–681.
25. Khan, M.U.; Aziz, S.; Malik, A.; Imtiaz, M.A. Detection of Myocardial Infarction using Pulse Plethysmograph Signals. In Proceedings of the 2019 International Conference on Frontiers of Information Technology (FIT), Islamabad, Pakistan, 16–18 December 2019; pp. 95–955.
26. Khan, M.U.; Aziz, S.; Amjad, F.; Mohsin, M. Detection of Dilated Cardiomyopathy using Pulse Plethysmographic Signal Analysis. In Proceedings of the 2019 22nd International Multitopic Conference (INMIC), Islamabad, Pakistan, 29–30 November 2019; pp. 1–5.

27. Khan, M.U.; Aziz, S.; Iqtidar, K.; Zainab, A.; Saud, A. Prediction of Acute Coronary Syndrome Using Pulse Plethysmograph. In Proceedings of the 2019 4th International Conference on Emerging Trends in Engineering, Sciences and Technology (ICEEST), Karachi, Pakistan, 10–11 December 2019; pp. 1–6.
28. Naqvi, S.Z.H.; Aziz, S.; Khan, M.U.; Asghar, N.; Rasool, G. Emotion Recognition System using Pulse Plethysmograph. In Proceedings of the 2020 International Conference on Emerging Trends in Smart Technologies (ICETST), Karachi, Pakistan, 26–27 March 2020; pp. 1–6.
29. Khan, M.U.; Aziz, S.; Naqvi, S.Z.H.; Zaib, A.; Maqsood, A. Pattern Analysis Towards Human Verification using Photoplethysmograph Signals. In Proceedings of the 2020 International Conference on Emerging Trends in Smart Technologies (ICETST), Karachi, Pakistan, 26–27 March 2020; pp. 1–6.
30. Zhang, G.; Si, Y.; Yang, W.; Wang, D. A Robust Multilevel DWT Densely Network for Cardiovascular Disease Classification. *Sensors* **2020**, *20*, 4777. [CrossRef] [PubMed]
31. Aileni, R.M.; Pasca, S.; Florescu, A. EEG-Brain Activity Monitoring and Predictive Analysis of Signals Using Artificial Neural Networks. *Sensors* **2020**, *20*, 3346. [CrossRef] [PubMed]
32. Alturki, F.A.; AlSharabi, K.; Abdurraqeeb, A.M.; Aljalal, M. EEG Signal Analysis for Diagnosing Neurological Disorders Using Discrete Wavelet Transform and Intelligent Techniques. *Sensors* **2020**, *20*, 2505. [CrossRef] [PubMed]
33. Gradolewski, D.; Redlarski, G. Wavelet-based denoising method for real phonocardiography signal recorded by mobile devices in noisy environment. *Comput. Biol. Med.* **2014**, *52*, 119–129. [CrossRef] [PubMed]
34. Ebrahim, M.P.; Heydari, F.; Redoute, J.M.; Yuce, M.R. Accurate heart rate detection from on-body continuous wave radar sensors using wavelet transform. In Proceedings of the 2018 IEEE SENSORS, New Delhi, India, 28–31 October 2018; pp. 1–4.
35. Ebrahim, M.P.; Heydari, F.; Walker, K.; Joe, K.; Redoute, J.M.; Yuce, M.R. Systolic Blood Pressure Estimation Using Wearable Radar and Photoplethysmogram Signals. In Proceedings of the 2019 IEEE International Conference on Systems, Man and Cybernetics (SMC), Bari, Italy, 6–9 October 2019; pp. 3878–3882.
36. Asghar, M.A.; Khan, M.J.; Rizwan, M.; Mehmood, R.M.; Kim, S.H. An Innovative Multi-Model Neural Network Approach for Feature Selection in Emotion Recognition Using Deep Feature Clustering. *Sensors* **2020**, *20*, 3765. [CrossRef] [PubMed]
37. Aziz, S.; Khan, M.U.; Alhaisoni, M.; Akram, T.; Altaf, M. Phonocardiogram Signal Processing for Automatic Diagnosis of Congenital Heart Disorders through Fusion of Temporal and Cepstral Features. *Sensors* **2020**, *20*, 3790. [CrossRef]
38. Tsimpiris, A.; Kugiumtzis, D. Feature selection for classification of oscillating time series. *Expert Syst.* **2012**, *29*, 456–477. [CrossRef]
39. Goyal, D.; Choudhary, A.; Pabla, B.; Dhami, S. Support vector machines based non-contact fault diagnosis system for bearings. *J. Intell. Manuf.* **2019**, *31*, 1275–1289. [CrossRef]
40. Fasil, O.; Rajesh, R. Time-domain exponential energy for epileptic EEG signal classification. *Neurosci. Lett.* **2019**, *694*, 1–8.
41. Banker, R.; Natarajan, R.; Zhang, D. Two-stage estimation of the impact of contextual variables in stochastic frontier production function models using data envelopment analysis: Second stage OLS versus bootstrap approaches. *Eur. J. Oper. Res.* **2019**, *278*, 368–384. [CrossRef]
42. Kumar, D.; Carvalho, P.; Antunes, M.; Paiva, R.; Henriques, J. Heart murmur classification with feature selection. In Proceedings of the 2010 Annual International Conference of the IEEE Engineering in Medicine and Biology, Buenos Aires, Argentina, 31 August–4 September 2010; pp. 4566–4569.
43. Dhindsa, I.S.; Agarwal, R.; Ryait, H.S. Performance evaluation of various classifiers for predicting knee angle from electromyography signals. *Expert Syst.* **2019**, *36*, e12381. [CrossRef]
44. Mukhopadhyay, A.K.; Samui, S. An experimental study on upper limb position invariant EMG signal classification based on deep neural network. *Biomed. Signal Process. Control.* **2020**, *55*, 101669. [CrossRef]
45. Phinyomark, A.; Phukpattaranont, P.; Limsakul, C. Feature reduction and selection for EMG signal classification. *Expert Syst. Appl.* **2012**, *39*, 7420–7431. [CrossRef]
46. Peeters, G. A large set of audio features for sound description (similarity and classification) in the CUIDADO project. *Cuid. IST Proj. Rep.* **2004**, *54*, 1–25.
47. Yadav, A.; Singh, A.; Dutta, M.K.; Travieso, C.M. Machine learning-based classification of cardiac diseases from PCG recorded heart sounds. *Neural Comput. Appl.* **2019**, 1–14. [CrossRef]
48. Amezquita-Sanchez, J.P.; Mammone, N.; Morabito, F.C.; Marino, S.; Adeli, H. A novel methodology for automated differential diagnosis of mild cognitive impairment and the Alzheimer's disease using EEG signals. *J. Neurosci. Methods* **2019**, *322*, 88–95. [CrossRef]
49. Peng, J.; Hao, D.; Yang, L.; Du, M.; Song, X.; Jiang, H.; Zhang, Y.; Zheng, D. Evaluation of electrohysterogram measured from different gestational weeks for recognizing preterm delivery: A preliminary study using random Forest. *Biocybern. Biomed. Eng.* **2020**, *40*, 352–362. [CrossRef]
50. Birajdar, G.K.; Patil, M.D. Speech/music classification using visual and spectral chromagram features. *J. Ambient. Intell. Humaniz. Comput.* **2020**, *11*, 329–347. [CrossRef]
51. Leite, J.P.R.; Moreno, R.L. Heartbeat classification with low computational cost using Hjorth parameters. *IET Signal Process.* **2017**, *12*, 431–438. [CrossRef]
52. Dua, M.; Aggarwal, R.K.; Biswas, M. GFCC based discriminatively trained noise robust continuous ASR system for Hindi language. *J. Ambient. Intell. Humaniz. Comput.* **2019**, *10*, 2301–2314. [CrossRef]

53. Adnan, S.M.; Irtaza, A.; Aziz, S.; Ullah, M.O.; Javed, A.; Mahmood, M.T. Fall detection through acoustic local ternary patterns. *Appl. Acoust.* **2018**, *140*, 296–300. [CrossRef]
54. Guyon, I. Practical feature selection: From correlation to causality. In Proceedings of the NATO Advanced Study Institute on Mining Massive Data Sets for Security (MMDSS 2007), Gazzada, Italy, 10–21 September 2007; pp. 27–43.
55. Li, X.; Ling, S.H.; Su, S. A Hybrid Feature Selection and Extraction Methods for Sleep Apnea Detection Using Bio-Signals. *Sensors* **2020**, *20*, 4323. [CrossRef]
56. Zhen, Z.; Zeng, X.; Wang, H.; Han, L. A global evaluation criterion for feature selection in text categorization using Kullback-Leibler divergence. In Proceedings of the 2011 International Conference of Soft Computing and Pattern Recognition (SoCPaR), Dalian, China, 14–16 October 2011; pp. 440–445.
57. Guorong, X.; Peiqi, C.; Minhui, W. Bhattacharyya distance feature selection. In Proceedings of the 13th International Conference on Pattern Recognition, Vienna, Austria, 25–29 August 1996; Volume 2, pp. 195–199.
58. Chowdhury, M.H.; Shuzan, M.N.I.; Chowdhury, M.E.; Mahbub, Z.B.; Uddin, M.M.; Khandakar, A.; Reaz, M.B.I. Estimating Blood Pressure from the Photoplethysmogram Signal and Demographic Features Using Machine Learning Techniques. *Sensors* **2020**, *20*, 3127. [CrossRef] [PubMed]
59. Ding, C.; Peng, H. Minimum redundancy feature selection from microarray gene expression data. *J. Bioinform. Comput. Biol.* **2005**, *3*, 185–205. [CrossRef] [PubMed]
60. Xu, Y.; Zhao, X.; Chen, Y.; Zhao, W. Research on a mixed gas recognition and concentration detection algorithm based on a metal oxide semiconductor olfactory system sensor array. *Sensors* **2018**, *18*, 3264. [CrossRef] [PubMed]
61. Schölkopf, B.; Smola, A.; Müller, K.R. Nonlinear component analysis as a kernel eigenvalue problem. *Neural Comput.* **1998**, *10*, 1299–1319. [CrossRef]
62. Khan, M.U.; Saad, M.; Aziz, S.; Ch, J.M.; Naqvi. S.Z.H.; Qasim, M.A. Electrocardiogram based Gender Classification. In Proceedings of the 2020 International Conference on Electrical, Communication, and Computer Engineering (ICECCE), Istanbul, Turkey, 12–13 June 2020; pp. 1–6.
63. Hoffmann, H. Kernel PCA for novelty detection. *Pattern Recognit.* **2007**, *40*, 863–874. [CrossRef]
64. Mohamed, M.; Deriche, M. An approach for ECG feature extraction using daubechies 4 (DB4) wavelet. *Int. J. Comput. Appl.* **2014**, *96*, 36–41.
65. Acharya, U.R.; Fujita, H.; Adam, M.; Lih, O.S.; Sudarshan, V.K.; Hong, T.J.; Koh, J.E.; Hagiwara, Y.; Chua, C.K.; Poo, C.K.; et al. Automated characterization and classification of coronary artery disease and myocardial infarction by decomposition of ECG signals: A comparative study. *Inf. Sci.* **2017**, *377*, 17–29. [CrossRef]
66. Labate, D.; La Foresta, F.; Occhiuto, G.; Morabito, F.C.; Lay-Ekuakille, A.; Vergallo, P. Empirical mode decomposition vs. wavelet decomposition for the extraction of respiratory signal from single-channel ECG: A comparison. *IEEE Sens. J.* **2013**, *13*, 2666–2674. [CrossRef]
67. Mills, K.T.; Bundy, J.D.; Kelly, T.N.; Reed, J.E.; Kearney, P.M.; Reynolds, K.; Chen, J.; He, J. Global disparities of hypertension prevalence and control: A systematic analysis of population-based studies from 90 countries. *Circulation* **2016**, *134*, 441–450. [CrossRef] [PubMed]

Article

A Bipolar-Channel Hybrid Brain-Computer Interface System for Home Automation Control Utilizing Steady-State Visually Evoked Potential and Eye-Blink Signals

Dalin Yang, **Trung-Hau Nguyen and Wan-Young Chung** *

Department of Electronic Engineering, Pukyong National University, Busan 48513, Korea; dalinyang@pukyong.ac.kr (D.Y.); haunguyen85@pukyong.ac.kr (T.-H.N.)

* Correspondence: wychung@pknu.ac.kr

Received: 27 July 2020; Accepted: 21 September 2020; Published: 24 September 2020

Abstract: The goal of this study was to develop and validate a hybrid brain-computer interface (BCI) system for home automation control. Over the past decade, BCIs represent a promising possibility in the field of medical (e.g., neuronal rehabilitation), educational, mind reading, and remote communication. However, BCI is still difficult to use in daily life because of the challenges of the unfriendly head device, lower classification accuracy, high cost, and complex operation. In this study, we propose a hybrid BCI system for home automation control with two brain signals acquiring electrodes and simple tasks, which only requires the subject to focus on the stimulus and eye blink. The stimulus is utilized to select commands by generating steady-state visually evoked potential (SSVEP). The single eye blinks (i.e., confirm the selection) and double eye blinks (i.e., deny and re-selection) are employed to calibrate the SSVEP command. Besides that, the short-time Fourier transform and convolution neural network algorithms are utilized for feature extraction and classification, respectively. The results show that the proposed system could provide 38 control commands with a 2 s time window and a good accuracy (i.e., 96.92%) using one bipolar electroencephalogram (EEG) channel. This work presents a novel BCI approach for the home automation application based on SSVEP and eye blink signals, which could be useful for the disabled. In addition, the provided strategy of this study—a friendly channel configuration (i.e., one bipolar EEG channel), high accuracy, multiple commands, and short response time—might also offer a reference for the other BCI controlled applications.

Keywords: hybrid brain-computer interface (BCI); home automation; electroencephalogram (EEG); steady-state visually evoked potential (SSVEP); eye blink; short-time Fourier transform (STFT); convolution neural network (CNN)

---

1. Introduction

A brain-computer interface (BCI) is a connection between a brain and a device that enables signals from the brain to direct various external activities without the participant of the peripheral nerve and muscles [1]. BCI is typically utilized by people with severe motor disabilities, such as patients with amyotrophic lateral sclerosis, brainstem strokes, or other neuromuscular diseases [2–4]. People can utilize BCI-based applications to control wheelchairs, prosthetics, toys, video games, and various computer applications. Additionally, a BCI-based home automation control system was recently proposed based on the promising advantage in the field of artificial intelligence. In contrast to other types of home automation systems (e.g., gesture and voice recognition), BCI-controlled home automation systems have no limitations in terms of lighting and noise conditions [5]. Also, it essentially allows a home automation system to be controlled based on user intentions directly.

To avoid the surgical procedures, there are four popular non-invasive measurement methods for acquiring the brain information: functional magnetic resonance imaging (fMRI) [6], magnetoencephalography (MEG) [7], electroencephalogram (EEG) [8], and functional near-infrared spectroscopy (fNIRS) [9,10]. Due to the advantage of the good temporal resolution, portable, and low cost, EEG shows unique usability superiority for the BCI-based control system in comparison to the other type of brain techniques (i.e., MEG, fMRI, and fNIRS) [11].

Traditionally, the EEG-based BCI system divided four various patterns, such as motor imagery (MI), Steady-state visual evoked potential (SSVEP), P300 potentials, and slow cortical potentials. Each category has its advantages and disadvantages [12]. One of novel literature [13] employed MI signals to control a system based on hand grasps, which cover only a small range of commands. Additionally, the MI model requires much training and suffers from significant accuracy problems due to the BCI illiteracy, which is sourced from the sensorimotor rhythms [14]. One of the research team [15] utilized eye blinks and eye movement as a control mode for a home lighting system and determined that eye blinking is more accurate compared to eye movement, which has a margin of error that may lead to unreliable results. In the comparison of MI and eye-blink signals, P300 and SSVEP signals showed an excellent performance (i.e., accuracy) and fast response times, which is the reason that most existing BCI-based home automation systems employ P300 [16–21], alpha rhythm [22,23], and SSVEP [24–26] signals for obtaining a faster performance and more accurate control. As the principle of P300 and SSVEP, both signals are based on external stimuli. P300 signals are excellent for multi-stimuli recognition (more than six), whereas SSVEP signals provide superior performance when the number of stimuli is less than six. Additionally, one reference [27] demonstrated that SSVEP signals yield faster responses to user mental activity and are less reliant on channel selection. The current study describes a hybrid BCI system that combines two/more than two brain techniques to control the external device, which could make up for the disadvantages of each individual technique. A review article [28] proved that hybrid BCI systems could provide more commands and provide the potential to increases the classification accuracy and information transfer rates. It was also determined by recent literature [29,30] on the hybrid BCI system, which used EEG (i.e., SSVEP and MI) and eye blink/movement signals as the input for a speller.

Most existing BCI-based home automation systems employ multiple channels to acquire EEG signals. However, multichannel data processing leads to considerable time consumption, and more electrodes cause higher costs and more complex experimental setups [31]. Also, the multiple channels configuration is the biggest challenge for daily usage, especially for disabled patients. Therefore, the selection of the proper EEG channels and the related brain cortex is significant. For SSVEP signal acquisition, one study [32] demonstrated that even though one can detect SSVEP signals from the scalp from non-hair-bearing regions, the occipital region provides satisfactory SSVEP signals. Also, a review article [33] states that the bipolar channel, based on the occipital region, can further enhance the signal to noise ratio of SSVEP. Additionally, eye closure induces a strong alpha wave, which can be detected via EEG or magnetoencephalography from the occipital region [34]. The above results reveal that both SSVEP and eye blink signals could be acquired from the occipital region, which further reduces the number of channels required for recording signals.

Among the studies published by BCI-based home automation researchers, many have been published based on multi-channel systems with numbers of commands ranging from 2–113. A summary of related studies is provided in Table 1. The highest accuracy of 94.17% was achieved by Goel [35] with a response time of 5.2 s to produce two commands using four EEG channels. As we mentioned above, two commands are not sufficient for system control in daily life. A more friendly system is essential with a lower number of channels configuration, a high classification accuracy and multiple control commands needs to be proposed.

Table 1. Summary of related works on BCI-based home automation systems.

Publication	Type	Category	Commands	Channel	Evaluation Criteria		
					Accuracy (%)	Time (s)	ITR (bits/min)
Aloise et al. [17]	Real	P-300	16	8	90.00	4–5.6	11.19
Holzner et al. [21]	Virtual	P-300	13	N/A	79.35	N/A	N/A
Karmali et al. [23]	Virtual	Alpha Rhythm	44	4	N/A	27.7	N/A
Kosmyna et al. [5]	Virtual	Conceptual imagery	8	16	77–81	N/A	N/A
Goel et al. [35]	Virtual	SSVEP + Eye blink	2	4	94.17	5.2	11.6
Lin et al. [22]	Virtual	Alpha Rhythm	2	1	81.40	36–37	N/A
Prateek et al. [25]	Virtual	SSVEP	5	8	84.80	15	N/A
Perego et al. [26]	Virtual	SSVEP	4	N/A	N/A	350	N/A
Corralejo et al. [20]	Virtual	P-300	113	8	75–95	10.2	20.1
Carabalona et al. [19]	Real	P-300	36	12	50–80	N/A	N/A
Sellers et al. [18]	Real	P-300	72	8	83.00	N/A	N/A
Our study	Virtual	SSVEP + Eye blink	38	1	96.92	2	146.67

This paper proposes a hybrid BCI-based home automation system utilizing SSVEP and eye blink signals to provide 38 commands (i.e., 6 × 6 SSVEP commands and two eye blink commands) for controlling daily life activities through a single bipolar channel. SSVEP signals provide selection functions, and single eye blinks provide the functionality to confirm selections. Resetting a selection requires one to perform a simple double eye blink.

The short-time Fourier transform (STFT) is applied to extract the feature. Moreover, the classification was conducted by a convolutional neural network (CNN). The offline/real-time results demonstrate that the proposed system could be used in daily life for home automation control with a robust classification accuracy and simple EEG headset structure by performing an easy task. The proposed system provides a novel strategy for a BCI controlled system. Also, this BCI application could offer the possibility for the disable people to utilize the home facility conveniently.

The rest of this paper is organized as follows: The materials and methodology regarding the brain signal based home automation system are introduced in Section 2, which includes the information of participants, device parameters, experimental protocol, and theoretical algorithm of brain signal processing. Section 3 illustrates the results regarding the optimal channel selection, performance comparison of different time windows, offline classification, and real-time evaluation. In Section 4, the proposed system is compared and discussed. Conclusions are presented in the last section.

2. Materials and Methods

2.1. System Architecture and Parameters

This study describes a hybrid BCI-based home automation system utilizing SSVEP signals and eye blinks. As shown in Figure 1, the proposed system consists of a SSVEP stimulus panel, BCI module, and visual home automation interface. The stimulus panel (Samsung, Seoul, Korea, 21.5″, 60 Hz refresh rate, 1920 × 1080 screen resolution) is utilized as the stimulus source. Each stimulus is a square with a side length of 4 cm, horizontal spacing between squares of 16.5 cm, and vertical spacing between squares of 9 cm. Six targets are presented in the BCI system with flicking frequencies of 6.6, 7.5, 8.57, 10, 11 and 12 Hz. The interface was designed in the C# platform, as shown in Figure 2, in which the text indicators were displayed along the flickers to make corresponding control by the user. A high-performance EEG device (Cognionics Inc., San Diego, CA, USA) was utilized to acquire eye blinking and SSVEP signals from the O-bipolar channel (i.e., O1 and O2) with the reference of the international 10–20 EEG system. HD-72 dry wireless EEG headset (Cognionics Inc., San Diego, CA, USA) is a commercial high-density EEG recording device, which contains the 64 EEG electrodes plus reference and ground. To utilize the low analog amplification and elimination of ac-coupling of the signal path, the 24-bit ADC is applied in the headset system. Furthermore, EEG signals are referenced to the right earlobe. The impedance of all electrodes was kept below 5 kΩ. The EEG data were amplified and digitalized with a sampling frequency of 500 Hz and a band-pass filter in the range of 1–50 Hz.

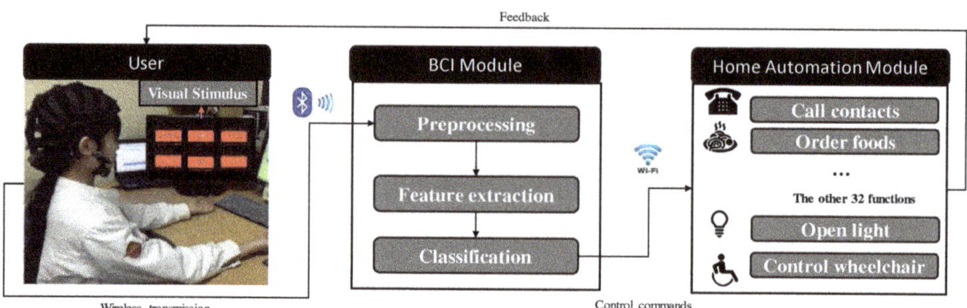

Figure 1. The control structure for our hybrid BCI system.

In this study, five healthy subjects (two females, three males, the median age of 24 years) with no prior brain-related or health issues participated after giving informed consent. The experiments were carried out following rules of the Declaration of Helsinki of 1975, revised in 2008. The identification code of approval is 1041386-202003-HR-11-02, approved on 3 March 2020, by the ethic committee of Pukyong National University. The subjects were asked to sit approximately 50 cm away from the monitor. They were then asked to focus their eyes on one of the six stimulus targets with flicking by the different frequencies (i.e., 6.6, 7.5, 8.57, 10, 11 and 12 Hz). EEG data were collected for 2 s for target identification, and a pop-up window would show the identification result (e.g., "Do you want to enter the sub-1 system?"). If the system detected a single-eye-blink signal, it meant the subject confirmed their selection. If a double-eye-blink signal was received, the system determined that the identified command was different from the subject's intention. The system would then return to the previous interface and resume gaze control for the stimulus targets. After choosing the correct command, the participants could then further control the home automation system. A flow chart for this process is presented in Figure 3.

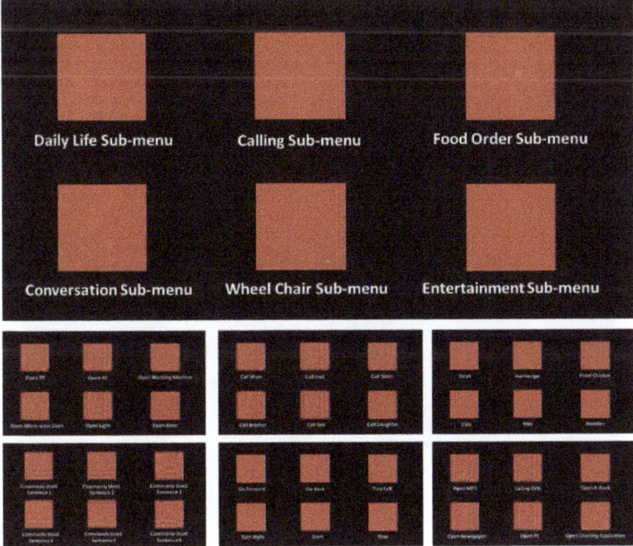

Figure 2. Interface for the selection menu.

2.2. *Setting Up the Interface*

The interface of the proposed BCI-based home automation system was set up based on a 6×6 categorical system. A user first chooses one of the six main categories. The categories were designed based on the most common daily life activities, which could make life easier and more comfortable for users. The main categories are presented in Figure 2 with six categories of daily life control, calling, food ordering, conversation control, wheelchair control, and entertainment. Each category is further divided into six subcategories, which contain the common tasks relevant to each subcategory. Users can select an option by gazing at a flicker and confirm one selection via the single eye blinking. To undo a selection, a double blink will return to the previous menu. The interface was constructed in a C#-based visual environment. Each selection result is presented to users through pop-up windows. During the experiment, all six stimulus buttons were displayed simultaneously. The users were instructed to select their menu options by shifting their gaze to the corresponding button.

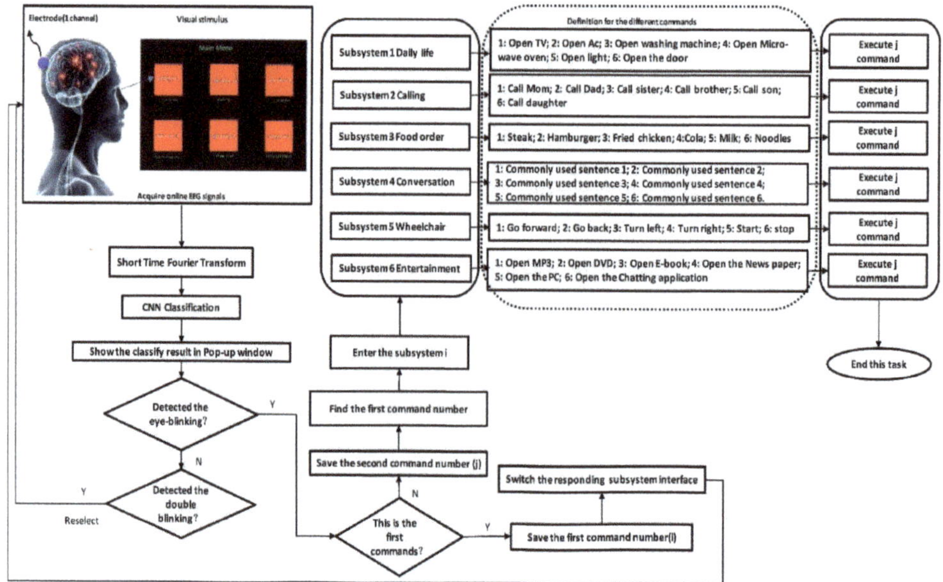

Figure 3. Control strategy for our hybrid BCI-based home automation system.

2.3. Experimental Protocol

Step 1: Setting up the headset on the user's head and ensure that the scalp and electrode have good contact with good signal quality. The real-time data acquisition software (Cognionics Inc., San Diego, CA, USA) with its interaction channel between it and C# programming environment. As a result, EEG signals can be captured via Bluetooth communication using a PC's serial port.

Step 2: Initiate the interface for the home automation system utilizing the Microsoft visual studio. A screen will appear on the monitor showing an interface with six main categories. To release the buffer pool of C#, once the buffer pool of C# cached more than 500 data samples of each channel, packaged MATLAB code was called for saving the EEG signals as the. mat file. After receiving two packages (i.e., 1000 data points), further analysis would be performed in MATLAB for generating the commands.

Step 3: To choose one of the six main categories, the user gazes at the corresponding category block. Each block is flickering with a different frequency (i.e., 6.6, 7.5, 8.57, 10, 11 and 12 Hz). The selected target would be encoded by the signal acquired from the occipital cortex. The subject then blinks their eyes to enter the subcategory menu (target block).

Step 4: A pop-up window will appear and ask the user if the window shows the correct selection. The user blinks their eyes again to confirm or blinks twice to return to the previous menu.

Step 5: A new window will appear, displaying the common tasks relevant to the selected category. The user can choose one term by blinking their eyes. Again, a pop-up window will appear to confirm the selection. They are blinking the eyes once will confirm the selection. Blinking twice will return to the previous menu.

2.4. Feature Extraction Protocol

A broad range of features has been implemented with the continuous development of BCIs to design BCI applications, such as the amplitude of EEG signals, band power, power spectral density, autoregressive models, and time-frequency features. In order to obtain effective output for BCI classification, it is necessary to understand clearly which features are accessible and how they are used. It is essential to select the relevant feature as the input for the classification [8]. As the literature [36] demonstrated that most brain activity patterns utilized to drive BCIs are defined by specific EEG time point and frequency band. Therefore, the time window of EEG signals should be considered as the important parameters during feature extraction Additionally, as the real-time analysis result indicated [37] that the EEG control range could facilitate to discover the beginning of alpha wave synchronization with low counts of false positives. Therefore, this study utilized the short-time Fourier transform (STFT) to simultaneously extract the features of the SSVEP signal and the eye-blink signal, which could contain the information in the time series and frequency band:

$$S(f,k) = \sum_{n=0}^{N-1} S(n)\left[w(n-k)e^{-\frac{j2\pi fn}{N}}\right]. \tag{1}$$

After recording the data from the selected EEG channel, the infinite pulse filter is applied with a cutting-off frequency ranging from 5–30 Hz. 2 s EEG signals after the onset of the task was selected for further analysis. The EEG power was determined by the STFT algorithm utilizing functions of *spectrogram* (MATLAB™) over a 1 s (i.e., 500 data point) *Kaiser* window. The length of each step is 1, and the overlap window length is set up as a value of 499. The detailed calculation is shown in Equation (1), where $S(n)$ is the original data in the time series, f is the frequency, the window function is represented by $W(n)$, k refers to the power. All features are derived from the time windows, and the oldest signal is eliminated from the active buffer when the new time series data reach. Then, the extracted feature is saved for the classification step. The procedure for signal processing is presented in Figure 4.

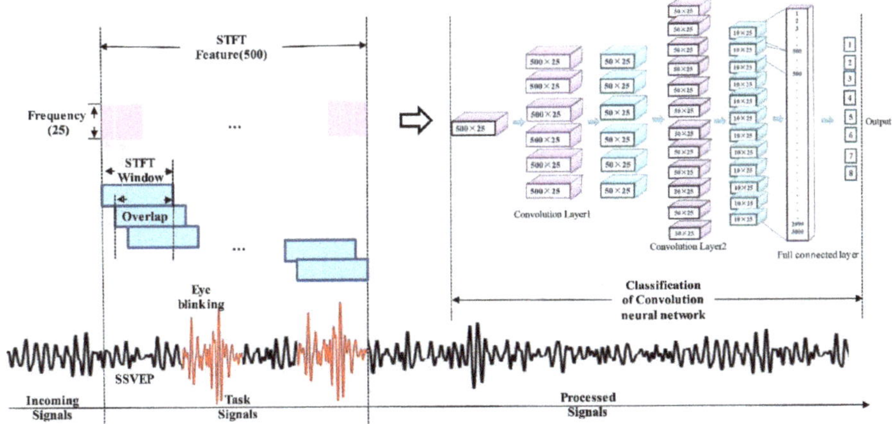

Figure 4. Procedure for signal acquisition, feature extraction, and classification.

2.5. Classification

Convolution neural networks (CNN) can be used efficiently for the identification of characters and produce outstanding outcomes for multiple datasets [38], such as the MNIST database. A CNN model can accommodate geometric deformation, and the receptive field/convolutional kernel can be readily understood, and the forms of high-level features to identify are detected [39]. Therefore, numerous studies [40–42] have employed CNNs as classifiers to identify EEG signals. Network topology is the crucial feature in a CNN algorithm. Our Network topology is shown in Figure 4. Our network is made up of four layers with one or more maps in each. The CNN model measures the forwarding propagation activation by using a rectified linear unit as an activation function:

$$Z(u,v) = \sum_{i=-\infty}^{\infty} \sum_{j=-\infty}^{\infty} s(f,k) \cdot N1 \cdot R(i,j) + \beta, \tag{2}$$

$$R(i,j) = \begin{cases} 1, & j < 2, 0 < i; \\ 0, & Others; \end{cases}, \tag{3}$$

$$a(u,v) = max(z(u,v)). \tag{4}$$

The normalization was performed for the extracted features with the frequency range of 5 to 30 Hz, which maintains the important information for the identification of the different features. The matrix of each input sample is 25 × 500. In this study, the size of the convolutional kernel (i.e., N) is 2 × 2. The bias is β. The output (i.e., Z) of the convolutional layer is calculated, as shown in Equation (2). Since the superiority of the fast speed of convergent, we applied the *Relu* function as the active function (i.e., shown in Equation (4)):

$$p(u,v) = w \cdot \sum_{i=-\infty}^{\infty} \sum_{j=-\infty}^{\infty} a(u,v) + \beta. \tag{5}$$

The pooling layer was used to reduce the size of the feature map. In this study, the max-pooling was conducted, which was employed to avoid the overfitting issue. In Equation (5), the weight is W, P represents the output of the convolutional layer. There were two fully connected layers (Layer 3 and Layer 4). This study employed the backpropagation to calculate the error term and gradient loss. The cost function is shown in Equation (6), as the input is given by Equation (7). Here, $h_{w,b}(^{(i)})$ is the desired values, and $y^{(i)}$ is the output value after the four-layers propagation:

$$J(w,b) = \frac{1}{m} \sum_{m=1}^{m} J\left(w,b;x^{(i)},y^{(i)}\right), \tag{6}$$

$$J\left(w,b;x^{(i)},y^{(i)}\right) = \frac{1}{2}\left(y^{(i)} - h_{w,b}\left(x^{(i)}\right)\right)^2, \tag{7}$$

After the calculation of each epoch, the unknown terms (i.e., w b) was updated with the negative lag direction. The algorithm is given by Equation (8) and Equation (9), respectively. The parameter α is the learning rate. After testing the trained CNN model, the error rate (ε = wrongly classified samples/total samples) for the testing sample was computed. The accuracy is calculated based on the equation: *accuracy* = (1 − ε) × 100%. Information transfer rate (ITR) is widely used om BCI filed [43], and it was calculated using the equations below:

$$w_{(i,j)}^{(l)} = w_{(i,j)}^{(l)} - \alpha \frac{\partial J}{\partial w_{(i,j)}^{(l)}}, \tag{8}$$

$$b_{(i,j)}^{(l)} = b_{(i,j)}^{(l)} - \alpha \frac{\partial J}{\partial b_{(i,j)}^{(l)}}, \tag{9}$$

where P is the probability of detecting correct commands (i.e., refer to the accuracy in this study), N is the number of the commands performed, and T (i.e., 2 s) is the time required to produce the number of commands.

$$\psi = \left(Plog_2P + (1-P)log_2\left(\frac{1-P}{N-1}\right) + log_2N\right), \tag{10}$$

$$ITR = \frac{No\,of\,commands \times \psi}{T}. \tag{11}$$

3. Results

3.1. Channel Selection

Applying a large number of EEG channels may result in noisy or redundant signals that degrade BCI performance and user convenience. As demonstrated in [21], the occipital region provides the best SSVEP signals. In this paper, we compared the performances of three occipital channels (i.e., O1, O2, and O-bipolar) and choose the best electrode for online testing. The subjects were asked to sit in front of the screen and execute each command in turn. To reduce the error caused by changes in the environment, we compared the data acquired from the same subjects and time windows. Each channel recorded 800 trials (i.e., five subjects × eight tasks × 20 trials) and then utilized the CNN algorithm for training the model (i.e., 80% × 800 = 640 samples). The results after the testing (i.e., 20% × 800 = 160 samples), as shown in Figure 5, reveal that the O-bipolar channel provides the best performance (average accuracy of 96.92%) for all the subjects among the three channels. And the performance of O1 and O2 are no significant differences.

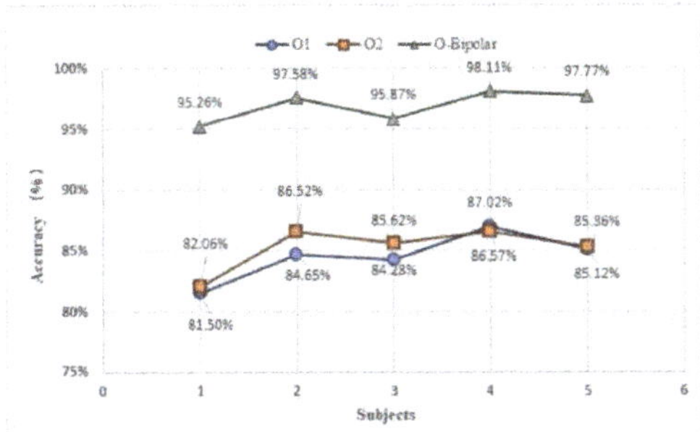

Figure 5. Performance comparison between different channels.

3.2. Time Window Selection

To calibrate the tradeoff between performance and time window duration, we trained the CNN models utilizing four different time-windows (1, 2, 3 and 4 s). The subjects were then asked to execute tasks based on cues on the screen. We created different segmentations for each of the time windows. Each time window contained 800 samples as input data. As shown in Figure 6, the 2 s time window provided good accuracy at 96.92%. Moreover, the 3 s and 4 s windows provided better performance by 97.28% and 98.51%, respectively. Considering the importance of time windows for the control system, the 2 s window is the selected for our system, since the 2 s time window could achieve the satisfied classification results. The classification performance (i.e., accuracy and loss) of different time windows are shown in Figure 7.

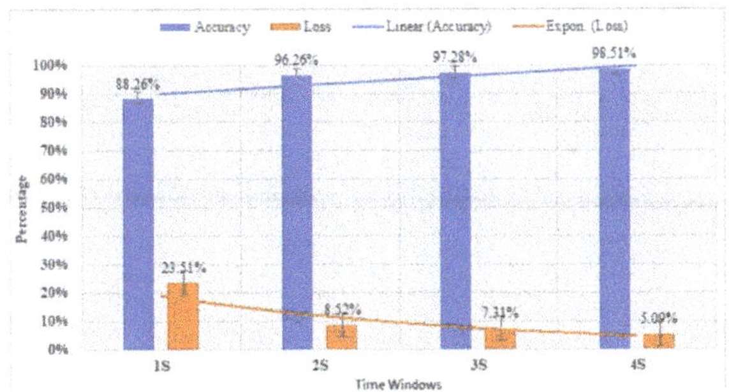

Figure 6. Performance comparison between different time windows.

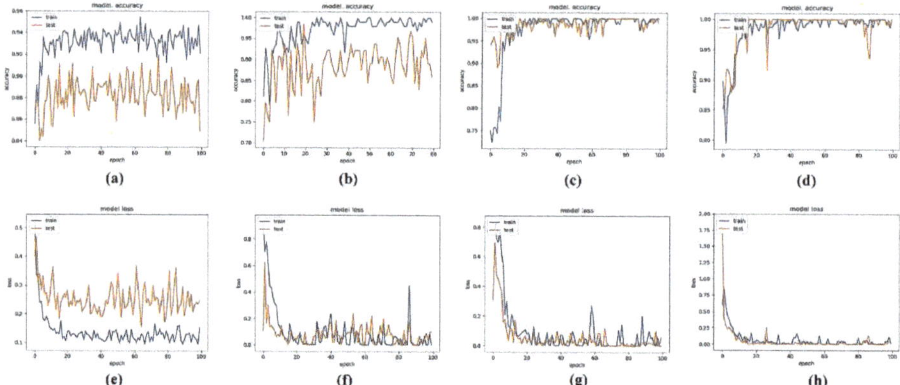

Figure 7. Performance results for different time windows, the accuracy results for the 1 s, 2 s, 3 s, and 4 s time windows are shown in (**a–d**), respectively. The corresponding training losses are shown in (**e–h**).

3.3. Feature Extraction

As shown in Figure 8, the extracted features were calculated by the STFT in the time (i.e., 2 s) and frequency domain. Totally, 800 feature maps (i.e., five subjects × eight tasks × 20 trials) were obtained from five subjects. Since the SSVEP frequency in subcategory interfaces are consistent with the main interface flickers, only six SSVEP features and two eye blinking features were extracted to train the model. In other words, once the selected feature was trained in the CNN model, the CNN model would recognize a similar pattern either during the selection in the subcategory interface or the main category interface.

In Figure 8, the power bar is shown on the right side. The feature of single eye blink present in Figure 8a; two power peaks were observed when the subjects performed double eye blink tasks, as shown in Figure 8b. The powerband for the SSVEP tasks with specific frequency are shown in Figure 8c–h, respectively, which refer to the brain features caused by gazing the different frequency flickering.

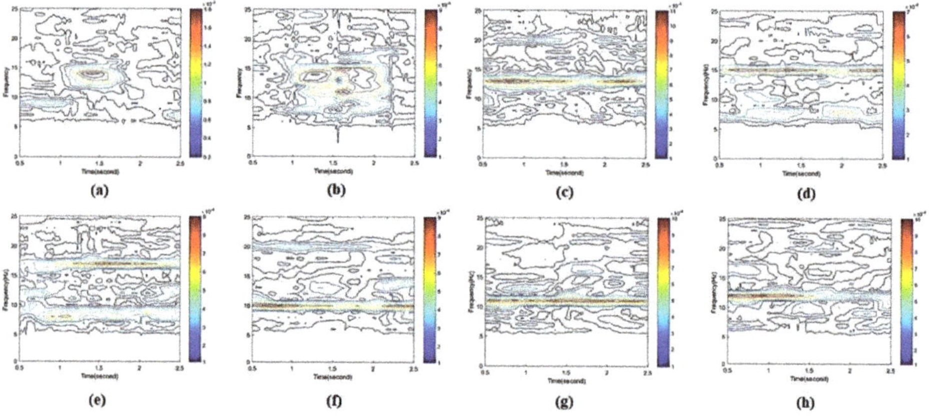

Figure 8. STFT features for (**a**) single eye blink, (**b**) double eye blink, (**c**) SSVEP at 6.6 Hz, (**d**) SSVEP at 7.5 Hz, (**e**) SSVEP at 8.57 Hz, (**f**) SSVEP at 10 Hz, (**g**) SSVEP at 11 Hz, and (**h**) SSVEP at 12 Hz.

3.4. Real-Time Evaluation

In the online section, each subject performs eight trials to test the proposed system. Each trial includes a selection from the main interface, confirmation via eye blinking, selecting a command from the subcategory interface, and confirmation of the second command in sequence. The targeted selection from 36 control commands was decided by the willingness of each participant. The results for three subjects in the online evaluation experiments are listed in Table 2. As shown in Table 2, the mean eye-blink time for a single blink was 1.3163 s, with the fastest time for a single blink recorded for subject S1. The shortest time for sending a control command was 1.2 s and the longest time was 1.425 s. For a double blink, the longest time was 1.652 s, and the shortest time was 1.576 s, with a mean value of 1.608 s for sending control commands. In addition, the correct identified command of the single eye blink case was 47/48. In the double blink case, the five commands were detected correctly. In comparison with the eye blink, the SSVEP shows the lower ability (43/48) for identification.

Table 2. Real-time results for three subjects controlling the virtual home automation system.

Subjects	Correct Command			Response Time		
	$N_{eye\text{-}blink}$	$N_{double\text{-}blink}$	N_{SSVEP}	$T_{eye\text{-}blink}$ (s)	$T_{double\text{-}blink}$ (s)	T_{SSVEP} (s)
S1	16/16	1/2	14/16	1.25	1.652	1.754
S2	15/16	2/2	15/16	1.425	1.576	1.953
S3	16/16	2/2	14/16	1.274	1.597	1.731
Total	47/48	5/6	43/48	1.316	1.608	1.813
Std.	0.036	0.289	0.036	0.095	0.039	0.122

N: Number; T: Time; Std: Standard Deviation.

4. Discussion

In this study, we designed a control mechanism for a BCI-based home automation system. The proposed system can identify 38 commands (i.e., 36 control commands and two calibration commands) utilizing single eye blinks, double eye blinks, and SSVEP signals recorded from a single bipolar channel with a classification accuracy of 96.92%. As the best of the authors' knowledge, this is the first study that utilizes only one bipolar EEG channel for a home automation control BCI system with good accuracy within a short time window. The proposed system could provide a novel way for physical disordered people to control external devices by gazing and eye blinks. It offers a possibility to conduct daily routine tasks using brain signals directly without any physical movement.

In comparison with the previous relevant research [29,30,44–47] (listed in Table 3), instead of using the eye-tracking to detect the eye movement, this study applied the eye blinking to be a trigger in this control system. In the eye-tracking system, the extra device is required to monitor the eye movement. As demonstrated in the hybrid eye-tracking and SSVEP system [30], the participants need to wear the extra video eye-tracking system, for which the threshold of velocity, acceleration, and minimum deflection was 30°/s, 8000°/s^2, and 0.1°, respectively. Since the eyeblink and the SSVEP signal can be acquired from the EEG device with the same channel, the unfriendly hardware burden and extra cost of the hybrid device (e.g., EEG and eye-tracking device) will be reduced. With the development of the eye-tracking, the pure eye-tracking system is going to be an alternative technique of the BCI. However, one reference reported that the EEG-based system is easiest to use, also, the SSVEP-based system shows better performance than the eye-tracking system [48].

Table 3. Related works regarding the hybrid EEG and eye blink BCI system.

Study	Type	Category	Commands	Channel	Accuracy (%)	Time (s)	ITR (bits/min)
Wang et al. [44]	Real	P-300 + Eye blink + MI	8	17	91.25	4	N/A
Wang et al. [45]	Virtual	SSVEP + Eye blink	12	4	40–100	0–20	N/A
Park et al. [46]	Virtual	SSVEP + AR + Eye blink	16	32	92.8	20	37.4
Duan et al. [47]	Real	MI + SSVEP + Eye blink	6	12	86.5	48	1.69
Malik et al. [30]	Virtual	SSVEP + Eye tracking	48	8	90.35	1	184.06
He et al. [29]	Virtual	MI + Eye blink	N/A	32	96.02	6.16	45.97
Our study	Virtual	SSVEP + Eye blink	38	1	96.92	2	146.67

Five subjects participated in an offline training experiment to evaluate the performance of different channels and time windows. The results demonstrate that the O-bipolar channel provides better performance compared to channels O1 and O2, as shown in Figure 5. Based on our results, we concluded that utilizing the O-bipolar channel significantly reduces the interference of noise. Response time plays a vital role in real-time systems. Therefore, four different time windows for data partitioning are utilized for the evaluation in this study. Figure 7 presents a comparison performance (i.e., classification accuracy and loss) between the different time windows. The 4 s time windows offer the best performance in terms of accuracy (i.e., 98.51%), as more information is included for the longer time window in comparison to the short time windows. This finding is consistent with the pioneer study [45], which could achieve 100% by conducting the task with a 20 s time window. With the development of deep learning, an improved classification may be achieved by utilizing the hybrid modality (i.e., EEG and fNIRS) [49,50] advanced machine learning algorithms, such as long-short team memory [51] and deep neural network [52]. In addition, in this study, we applied red squares with the text indicators to guide the participants to select the corresponding control commands. A more intuitive display method might provide a more friendly interface (e.g., pictures, or different color squares, or various shapes) for the user and we will consider this in our future work.

For the time windows selection, we applied four different time windows (i.e., 1, 2, 3, and 4 s) to assess their performance. As shown in Figure 6, the time window of 4 s could achieve the highest accuracy compared with the other time windows (e.g., 1, 2, and 3 s). However, to reduce the time consumed of this proposed control system, we choose 2 s as the time window to obtain a satisfactory result (i.e., 96.26%). In the online experiment, three subjects were recruited to control the proposed system in real-time. Each participant performed eight trials, which were conducted in sequence: (i) selection in the main interface, (ii) calibration via eye blinking, (iii) selection from the subcategory interface, and (iv) calibration for the second selection. Before the experiment, the participants were informed to select all the flickers in the main interface. The decision of selection in each subcategory interface was made by the subjects randomly. Thus, part of the commands from the 36 functions was evaluated. As shown in Table 2, the initial investigation results indicated the feasibility of the proposed system. In future work, we will develop our own simplified device (i.e., two electrodes) and examine the real-time system with a total of 36 functions.

Although the offline and real-time virtual results showed good performance (i.e., single bipolar channel, good accuracy, and short time window) for the proposed home automation control system, some limitations need to be mentioned. First, the system is simulated in the virtual environment. The real home automation application results may lead to less accuracy due to the lousy human mental state (e.g., distracted and motor artifact, etc.) and the signal transmission problem between the home automation application and the EEG device. Secondly, eye blink is typically considered an undesirable electrical potential. With the advantage of high amplitude and analytic features, voluntary eye blinks are widely employed as an input or control command in BCI areas. Therefore, one needs to pay attention during the analysis to noise reduction. Also, the control of the eye blink should follow the cue while performing the task. In addition, the idle state/resting state was not considered for state identification. Non-detection of the idle state may lead to misclassification during long duration experiments. For a future study it was suggested to add an extra idle state detection to avoid this issue. Lastly, the objective of this study was to investigate the feasibility of home automation control application with a single bipolar channel by using the hybrid SSVEP and eye blink signals. In this pilot study, only a few participants were considered (i.e., a total of 800 trials was conducted to increase the dataset of the CNN classification subjects to assess the repeatability and stability). In future work, more participants need to be investigated for comprehensive analysis in a real-time environment.

5. Conclusions

This study proposed a hybrid BCI-based home automation system utilizing SSVEP and eye-blink signals with a single bipolar channel for multiple comment control (i.e., 38 commands). SSVEP signals are utilized to select desired commands, and eye-blink signals are utilized to calibrate command selections. Both signals are obtained from the same bipolar channel and classified by the same CNN model. Our experiments included two modules for processing and analyzing EEG signals. An offline module was employed to assess general model performance (e.g., channel selection, time window selection, feature extraction, and CNN model training). Five subjects participated in offline experiments. The results demonstrated that 38 daily task commands could be identified with an accuracy of 96.92% based on a 2 s time-window using the signal acquired from the O-bipolar channel. Three subjects participated in a real-time experiment, and the results demonstrated that changes in brain intentions could automatically control the proposed system. As the best knowledge of the authors, this is the first work to utilize the combination of SSVEP and eye-blinks to perform home automation control. This study demonstrated that it is possible to achieve multidimensional control with good performance using SSVEP and eye blink signals from only one single bipolar channel. Also, the proposed system could be applied to home automation control system, which could be a helpful assistant for disabled/healthy people.

Author Contributions: Conceptualization, W.-Y.C.; methodology, D.Y., T.-H.N.; data curation, D.Y; writing—original draft preparation, D.Y; supervision, W.-Y.C. All authors have read and agreed to the published version of the manuscript.

Funding: This work was supported by the National Research Foundation of Korea (NRF) grant funded by the Korea Government (MSIT) (2019R1A2C1089139).

Conflicts of Interest: The authors declare no conflict of interest.

References

1. Lance, B.J.; Kerick, S.E.; Ries, A.J.; Oie, K.S.; McDowell, K. Brain–Computer Interface Technologies in the Coming Decades. *Proc. IEEE* **2012**, *100*, 1585–1599. [CrossRef]
2. Leuthardt, E.C.; Schalk, G.; Wolpaw, J.R.; Ojemann, J.G.; Moran, D.W. A brain–computer interface using electrocorticographic signals in humans. *J. Neural Eng.* **2004**, *1*, 63–71. [CrossRef] [PubMed]
3. Nguyen, T.-H.; Yang, D.; Chung, W.-Y. A High-Rate BCI Speller Based on Eye-Closed EEG Signal. *IEEE Access* **2018**, *6*, 33995–34003. [CrossRef]
4. Yang, D.; Huang, R.; Yoo, S.-H.; Shin, M.-J.; Yoon, J.A.; Shin, Y.-I.; Hong, K.-S. Detection of Mild Cognitive Impairment Using Convolutional Neural Network: Temporal-Feature Maps of Functional Near-Infrared Spectroscopy. *Front. Aging Neurosci.* **2020**, *12*. [CrossRef]
5. Kosmyna, N.; Tarpin-Bernard, F.; Bonnefond, N.; Rivet, B. Feasibility of BCI Control in a Realistic Smart Home Environment. *Front. Hum. Neurosci.* **2016**, *10*, 1–10. [CrossRef]
6. Birn, R.M. The role of physiological noise in resting-state functional connectivity. *NeuroImage* **2012**, *62*, 864–870. [CrossRef]
7. Tanaka, M. Effects of Mental Fatigue on Brain Activity and Cognitive Performance: A Magnetoencephalography Study. *Anat. Physiol.* **2015**, *4*, 1–5. [CrossRef]
8. Lotte, F.; Congedo, M.; Lecuyer, A.; Lamarche, F.; Arnaldi, B. A review of classification algorithms for EEG-based brain–computer interfaces. *J. Neural Eng.* **2007**, *4*, R1–R13. [CrossRef]
9. Naseer, N.; Hong, K.-S. fNIRS-based brain-computer interfaces: A review. *Front. Hum. Neurosci.* **2015**, *9*. [CrossRef]
10. Yang, D.; Hong, K.-S.; Yoo, S.-H.; Kim, C.-S. Evaluation of Neural Degeneration Biomarkers in the Prefrontal Cortex for Early Identification of Patients With Mild Cognitive Impairment: An fNIRS Study. *Front. Hum. Neurosci.* **2019**, *13*, 317. [CrossRef]
11. Abdulkader, S.N.; Atia, A.; Mostafa, M.-S.M. Brain computer interfacing: Applications and challenges. *Egypt. Inform. J.* **2015**, *16*, 213–230. [CrossRef]
12. Zhang, W.; Tan, C.; Sun, F.; Wu, H.; Zhang, B. A review of EEG-based brain-computer interface systems design. *Brain Sci. Adv.* **2018**, *4*, 156–167. [CrossRef]
13. Zakaria, M.H.F.; Mansor, W.; Lee, K.Y. Time-frequency analysis of executed and imagined motor movement EEG signals for neuro-based home appliance system. In Proceedings of the TENCON 2017—2017 IEEE Region 10 Conference, Penang, Malaysia, 5–8 November 2017; pp. 1657–1660. [CrossRef]

14. Vidaurre, C.; Blankertz, B. Towards a Cure for BCI Illiteracy. *Brain Topogr.* **2009**, *23*, 194–198. [CrossRef] [PubMed]
15. Rani, M.S.B.A.; Mansor, W.B. Detection of eye blinks from EEG signals for home lighting system activation. In Proceedings of the 2009 6th International Symposium on Mechatronics and its Applications, Sharjah, UAE, 23–26 March 2009; pp. 1–4. [CrossRef]
16. Sajda, P.; Pohlmeyer, E.; Wang, J.; Parra, L.; Christoforou, C.; Dmochowski, J.; Hanna, B.; Bahlmann, C.; Singh, M.; Chang, S.-F. In a Blink of an Eye and a Switch of a Transistor: Cortically Coupled Computer Vision. *Proc. IEEE* **2010**, *98*, 462–478. [CrossRef]
17. Aloise, F.; Schettini, F.; Aricò, P.; Leotta, F.; Salinari, S.; Mattia, D.; Babiloni, F.; Cincotti, F. P300-based brain–computer interface for environmental control: An asynchronous approach. *J. Neural Eng.* **2011**, *8*, 25025. [CrossRef] [PubMed]
18. Sellers, E.W.; Vaughan, T.M.; Wolpaw, J.R. A brain-computer interface for long-term independent home use. *Amyotroph. Lateral Scler.* **2010**, *11*, 449–455. [CrossRef]
19. Carabalona, R.; Grossi, F.; Tessadri, A.; Castiglioni, P.; Caracciolo, A.; De Munari, I. Light on! Real world evaluation of a P300-based brain–computer interface (BCI) for environment control in a smart home. *Ergonomics* **2012**, *55*, 552–563. [CrossRef]
20. Corralejo, R.; Nicolas-Alonso, L.F.; Alvarez, D.; Hornero, R. A P300-based brain–computer interface aimed at operating electronic devices at home for severely disabled people. *Med. Biol. Eng.* **2014**, *52*, 861–872. [CrossRef]
21. Holzner, C.; Guger, C.; Edlinger, G.; Gronegress, C.; Slater, M. Virtual Smart Home Controlled by Thoughts. In Proceedings of the 2009 18th IEEE International Workshops on Enabling Technologies: Infrastructures for Collaborative Enterprises, Groningen, The Netherlands, 29 June–1 July 2009; pp. 236–239. [CrossRef]
22. Lin, C.-T.; Lin, B.-S.; Lin, F.-C.; Chang, C.-J. Brain Computer Interface-Based Smart Living Environmental Auto-Adjustment Control System in UPnP Home Networking. *IEEE Syst. J.* **2012**, *8*, 363–370. [CrossRef]
23. Karmali, F.; Polak, M.; Kostov, A. Environmental control by a brain-computer interface. In Proceedings of the 22nd Annual International Conference of the IEEE Engineering in Medicine and Biology Society (Cat No 00CH37143) IEMBS-00, Chicago, IL, USA, 23–28 July 2002; Volume 4, pp. 2990–2992. [CrossRef]
24. Piccini, L.; Parini, S.; Maggi, L.; Andreoni, G. A Wearable Home BCI system: Preliminary results with SSVEP protocol. In Proceedings of the 2005 IEEE Engineering in Medicine and Biology 27th Annual Conference, Shanghai, China, 17–18 January 2006; Volume 5, pp. 5384–5387. [CrossRef]
25. Prateek, V.; Poonam, S.; Preeti, K. Home Automation Control System Implementation using SSVEP based Brain-Computer Interface. In Proceedings of the IEEE ICICI—2017, Coimbatore, India, 23–24 November 2017; pp. 1068–1073.
26. Perego, P.; Maggi, L.; Andreoni, G.; Parini, S. A Home Automation Interface for BCI application validated with SSVEP protocol. In Proceedings of the 4th International BrainComputer Interface Workshop and Training Course, Graz, Austria, 18–21 September 2008.
27. Zhao, J.; Li, W.; Li, M. Comparative Study of SSVEP- and P300-Based Models for the Telepresence Control of Humanoid Robots. *PLoS ONE* **2015**, *10*, e0142168. [CrossRef]
28. Amiri, S.; Fazel-Rezai, R.; Asadpour, V. A Review of Hybrid Brain-Computer Interface Systems. *Adv. Hum. Comput. Interact.* **2013**, *2013*, 187024. [CrossRef]
29. He, S.; Zhou, Y.; Yu, T.; Zhang, R.; Huang, Q.; Chuai, L.; Mustafa, M.-U.; Gu, Z.; Yu, Z.L.; Tan, H.; et al. EEG- and EOG-Based Asynchronous Hybrid BCI: A System Integrating a Speller, a Web Browser, an E-Mail Client, and a File Explorer. *IEEE Trans. Neural Syst. Rehabil. Eng.* **2020**, *28*, 519–530. [CrossRef] [PubMed]
30. Mannan, M.M.N.; Kamran, M.A.; Kang, S.; Choi, H.S.; Jeong, M.Y. A Hybrid Speller Design Using Eye Tracking and SSVEP Brain–Computer Interface. *Sensors* **2020**, *20*, 891. [CrossRef] [PubMed]
31. Lau, T.M.; Gwin, J.T.; Ferris, D.P. How Many Electrodes Are Really Needed for EEG-Based Mobile Brain Imaging? *J. Behav. Brain Sci.* **2012**, *2*, 387–393. [CrossRef]
32. Wang, Y.-T.; Wang, Y.; Cheng, C.-K.; Jung, T.-P. Measuring Steady-State Visual Evoked Potentials from non-hair-bearing areas. In Proceedings of the Annual International Conference of the IEEE Engineering in Medicine and Biology Society, San Diego, CA, USA, 28 August–1 September 2012; Volume 2012, pp. 1806–1809. [CrossRef]
33. Zhu, D.; Bieger, J.; Molina, G.G.; Aarts, R.M. A Survey of Stimulation Methods Used in SSVEP-Based BCIs. *Comput. Intell. Neurosci.* **2010**, *2010*, 702357. [CrossRef]

34. Isa, I.S.; Zainuddin, B.S.; Hussain, Z.; Sulaiman, S.N. Preliminary Study on Analyzing EEG Alpha Brainwave Signal Activities Based on Visual Stimulation. *Procedia Comput. Sci.* **2014**, *42*, 85–92. [CrossRef]
35. Goel, K.; Vohra, R.; Kamath, A.; Baths, V. Home automation using SSVEP & eye-blink detection based brain-computer interface. In Proceedings of the IEEE International Conference on Systems, Man, and Cybernetics (SMC), San Diego, CA, USA, 5–8 October 2014; pp. 4035–4036. [CrossRef]
36. Wolpaw, J.R.; Birbaumer, N.; McFarland, D.J.; Pfurtscheller, G.; Vaughan, T.M. Brain–computer interfaces for communication and control. *Clin. Neurophysiol.* **2002**, *113*, 767–791. [CrossRef]
37. Kirkup, L.; Searle, A.; Craig, A.; McIsaac, P.; Larsen, G. Three methods compared for detecting the onset of alpha wave synchronization following eye closure. *Physiol. Meas.* **1998**, *19*, 213–224. [CrossRef]
38. Simard, P.Y.; Steinkraus, D.; Platt, J.C. *Best Practices for Convolutional Neural Networks Applied to Visual Document Analysis*; Icdar: Edinburgh, UK, 2003.
39. Cecotti, H.; Graser, A. Convolutional Neural Networks for P300 Detection with Application to Brain-Computer Interfaces. *IEEE Trans. Pattern Anal. Mach. Intell.* **2010**, *33*, 433–445. [CrossRef]
40. Ieracitano, C.; Mammone, N.; Bramanti, A.; Hussain, A.; Morabito, F.C. A Convolutional Neural Network approach for classification of dementia stages based on 2D-spectral representation of EEG recordings. *Neurocomputing* **2019**, *323*, 96–107. [CrossRef]
41. Trachtman, J.P. The Limits of PTAs. *Prefer. Trade Agreem.* **2011**, 115–149. [CrossRef]
42. Lawhern, V.J.; Solon, A.J.; Waytowich, N.R.; Gordon, S.; Hung, C.P.; Lance, B.J. EEGNet: A compact convolutional neural network for EEG-based brain–computer interfaces. *J. Neural Eng.* **2018**, *15*, 056013. [CrossRef] [PubMed]
43. Wolpaw, J.R.; Birbaumer, N.; Heetderks, W.; McFarland, D.; Peckham, P.; Schalk, G.; Donchin, E.; Quatrano, L.; Robinson, C.; Vaughan, T. Brain-computer interface technology: A review of the first international meeting. *IEEE Trans. Rehabil. Eng.* **2000**, *8*, 164–173. [CrossRef] [PubMed]
44. Wang, H.; Li, Y.; Long, J.; Yu, T.; Gu, Z. An asynchronous wheelchair control by hybrid EEG–EOG brain–computer interface. *Cogn. Neurodyn.* **2014**, *8*, 399–409. [CrossRef]
45. Wang, M.; Qu, W.; Chen, W.-Y. Hybrid sensing and encoding using pad phone for home robot control. *Multimed. Tools Appl.* **2017**, *77*, 10773–10786. [CrossRef]
46. Park, S.; Cha, H.-S.; Im, C.-H. Development of an Online Home Appliance Control System Using Augmented Reality and an SSVEP-Based Brain–Computer Interface. *IEEE Access* **2019**, *7*, 163604–163614. [CrossRef]
47. Duan, X.; Xie, S.; Xie, X.; Meng, Y.; Xu, Z. Quadcopter Flight Control Using a Non-invasive Multi-Modal Brain Computer Interface. *Front. Neurorobot.* **2019**, *13*. [CrossRef]
48. Suefusa, K.; Tanaka, T. A comparison study of visually stimulated brain–computer and eye-tracking interfaces. *J. Neural Eng.* **2017**, *14*, 036009. [CrossRef]
49. Hong, K.-S.; Naseer, N. Reduction of Delay in Detecting Initial Dips from Functional Near-Infrared Spectroscopy Signals Using Vector-Based Phase Analysis. *Int. J. Neural Syst.* **2016**, *26*, 1650012. [CrossRef]
50. Khan, M.J.; Hong, K.-S. Hybrid EEG–fNIRS-Based Eight-Command Decoding for BCI: Application to Quadcopter Control. *Front. Neurorobot.* **2017**, *11*. [CrossRef]
51. Tanveer, M.A.; Khan, M.J.; Qureshi, M.J.; Naseer, N.; Hong, K.-S. Enhanced Drowsiness Detection Using Deep Learning: An fNIRS Study. *IEEE Access* **2019**, *7*, 137920–137929. [CrossRef]
52. Asgher, U.; Khalil, K.; Khan, M.J.; Ahmad, R.; Butt, S.I.; Ayaz, Y.; Naseer, N.; Nazir, S. Enhanced Accuracy for Multiclass Mental Workload Detection Using Long Short-Term Memory for Brain–Computer Interface. *Front. Mol. Neurosci.* **2020**, *14*. [CrossRef]

Article

Identification of Motor and Mental Imagery EEG in Two and Multiclass Subject-Dependent Tasks Using Successive Decomposition Index

Muhammad Tariq Sadiq †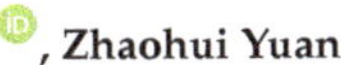, **Xiaojun Yu** †, **Zhaohui Yuan** * and **Muhammad Zulkifal Aziz** 

School of Automation, Northwestern Polytechnical University, 127 West Youyi Road, Xi'an 710072, China; tariq.sadiq@mail.nwpu.edu.cn (M.T.S.); XJYU@nwpu.edu.cn (X.Y.); zulkifalaziz@mail.nwpu.edu.cn (M.Z.A.)

* Correspondence: yuanzhh@nwpu.edu.cn

† Muhammad Tariq Sadiq and Xiaojun Yu are co-first author.

Received: 29 July 2020; Accepted: 10 September 2020; Published: 16 September 2020

Abstract: The development of fast and robust brain–computer interface (BCI) systems requires non-complex and efficient computational tools. The modern procedures adopted for this purpose are complex which limits their use in practical applications. In this study, for the first time, and to the best of our knowledge, a successive decomposition index (SDI)-based feature extraction approach is utilized for the classification of motor and mental imagery electroencephalography (EEG) tasks. First of all, the public datasets IVa, IVb, and V from BCI competition III were denoised using multiscale principal analysis (MSPCA), and then a SDI feature was calculated corresponding to each trial of the data. Finally, six benchmark machine learning and neural network classifiers were used to evaluate the performance of the proposed method. All the experiments were performed for motor and mental imagery datasets in binary and multiclass applications using a 10-fold cross-validation method. Furthermore, computerized automatic detection of motor and mental imagery using SDI (CADMMI-SDI) is developed to describe the proposed approach practically. The experimental results suggest that the highest classification accuracy of 97.46% (Dataset IVa), 99.52% (Dataset IVb), and 99.33% (Dataset V) was obtained using feedforward neural network classifier. Moreover, a series of experiments, namely, statistical analysis, channels variation, classifier parameters variation, processed and unprocessed data, and computational complexity, were performed and it was concluded that SDI is robust for noise, and a non-complex and efficient biomarker for the development of fast and accurate motor and mental imagery BCI systems.

Keywords: electroencephalography; Brain-Computer Interface; multiscale principal component analysis; successive decomposition index; motor imagery; mental imagery; neurorehabilitation; classification

---

1. Introduction

With the rampant growth in automated systems, computer-aided physical systems, and artificial intelligence, brain–computer interface (BCI) has gained significant attention from researchers as it can bind a human mind to the computer and operate complex physical applications. The healthcare realm has been overwhelmed by the development of computer-aided brain devices, namely, prosthetic arms, brain-controlled wheelchairs, mind-controlled home automation, etc., for physically impaired people [1–7]. The fundamental source of BCI is the low-key signal generated on the surface of the human scalp as a result of neural activity and it acts as a watershed for the plethora of brain-controlled applications.

The common practices involved to retrieve such signals are invasive and noninvasive methods. Invasive methods, as the name implies, record signals from the inside of the human brain which results in artifact-free data. On the other hand, noninvasive techniques accumulate noise artifacts that degrade the performance of BCI systems. Electroencephalography (EEG) is the commonly employed technology for the development of practical BCI systems.

Motor and mental imagery are subdomains in BCI which deal with the simulation of motor and mental activities in the brain without performing any activity in real. The inherent nature of motor and mental imagery suggests an economical, noninvasive, portable, and high temporal resolution mode of acquiring signals and the best choice is electroencephalography (EEG) [3,8]. After acquiring the signal, the subsequent process is to correctly wring out useful information from it [9,10].

The analysis of any signal processing problem comprises at least three basic procedures: preprocessing (data preparation and artifacts removing), feature extraction (identifies the most significant characteristics in signals), and classification (segregating classes between features). As noninvasive mode of signal acquisition heavily accumulates noise artifacts and it is crucial to filter out alienated signals without disturbing the original content. Recent studies proposed independent component analysis (ICA) [11], principal component analysis, and canonical correlation analysis [12] for the noise removal of EEG signals; however, these methods are not very effective for the analysis of non-stationary signals [12]. Another hybrid algorithm namely multiscale principal component analysis (MSPCA) is recently proposed and studies [13,14] revealed its robustness in denoising non-stationary and nonlinear signals.

After preprocessing the data, the subsequent steps are features estimation and classification [15]. In EEG signal processing, some widely adopted feature extraction methods are categorized as Fourier transform (FT) [16], power spectral density (PSD) [17], common spatial patterns (CSP) [18,19], autoregressive (AR) [20,21], sparse representation, and signal decomposition (SD) [22–24] based methods. All of these methods have their associated demerits and complications, for example, FT-based features only preserve the spectral resolution of the signal and completely loses the temporal information, PSD-based methods are susceptible to electrodes locality, AR-based techniques are sensitive to noise content, etc.

Chattarjee et al. [25] does a comparative analysis for a different time, energy, entropy, and statistical features using a different machine and deep learning classifiers for EEG signals. The maximum classification outcome of 85% was observed for energy and entropy-based features using the support vector machine (SVM) classifier. Wang et al. [26] amalgamates empirical mode decomposition (EMD) with Hilbert spectral analysis for motor imagery EEG signals and backpropagation neural network for classification purposes. The maximum recorded accuracy was 93.8%. Gupta et al. [17] extracted PSD features for EEG signals and did a comparative analysis for different univariate and multivariate features selection methods using different classifiers. A maximum classification accuracy of 85% was obtained for the combination of the Burg and linear regression features selection method using the linear discriminant analysis classifier.

Jasmine et al. [22] presents a comparative analysis for three signal decomposition techniques, i.e., EMD, discrete wavelet transform, and wavelet packet decomposition (WPD), using motor imagery EEG datasets. The highest accuracy of 92.8% was attained for higher-order statistical features extracted from WPD using K-nearest neighbors classifier. Chaudhary et al. [27] combines the non-dyadic wavelet decomposition method and CSP features extraction method for the classification of motor imagery EEG signals. Maximum classification accuracy of 85.6% was obtained for decision tree classifier. Jiacan et al. [28] presents a deep multi-view feature learning process for the classification of motor imagery EEG tasks. First, many multidomain features (time, frequency, time-frequency, and spatial) were extracted, and then a restricted Boltzmann machine network improved by t-distributed stochastic neighbor embedding (t-SNE) is employed for features learning. An average classification accuracy of 78.5% was obtained using the SVM classifier. Chen et al. [29] develops an NAO robot walking control system based on motor imagery by utilizing CSP and local characteristic scale decomposition (LCD). The experimental results yielded a classification accuracy of 87.5%.

Our study [23] proposed an instantaneous amplitude and instantaneous frequency component-based features. First, the empirical wavelet transform (EWT) was employed to decompose an EEG signal into representative modes, then the Welch PSD method was adopted for modes selection. The last step was to calculate the instantaneous components of each selected mode and classify the features with seven machine learning classifiers. The maximum accuracy achieved was 95.2% for the proposed mechanism. Our second study [24] on motor imagery EEG proposed a multivariate empirical wavelet transform (MEWT) for signal decomposition. By selecting features with correlation-based method and classifying them with three benchmark classifiers, we obtained 98% classification outcomes for the least square version of SVM classifier. All the methods discussed above, either utilized complex signal decomposition methods in combination with features selection methods or used complex features extraction methods, which are both impractical for the realization of functional BCI system. Raghu et al. [30] proposed the successive decomposition index (SDI) method for the classification of epileptic seizures. The classification outcomes suggested that SDI is a successful feature extraction method for epileptic seizures and it can be extended to other EEG domains.

Many different studies have built graphical user interface (GUI) systems for the visual implementation of their proposed approaches. EPILAB GUI was developed by Teixeirra et al. [31] for the analysis and classification of epileptic seizures. EEGLAB developed by Delorme et al. [32] presented an ICA-based EEG signal denoising method, time-frequency analysis, and visual representation of EEG signals. Moreover, Oostenveld et al. [33] reviews a MATLAB open source toolbox named FieldTrip, which does the time-frequency analysis, non-parametrical statistical tests, and reconstruction using dipoles and distributed sources of EEG and magnetoencephalography (MEG) signals. Each of these methods analyzes multidomain EEG signals, but a specialized GUI for motor and mental imagery is lagging.

For the robust, efficient, and non-complex analysis and classification of motor and mental imagery EEG signals, this article for the first time to the best of our knowledge and understanding, makes use of successive decomposition index (SDI) for feature extraction. This research attests the performance of SDI feature using six benchmark machine learning and neural network classifiers and different case studies confirms the effectiveness of proposed method. The main contributions of this study are listed as follows:

1. Successive decomposition index is proposed for the decoding of different motor and mental imagery activities in development of the BCI system.
2. Statistical analysis and novel performance evaluation criteria named polygon area metrics (PAM) are performed to confirm the efficacy of the SDI feature as a biomarker.
3. Four different channel selection schemes are employed to validate the performance of SDI features corresponding to the number of channels.
4. Classifier parameters are varied to investigate its fallouts on the proposed method.
5. A comparison was undertaken for denoised and noisy data to confirm the robustness of SDI features against noise artifacts.
6. Validate the performance of the proposed approach for multiclass mental imagery data.
7. Developed a computerized automatic detection of motor and mental imagery-successive decomposition index (CADMMI-SDI) application for the visual and practical implementation of SDI features.

The rest of paper is organized as follows. Sections 2 and 3 deal with the datasets and the description of methods employed during the study, Section 4 describes the performance measures, Section 5 presents experimental set-up, Section 6 provides the results and discussion of the experimental outcomes, and, finally, Section 7 summarizes the study.

2. Materials

This study makes use of three motor and mental imagery publicly available datasets: IVa, IVb, and V from BCI competition III. Dataset IVa is a motor imagery dataset with two tasks right hand

(RH) (Class 1) and right foot (RF) (Class 2). Five normal subjects or participants ("aa", "al", "av", "aw", and "ay") participated for the collection of datasets. The global 10-20 system was used for the placement of 118 electrodes on the scalp. All the participants were shown a visual sign for 3.5 s and a total number of 280 trials (140 trials for each class) were recorded for an individual participant and the data were sampled at 1000 Hz. Similarly, dataset IVb is another single participant binary class motor imagery dataset with tasks left hand (LH) (Class 1) and right foot (RF) (Class 2). The data acquisition parameters for dataset IVb are similar to dataset IVa. Dataset V is a data collection of 3 individuals with imaginative roles of LH movement, RH movement, and random word (RW) production. These tasks are named as Class 1, Class 2, and Class 3, respectively. Data was collected in three cycles from 3 individuals with 32 electrodes and sampling frequency of 512 Hz. Further information for data sets is presented online at http://www.bbci.de/competition/iii/.

3. Methods

The study proposed a SDI-based framework for automated classification of two and multi-category motor and mental imagery EEG tasks in the development of computer-aided BCI systems. Figure 1 shows a clear presentation of the proposed strategy. First, the MSPCA process is used to separate noise from the raw EEG signal. Afterward, SDI is employed, that is, an inspirational case of discrete wavelet transform where a time series is pass through n levels of low-pass and high-pass filters and the coefficient at each step is used as a feature, and at last the extracted features are used as the inputs to the several machine leaning and neural network classifiers. Moreover, this study built up a layout for the realistic implementation of proposed platform for identifying motor and mental imagery EEG signals known as computerized automated detection of motor and mental imagery successive decomposition index (CADMMI-SDI). The subsequent subsections describe the details of the proposed automated framework.

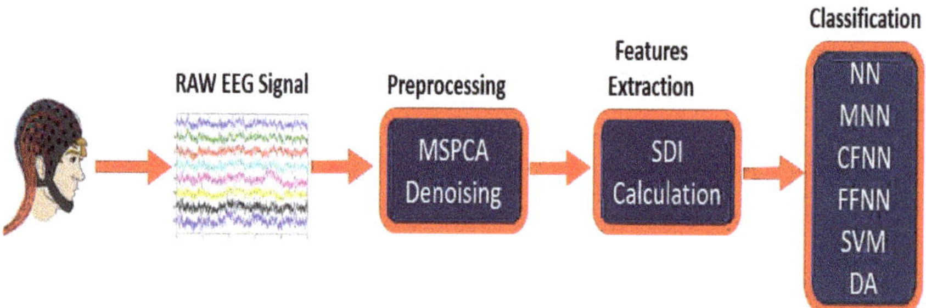

Figure 1. Block diagram of the successive decomposition index for identification of motor and mental imagery activities.

3.1. Module 1: MSPCA Denoising

EEG is a noninvasive method of signal retrieval from the subject that inherits different types of noise artifacts, i.e., systematic noise, blink signal noise, cardiac signals noise, thermal noise, etc. A mathematical model of the crude form signal can be described as follows [34],

$$X = X_{EEG} + X_N \tag{1}$$

where X_{EEG} is the desired EEG signal and X_N is the supplemental noise artifact added to the original signal. The objective is to model a system that can effectively remove noise from the raw signal without influencing the content of X_{EEG}. Principal Component Analysis (PCA) is conventionally adopted for determining the linear relationship between correlated data points. Furthermore, the nonlinear and non-stationary nature of the EEG signal demands a time-frequency resolution. Therefore, wavelet transform is commonly adopted and its significance is widely tested for non-stationary and nonlinear signals. A hybrid signal denoising algorithm called multiscale principal component analysis (MSPCA) is formulated by combining the properties of PCA and wavelet transform [24]. The workflow of MSPCA is given in Figure 2. We can define the procedure as follows.

1. Take a matrix A with dimensions $n \times m$, where n is the length of each signal and m is the number of channels. Decompose each channel into B levels using wavelet transform.
2. Formulate a detailed matrix A_jA and approximation matrix X_iA, and calculate PCA for all B decompositions and m channels. As the Kaiser rule suggests, select principal components with eigenvalues greater than the mean of collective eigenvalues.
3. Compute the inverse wavelet transform of the selected principal components.
4. A denoised signals matrix can be obtained by taking the PCA of the results obtained in step 3.

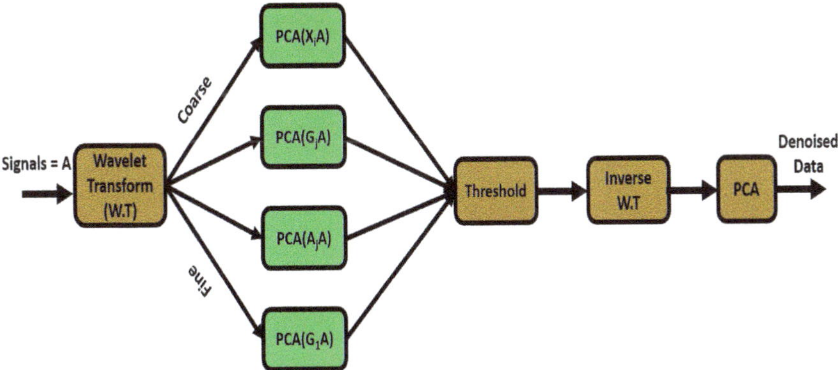

Figure 2. Multiscale principal component analysis (MSPCA) for denoising.

3.2. Module 2: Successive Decomposition Index Based Feature Extraction

In the past, a large number of studies [22,24,35,36] investigated the effectiveness of wavelet and signal decomposition-based methods for motor and mental imagery EEG signals using different mother wavelets and decomposition levels. The drawbacks of such methods are the selection of suitable mother wavelets and the number of decomposition levels which requires a thorough investigation in terms of classification outcomes and time complexity. The basic requirements of a practical BCI system are robustness, non-complexity and efficiency that are lagging in current researches. To overcome the aforementioned limitations a successive decomposition index (SDI) method is employed.

The proposed SDI method is an inspiration of discrete wavelet transform (DWT). In the first level of DWT, a time signal of length n is passed through a low and high pass filter. In the next level, the output of low pass filter is again passed through a high and low pass filter and this process is iterated for a specific number of decomposition levels. Finally, the coefficients from each decomposition level are used to extract features. The basic difference between DWT and SDI is that the former has to have a predefined number of decomposition levels whether the later has no predefined decomposition levels and the coefficient from the last level is considered for further analysis. The mathematical formulation of the SDI feature is described in following steps [30].

1. Consider an EEG signal $s = \{s_1, s_2, s_3, \ldots\ldots\ldots, s_n\}$, where n is the length of the signal. The first step is to compute the average of absolute values (S^+) of the EEG signal is as follows.

$$S^+ = \frac{1}{n}\sum_{i=1}^{n}|s_i| \tag{2}$$

2. The next step is to compute the average difference (S^-) of the signal and it can be calculated by the successive difference mean of non-overlapping pairs of time signal. It can mathematically represented as follows,

$$s^{(1)} = \left\{ \frac{s_1 - s_2}{2}, \frac{s_3 - s_4}{2} \ldots\ldots, \frac{s_{n-3} - s_{n-2}}{2}, \frac{s_{n-1} - s_n}{2} \right\} \tag{3}$$

where the length of $s^{(1)}$ is $n/2$. Similarly, $s^{(2)}$ can be calculated as

$$s^{(2)} = \left\{ \frac{s_1^{(1)} - s_2^{(1)}}{2}, \frac{s_3^{(1)} - s_4^{(1)}}{2} \ldots\ldots, \frac{s_{n/2-3}^{(1)} - s_{n/2-2}^{(1)}}{2}, \frac{s_{n/2-1}^{(1)} - s_{n/2}^{(1)}}{2} \right\} \tag{4}$$

The process of calculating $s(k)$ (where k is the number of iterations) continues until we get a single coefficient and that final coefficient is the average difference term S^-. The number of iterations required to calculate S^- can be determined as $k = 3.33\log10(n)$ and the total number of coefficients at each step are $n/2^k$. The next step is to calculate two new terms S^{++} and S^{--} as follows.

$$S^{++} = \frac{S^+ + S^-}{2} \tag{5}$$

$$S^{--} = \frac{S^+ - S^-}{2} \tag{6}$$

The terms S^{++} and S^{--} gives the relation between S^+ and S^-. In addition, a square matrix Z is formed from the four coefficients as follows.

$$Z = \begin{bmatrix} S^+ & S^{--} \\ S^- & S^{++} \end{bmatrix} \tag{7}$$

3. The final step is to calculate the determinant of matrix Z multiplied by a scalar n/k followed by log10.

$$SDI = \log_{10}\left(\frac{n}{k}\left(S^+S^{++} - S^-S^{--}\right)\right) \tag{8}$$

The resultant SDI is a single value bio marker for an EEG signal of length n. The significance of SDI is that it measures the variations of EEG signal successively with respect of time and packs it into a single representative value. In addition, unlike other wavelet and signal decomposition-based methods, there is no need to select a suitable mother wavelet and define the number of decomposition levels rather the process of calculating SDI is linear and non-complex, which makes it a suitable choice for the development of practical motor and mental imagery BCI systems.

3.3. Module 3: Classification

To segregate the motor and metal imagery tasks, we have utilized six widely used machine learning and neural network classifiers. Their description and parameters of classifiers utilized in this study are discussed as follows.

3.3.1. Support Vector Machine

A support vector machine (SVM) is a supervised learning classifier that formulates a hyperplane to maximize the separability between two classes. For nonlinear feature sets, different kernel functions are utilized to transform it into a linear problem at the cost of augmented dimensionality. The selection of SVM in this study is based on its robustness and reliability for motor imagery tasks discussed in [37,38]. In this study, we have utilized the radial basis function, linear function, and polynomial function as kernels and the default MATLAB toolbox hyperparameters were availed for each kernel.

3.3.2. Discriminant Analysis

Discriminative analysis (DA) is a supervised learning algorithm that formulates a predictive model during the learning phase that can be applied to test data for labeling them. DA can use lines, planes, and hyperplanes to segregate the normally distributed samples and thus it can classify multidimensional data robustly. To build a DA model, we have to compute the class probability, mean, and covariance matrix along with a suitable kernel function. In this study, we have utilized three kernels: linear, pseudo-linear, and pseudo-quadratic. The effectiveness of DA for motor imagery tasks has been accredited in [39,40].

3.3.3. Multilayer Perceptron with One Hidden Layers

A multilayer perceptron with single hidden layer (ANN) is the building block of deep learning classifiers and is robust in approximating linear, nonlinear functions and pattern recognition effectively. ANN has a three-layered structure consisting of input, hidden, and output layers. The number of input nodes is same as the number of features while the number of output nodes is equal to the number of classes. The number of hidden nodes is variable and depends primarily on classification outcomes. ANN propagates the input signal from first to last layer and the backpropagation algorithm tunes the hyperparameters of the network during training phase. The studies [41,42] attests the robustness of ANN for motor imagery EEG.

3.3.4. Multilayer Perceptron with Two Hidden Layers

A multilayer perceptron with two hidden layers (MNN) is an extension of ANN. The basic difference between both algorithms is that MNN has 2 to M hidden layers depending upon the classification results while ANN has only one hidden layer. The advantage of using MNN is that it has more parameters and hence it has an extra degree of freedom to approximate a nonlinear function or recognize a pattern. The disadvantage is that, because of the large number of hyperparameters, the training and testing time exceeds ANN and hence there is a trade-off between computational time and classification outcomes.

3.3.5. Cascade Feedforward Neural Network

The architecture of cascade feedforward neural network (CFNN) resembles ANN. The core difference between both classifiers is that CFNN has a connection from the output layer to the input layer that ANN lags in its structure. This extra connection gives CFNN the ability to memorize previous inputs and their outcomes and thus it is essential in learning sequential data. The authors of [43] utilized CFNN for the classification of motor imagery tasks.

3.3.6. Feed-Forward Neural Network

Throughout the feed-forward neural network (FFNN), a multilayered structure is used with each layer containing variable number of neurons. The signal is propagated from input to output across the network and an error is computed using a cost function. This error is then repropagated across the network and each parameter is tuned. In our research, tan sigmoid was used as an activation feature. The Levenberg–Marquardt algorithm was used for fast learning [43].

There is no structural difference between ANN and FFNN. In the present study, we utilized two different MATLAB functions named "patternenet()" for ANN and "feedforwardnet()" for FFNN. The basic difference between these two functions is that ANN uses "glorot" weights and biases initializer while FFNN uses "orthogonal" initializer. The "glorot" initializer takes random samples from a normal distribution where mean is zero and variance is 2/(size of inputs + size of outputs), while the orthogonal initializer takes a matrix from a unit uniform distribution and initializes the weights and biases with Q obtained from a QR decomposition [44,45].

4. Performance Parameters

This study utilizes a 10-fold cross-validation method to fairly evaluate the classification results. For this purpose, the feature matrix containing Class 1 and Class 2 features is divided into 10 equal parts, out of that 9 parts were used for training purposes and 1 part was used for validation. In this way, each trial of the feature set is being trained upon as well as validated. To evaluate the classification outcomes, we made use of 10-fold cross-validation method with different performance metrics, namely, classification accuracy (Acc), Sensitivity (Sen), Specificity (Spe), Kappa, and F1-Score. Their mathematical expressions are given respectively as follows,

$$A_{cc} = \frac{TP + TN}{TP + TN + FP + FN} \tag{9}$$

$$S_{en} = \frac{TP}{TP + FN} \tag{10}$$

$$S_{pe} = \frac{TN}{TN + FP} \tag{11}$$

$$Kappa = \frac{TP \times TN - FP \times FN}{\sqrt{((TP + FP)(TP + FN)(TN + FP)(TN + FN))}} \tag{12}$$

$$F1\_Score = 2 \times \frac{\text{Prec} \times S_{en}}{\text{Prec} + S_{en}} \tag{13}$$

where TP (True positive) is the amount of adequately identify Class 1 labels, TN (True negative) is the amount of adequately identify Class 2 labels, FP (False positive) is the number of inadequately classified Class 1 labels. and FN (False negative) is the number of inadequately identified Class 2 labels.

Apart from the above mentioned five performance parameters, we utilized a novel performance evaluation criteria named polygon area metrics (PAM) [46] for the very first time for motor and mental imagery EEG classification evaluation. The PAM constructs a hexagon with six performance parameters (F measure, Jaccard Index, Classification accuracy, Area under the curve, Sensitivity and Specificity) on each edge. The performance in this case is evaluated by the area of the polygon. The greater the area occupied by the polygon, the better the performance of the classifier and vice versa.

5. Experimental Setup

All experiments and simulations in this study were performed using MATLAB R2019b on an Intel(R) Core (TM) M-5Y10c CPU @0.80GHz cpu, Windows 10 64-bit operating system, and 8 GB RAM with WEKA 3.8.4.

Numerous studies have been performed in the past for effectively classifying motor and mental imagery tasks as detailed in Section 1. Most of them utilized complex signal processing techniques that make those unfeasible for the practical implementation and it also gets difficult for physicians to understand complex signal processing tools without having a piece of proper knowledge about the field. To cope up with such challenges, we have utilized a single non-complex feature that uses iterative signal decomposition coefficients to construct a representative feature with the least computational complexity and effective classification results.

Figure 1 shows the block diagram of the proposed methodology. At first, the raw data is passed through an MSPCA filter that suppresses the noise content from the signals. Then the data is divided into individual trials. In the case of dataset IVa and IVb, the single trial dimension is 400 $\times$ 118, where 400 is the signal length and 118 is the number of channels. For dataset V, the single-trial dimension is 512 $\times$ 32, where 512 is the signal length and 32 is the number of channels. Next, each trial is given to an SDI computational function which calculates features for that trial. In the case of dataset

IVa and IVb, we get 118 features for a single trial while for dataset V, we get 32 features for a single trial. In this way, a features matrix is formed with dimensions n × m, where *n* is the number of trials and *m* is the number of features (indirectly the number of channels) per trial. Last, the feature matrices of various classes are given to six benchmark classifiers to evaluate the performance of SDI features in estimating motor and mental imagery tasks.

6. Results and Discussion

6.1. Statistical Analysis

To analyze how the SDI feature segregates motor imagery tasks, we have performed a statistical analysis in this section. Figure 3 presents the SDI feature distribution for Class 1 and Class 2 tasks by utilizing channel C3 from all subjects of dataset IVa and IVb. Figure 3 suggests that subjects "aa", "al", "av", "aw", "ay", and dataset IVb have a highly nonlinear relationship between both task features and it is imperative to use a nonlinear classifier to trace the pattern between both classes. It can be seen in the Figure 3 that SDI feature has significantly singled out tasks for small training samples subject "ay" and later in this study we will see that subject "ay" is the best performant among all other subjects in terms of classification outcomes.

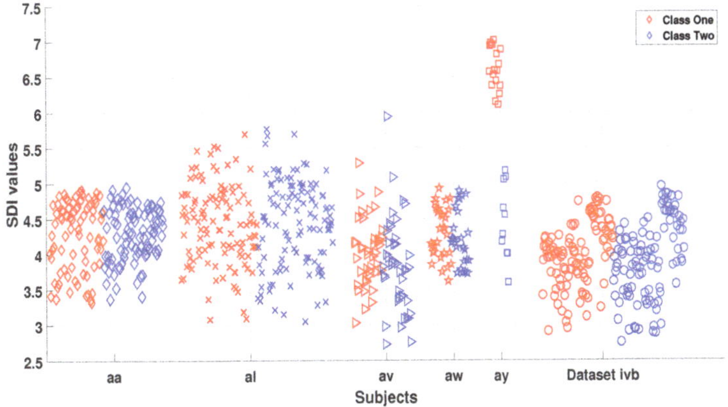

Figure 3. Scatter plot of SDI features for dataset IVa and IVb subjects.

In addition, a descriptive statistical analysis in terms of mean, standard deviation, median, and Kruskal–Wallis probability (p) values (KW test) of SDI features was performed for single trial cases of each subject. The results presented in Table 1 suggest that the mean and median values of subject "aa", "al", "av", "aw", "ay", and dataset IVb are higher for Class 2 cases than Class 1. For the subject "ay", the mean and median values for Class 1 are higher and this trend was consistent for all trials. Moreover, the KW p values for single-trial cases of all subjects are less than 0.05 which suggests the significance of SDI features for motor imagery tasks and the high discrimination ability of extracted features between two classes.

Table 1. Statistical analysis.

Participants	MI Tasks	Mean	Std	Median	KW p Values
"aa"	"Class 1 (RH)" "Class 2 (RF)"	4.071 4.633	0.506 0.175	4.088 4.626	2.16×10^{-19}
"al"	"Class 1 (RH)" "Class 2 (RF)"	3.994 4.333	0.498 0.931	3.986 4.178	0.06112
"av"	"Class 1 (RH)" "Class 2 (RF)"	3.655 5.961	0.516 0.139	3.698 5.979	3.27×10^{-40}
"aw"	"Class 1 (RH)" "Class 2 (RF)"	3.811 3.966	0.343 0.323	3.737 3.925	0.0001927
"ay"	"Class 1 (RH)" "Class 2 (RF)"	5.766 3.948	0.535 0.660	5.725 4.058	5.81×10^{-39}
"IVb"	"Class 1 (LH)" "Class 2 (RF)"	3.716 3.993	0.440 0.524	3.734 3.979	4.09×10^{-5}

6.2. Results by Selecting Different Number of Channels

Siuly et al. [47] conducted a comparative analysis for 18 and 118 channels motor imagery dataset IVa and IVb using two classification algorithms. Their study concludes that 118 channel results outperform 18 channels in terms of classification outcomes. In this section, a similar type of comparison is presented for dataset IVa and IVb with 18 channels, three channels and three channels selected with automated channel selection criteria. The 18 and three channels are widely adopted motor cortex channels while three-channel selection with automated channel selection criterion was proposed in our previous study [24]. The list of automated channels for each subject is given in Table 2. As motor imagery EEG signals are highly dependent upon subject physical and mental nature so for each subject, different channels are selected by the automated channel selection criteria. Figure 4 shows a visual representation of four channels selection schemes for best and worst-performing classifiers. The worst classifier is characterized in terms of least gain in accuracy while the best classifier symbolizes maximum gain in classification accuracy. This study made use of six machine learning and neural network classifiers (NN, MNN, CFNN, FFNN, SVM, and DA) out of which FFNN was the best performing classifier and SVM was the worst performer. The rest of the analysis is given as follows.

Table 2. List of 3 channels automated criteria.

Subjects	Selected Channels
aa	CCP5, CP5, CP6
al	C3, FFC7, CCP3
av	FT9, P8, PPO8
aw	C4, CCP6, CP6
ay	CCP5, C3, CFC5
IVb	C3, CCP5, C4

1. It is inferred from Figure 4 that 118 channels give the highest classification accuracy for dataset IVa as compared to other channel combinations. The average classification accuracy obtained using 118 channels with FFNN classifier is 97.46%. Similarly, the average accuracy for SVM classifier using 118 channels is 93.05%. Moving on to the 18-channel combination, it is observed that the average classification accuracy for FFNN and SVM classifier is 94.28% and 77.96% respectively. Furthermore, the 3-channel scheme resulted in a mean accuracy of 77.6% and 60.1% for FFNN and SVM classifiers respectively. Finally, the 3-channel automated scheme has the least average results as compared to other channels combination. The average results obtained for 3-channel automated criteria are 73.74% and 60.46% for FFNN and SVM classifiers, respectively.

2. For dataset IVb, a similar trend of classification accuracies is observed for varying channel combinations. Figure 4 shows that FFNN and SVM resulted in 99.5% accuracy each using 118 channels for dataset IVb and this is the highest among other channel selection schemes. For the combination of 18 channels, the FFNN classifier resulted in 96.2% accuracy and SVM yielded 91% classification certainty. Moving forward to the 3-channel strategy, it is noticed that FFNN turnout 82.9% accuracy and 52.4% for SVM. Likewise, 3-channel automated scheme resulted in similar pattern of results with 83.6% for FFNN and 54.6% for SVM classifier.
3. It is observed that the 118-channel combination has a maximum gain of 23.72% and 32.59% for FFNN and SVM classifiers, respectively using dataset IVa. Dataset IVb has a maximum gain of 16.6% and 47.1% for FFNN and SVM classifiers, respectively. It accredits the significance of using 118 channels for SDI features and advocates the channel comparison study performed by Siuly et al. [47].
4. One interesting observation is made that subject "ay" of dataset IVa has above 90% classification accuracy for all channel combinations and classifiers. As mentioned in the descriptive analysis section, the SDI features for subject "ay" tasks are well separated and distinguishable. We conclude that SDI feature extraction is more significant for subject with small training samples as compared to large one and this property makes it feasible for the development of practical BCI systems as disabled patients need small training to train a device.

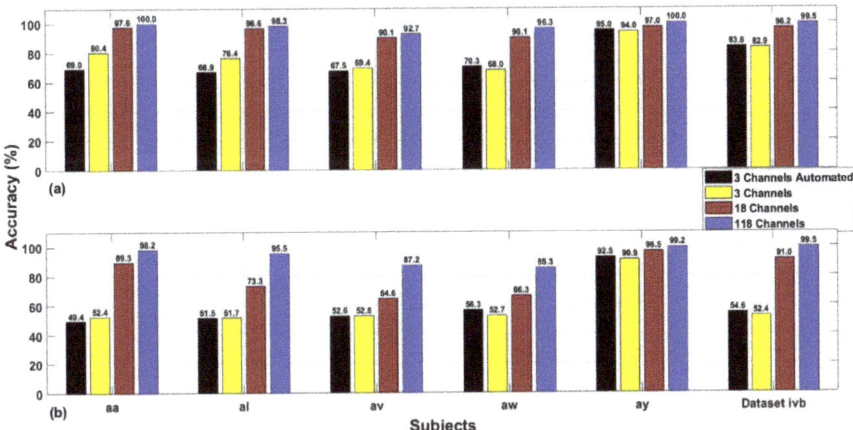

Figure 4. Bar plots for the comparison of 3-channel automated, 3-channel, 18-channel, and 118-channel results: (**a**) FFNN classifier and (**b**) SVM classifier.

6.3. Analysis with Sensitivity, Specificity, Kappa, F1-Score and PAM

In this section, we explain the effect of other performance measures namely sensitivity, specificity, kappa, F1-Score and most importantly a unified novel performance measure, the polygon area metric (PAM). Figure 5 shows the sensitivity, specificity, kappa, and F1-score values for FFNN and SVM classifiers using 118 channels with 10-fold cross-validation strategy. Figure 5a,e show the sensitivity values for FFNN and SVM classifier, respectively. The average sensitivity values are 98.8% and 94.8% accordingly for individual classifiers which suggests that FFNN correctly identified Class 1 instances 98 times and SVM classified them correctly 94 times. Similarly, Figure 5b,f show the specificity values for FFNN and SVM classifiers respectively. The average 10-fold specificity values are 98.25% and 95.57%, respectively, for each classifier, which indicates that FFNN classified Class 2 instances effectively 98 times and SVM classified them positively 95 times. Figure 5c,g presents the kappa scores for the aforementioned classifiers. It is noted that the average kappa for FFNN classifier is 96.93% with slight variations for subject "aw". The average kappa for SVM classifier is 91.5% with major variations in subject "av" and "aw". Hence, we conclude that FFNN is more stable and unbiased in classifying Class 1 and Class 2 tasks. Finally, Figure 5d,h show the F1-Score for each classifier, respectively, and the average F1-Score for individual classifier is 98.07% and 93.83% accordingly. The high value of F1-Score for FFNN classifier illustrates the high precision and recall measures.

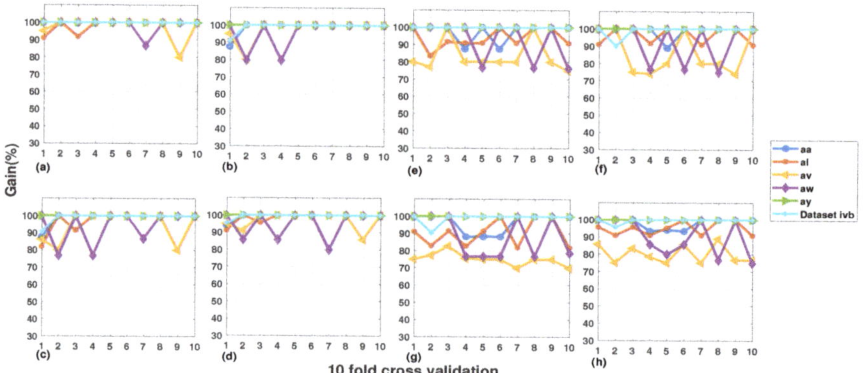

Figure 5. (**a–d**) 10-fold Sensitivity, Specificity, Kappa, F1-Score for FFNN Classifier. (**e–h**) Ten-fold Sensitivity, Specificity, Kappa, and F1-Score for SVM Classifier.

Figure 6 shows the PAM graphs for dataset IVa all subjects and dataset IVb using FFNN and SVM classifiers for 118 channels scheme. Figure 6a–f presents the PAM graphs for FFNN classifier and Figure 6g–l shows PAM graphs for SVM classifier. It can be seen that subject "aa" and "ay" have an area of 1 unit while subjects "al", "av", and "a" have areas of 0.95, 0.78, and 0.85 units for FFNN classifier, respectively. Dataset IVb has an area of 0.98 units for FFNN classifier. All of these results are consistent with the above-mentioned accuracy and other performance measures outcomes. Moreover, in the case of SVM classifier, subject "aa", "ay" and dataset IVb has an area of 0.98 units each, subject "al", "av", "aw", and "ay" has an area 0.95, 0.81, and 0.79 units, respectively. The key benefit of using PAM graph is that complete classification performance is represented in a single graph with several measures instead of looking into lengthy tables.

6.4. Results by Selecting Different Parameters of Classifiers

To investigate the fallouts of classifier parameters on the proposed approach, we compared the classification accuracies for varying classifier parameters of all classifiers. Table 3 shows the averaged 10-fold accuracies of all classifiers with varying parameters for the 118-channel scheme using dataset IVa individual subjects and dataset IVb. For neural network (NN) classifiers, the number of hidden layer neurons was varied and its effect was observed accordingly. For SVM classifier, three different kernels namely radial basis function (RBF), linear kernel and the polynomial kernel were utilized, for DA classifier, linear, pseudo quadratic and pseudo linear kernels were adopted and their performance was evaluated for both datasets individually. The findings are as following:

1. The experimental results suggest that NN classifiers have no significant effect on average accuracy by varying the number of hidden neurons. For NN classifiers, the maximum mean classification accuracy was recorded for 40 neurons with 94.6% and 99.05% results for dataset IVa and IVb, respectively. For MNN classifier, the best case mean classification outcome was obtained for 30 neurons for dataset IVa with an accuracy of 91.2% and 40 neurons for dataset IVb resulting in 94.29% mean accuracy. Moving on to CFNN classifier, it is noted that each number of neurons yields 97% accuracy for dataset IVa and 99% outcomes for dataset IVb. Lastly, FFNN classifier

turned out the maximum mean accuracy of 98.27% using dataset IVa and 99.52% using dataset IVb for 30 neurons each.

2. Amid the three kernels of the SVM classifier, it is observed that polynomial kernel is the best performant among others with the mean classification accuracy of 93.08% and 99.52% for datasets IVa and IVb, respectively. The linear and RBF kernels are ranked second and third accordingly with mean accuracies of 89.64% and 70.24% for dataset IVa and 96.19% and 83.33% for dataset IVb respectively.
3. In case of the DA classifier, we observe that the linear kernel is ranked highest as compared to pseudo-quadratic and pseudo-linear kernels. The average classification accuracy for linear kernel is 95.29% and 98.1% for dataset IVa and IVb each. Moving forward, the pseudo-linear kernel is ranked second with a mean classification accuracy of 93.29% and 95.1%, respectively, for datasets IVa and IVb. Last, the pseudo quadratic-kernel results in an average classification result of 92.63% and 92.24% accordingly for each dataset.
4. The thorough investigation of varying classifiers parameters suggests that polynomial and linear kernels are the best performers for SVM and DA classifiers respectively. Similarly, for NN, MNN, CFNN, and FFNN classifiers, 40 number of hidden neurons were chosen as the best parameter setting and these parameters are employed throughout this study.

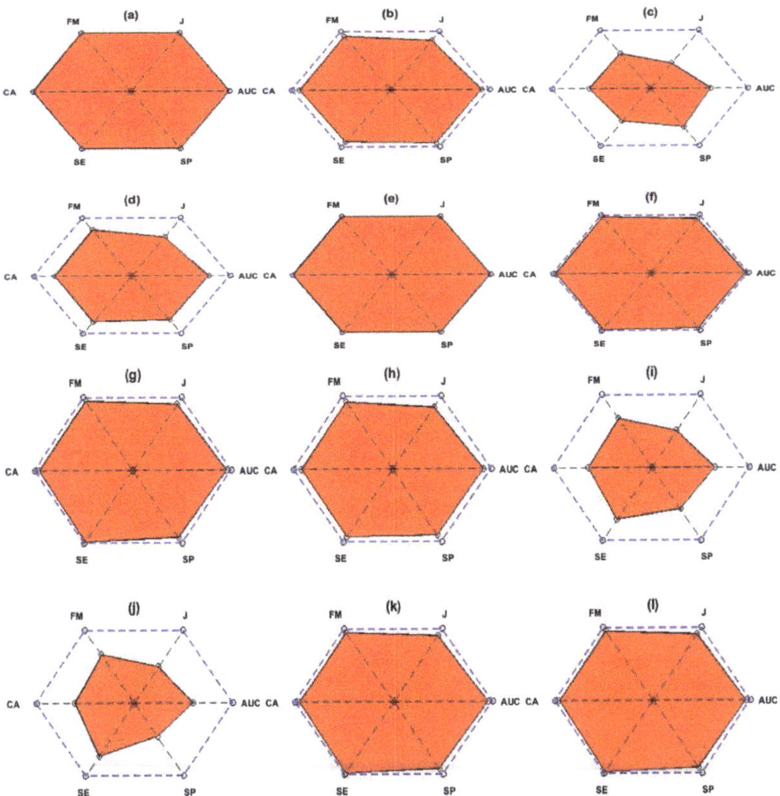

Figure 6. **(a–f)** PAM for Subjects "aa", "al", "av", "aw", "ay" and "Dataset IVb" respectively using FFNN classifier. **(g–l)** PAM for Subjects "aa", "al", "av", "aw", "ay" and "Dataset IVb" respectively for SVM classifier.

5. Figure 7 shows the average accuracies of 10 times repeated 10-fold experiments for best (FFNN) and worst (SVM) case classifiers and each subject of dataset IVa and IVb. It is noted that the average results obtained for both classes results in slight variations of ±1.5%. In case of "av" subject with the FFNN and subject "aw" with the SVM, the variations are larger than 10%, which is due to the outliers caused by classifiers in some fold results but the mean results are more or less the same as calculated previously. The extensive experimentation results obtained confirms the robustness and stability of SDI features in estimating motor imagery tasks.

Table 3. Classification (%) results for different parameters of the classifier.

"Classifiers"	"Variations in Parameters"	"aa"	"al"	"av"	"aw"	"ay"	"Dataset IVb"
"NN"	"5 Neurons"	97.61	93.30	77.78	83.67	95.43	99.52
	"10 Neurons"	95.00	91.05	86.67	90.67	97.33	99.05
	"20 Neurons"	97.61	98.20	87.22	89.33	98.23	97.62
	"30 Neurons"	97.61	97.33	76.94	93.00	99.24	98.57
	"40 Neurons"	99.41	95.49	85.97	93.00	99.10	99.05
"MNN"	"5 Neurons"	87.32	84.55	81.25	81.00	97.55	98.10
	"10 Neurons"	89.71	88.30	72.78	63.00	98.33	92.38
	"20 Neurons"	89.71	86.68	74.86	80.67	99.12	97.14
	"30 Neurons"	95.74	94.58	78.33	91.00	96.34	93.33
	"40 Neurons"	96.99	96.42	79.72	81.67	99.56	94.29
"CFNN"	"5 Neurons"	99.38	98.66	94.03	98.33	98.12	99.05
	"10 Neurons"	99.41	99.55	93.06	98.33	97.24	99.52
	"20 Neurons"	100.00	97.77	98.75	91.00	99.10	99.23
	"30 Neurons"	99.41	99.55	95.28	91.33	100	99.52
	"40 Neurons"	100.00	100.00	97.64	94.33	95.00	99.05
"FFNN"	"5 Neurons"	99.38	99.11	94.03	96.33	97.11	98.10
	"10 Neurons"	98.75	98.24	91.81	98.00	98.67	98.57
	"20 Neurons"	100.00	97.31	90.69	96.33	97.98	99.52
	"30 Neurons"	99.41	98.66	95.28	98.33	99.65	99.52
	"40 Neurons"	100	98.33	92.7	96.7	100	99.5
"SVM"	"RBF"	63.57	62.92	64.58	69.00	91.12	83.33
	"Linear"	94.08	88.38	85.42	84.67	95.65	96.19
	"Polynomial"	98.20	95.53	87.22	85.33	99.12	99.52
"DA"	"Linear"	98.82	99.09	86.53	94.67	97.33	98.10
	"Pseudo Quadratic"	94.12	97.81	84.58	86.67	100.00	93.24
	"Pseudo Linear"	98.82	99.09	85.42	84.00	99.12	95.10

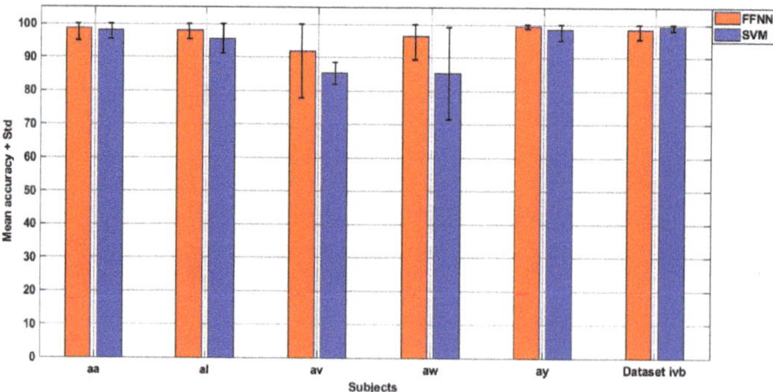

Figure 7. Results obtained with 10-fold 10 times.

6.5. Results with Raw EEG and Noise-Free EEG Signals

We discussed earlier that EEG is a noninvasive mode of signal retrieval and it inherits noise artifacts while recording the data. In this section, a comparative analysis for MSPCA denoised and unprocessed (noisy) data is performed and validated if SDI feature is being affected by noise artifacts or not.

Figure 8 shows the classification accuracy for MSPCA denoised and noisy data of dataset IVa and IVb. The classification results are calculated for best-case FFNN classifier. As observed from Figure 8, the classification accuracies for noisy data are 83.1%, 84.4%, 82.5%, 85%, 92.4%, and 81.4% for subjects "aa", "al", "av", "aw", "ay", and dataset IVb, respectively. The average results are 85.5% and 81.4% for dataset IVa and IVb respectively. We observe a significant improvement in individual and average classification results after denoising the data. The results after denoising with MSPCA are 100%, 97.3%, 90.6%, 96.3%, 100% and 99.52% for subjects "aa", "al", "av", "aw", "ay", and dataset IVb, respectively. The average accuracies for datasets IVa and IVb are 96.8% and 99.52%, respectively. By looking at the results obtained from two case scenarios, we observe an increase of 11.3% and 18.12% in accuracy for dataset IVa and IVb jointly. A similar trend of accuracy enhancement for denoised data was observed for other classifiers and hence it is concluded that the proposed SDI based feature extraction framework is robust against noise artifacts.

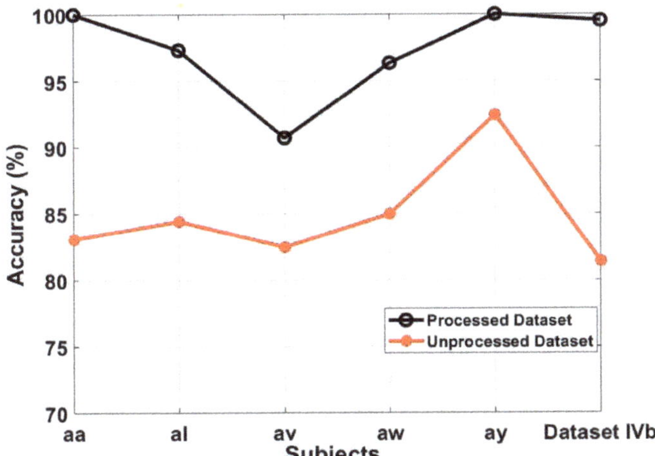

Figure 8. Comparison between denoised and noisy datasets.

It is important to note that we have also checked numerous conventional methods including such band pass filters, temporal filtering, and spatial filtering for meticulous selection of a suitable strategy in the preprocessing module and identified that MSPCA produces the best findings for the proposed SDI feature extraction approach.

6.6. Classification Performance (%) with Dataset

This section deals with the experimental results of multiclass mental imagery dataset V. At first, the dataset was denoised with MSPCA and rearranged into individual trials with dimensions 512×32 (where 512 is the signal length and 32 is the number of channels) for each trial. We have rearranged the multiclass problem into 3 binary class experiments for each subject. The number of cases are given in Table 4. Here cases 1 to 3 are dedicated for participant 1 (P1), cases 4 to 6 corresponds to the participant 2 (P2). and cases 7 to 9 are formed for the participant 3 (P3). Next, the SDI feature is calculated for all trials and fed into six classifiers. The classification outcomes in terms of accuracies are given in Table 5.

Table 4. Different cases consider for SDI experimental work by employing dataset V.

Case 1:	"Class 1 (LH)" vs. "Class 2 (RH)"	**Case 2:**	"Class 1 (LH)" vs. "Class 3 (RW)"	**Case 3:**	Class 2 (RH) vs. "Class 3 (RW)"
Case 4:	"Class 1 (LH)" vs. "Class 2 (RH)"	**Case 5:**	"Class 1 (LH)" vs. "Class 3 (RW)"	**Case 6:**	"Class 2 (RH)" vs. "Class 3 (RW)"
Case 7:	"Class 1 (LH)" vs. "Class 2 (RH)"	**Case 8:**	"Class 1 (LH)" vs. "Class 3 (RW)"	**Case 9:**	"Class 2 (RH)" vs. "Class 3 (RW)"

It is observed from Table 5 that all classifiers achieved an average accuracy of above 90% for each subject. Moreover, the average individual classification accuracy for NN, MNN, CFNN and FFNN is above 95% which shows the effectiveness of NN classifiers in segregating mental imagery tasks. The best-case scenario was observed in for FFNN classifier with an average accuracy of 99.07%, 98.16%. and 98.38% for participants 1, 2, and 3, respectively. It should be noted that FFNN was the best performer for motor imagery tasks and now it again gives the best results for mental imagery dataset. The worst-case scenario was observed for SVM classifier with accuracies 91.84%, 90.36%, and 93.81%, respectively, for first participant, second participant, and third participant. As per the experimental results, it is concluded that NN classifiers, especially FFNN classifier is intelligent in estimating mental imagery tasks.

Table 5. Classification accuracies (%) obtained with different cases by employing dataset V.

Classifiers	Cases	"P1"	"P2"	"P3"
"NN"	Case 1	100.00	98.22	96.67
	Case 2	93.21	97.98	97.44
	Case 3	100.00	97.12	98.43
	Average	**97.74**	**97.77**	**97.51**
"MNN"	Case 1	100.00	99.88	93.08
	Case 2	98.43	95.16	98.30
	Case 3	93.21	90.34	94.12
	Average	**97.21**	**95.13**	**95.16**
"CFNN"	Case 1	98.44	99.12	97.12
	Case 2	98.49	99.12	98.45
	Case 3	96.34	99.34	96.34
	Average	**97.76**	**99.19**	**97.30**
"FFNN"	Case 1	99.12	98.24	99.89
	Case 2	100.00	99.13	97.12
	Case 3	98.09	97.12	98.12
	Average	**99.07**	**98.16**	**98.38**
"SVM"	Case 1	95.23	94.32	99.12
	Case 2	94.74	95.23	94.98
	Case 3	85.54	81.52	87.34
	Average	**91.84**	**90.36**	**93.81**
"DA"	Case 1	93.45	94.55	93.19
	Case 2	94.14	96.34	95.83
	Case 3	88.32	91.78	89.15
	Average	**91.97**	**94.22**	**92.72**

Figure 9 shows the classification performance of SDI feature for dataset V in terms of four performance parameters (Sensitivity, Specificity, Kappa, and F1-Score). The performance parameters are shown for the best classifier which is FFNN in our case. It can be inferred from Figure 9 that the sensitivity and specificity values for all cases in each subject are above 95% and in some cases, it is 100% which shows the greatness of FFNN classifier in predicting Class 1, Class 2, and Class 3 tasks. It can also be seen that the kappa and F1-measures are above 95% in all cases which depict the stabilization and unbiased nature of FFNN classifier. Overall it can be concluded that SDI features are not only specific for motor imagery tasks but equally essential and significant for mental imagery tasks as well.

6.7. CADMMI-SDI Application

Apart from the theoretical analysis, we have developed a computerized automatic detection of motor and mental imagery using SDI (CADMMI-SDI) graphical user interface to assist physicians and laymen to utilize SDI method for their purpose without having to implement it their self. Table 6 presents the description of individual components present in the GUI while Figure 10 shows the detailed interface of our developed CADMMI-SDI. Some interesting features of the developed application are detailed in Table 6.

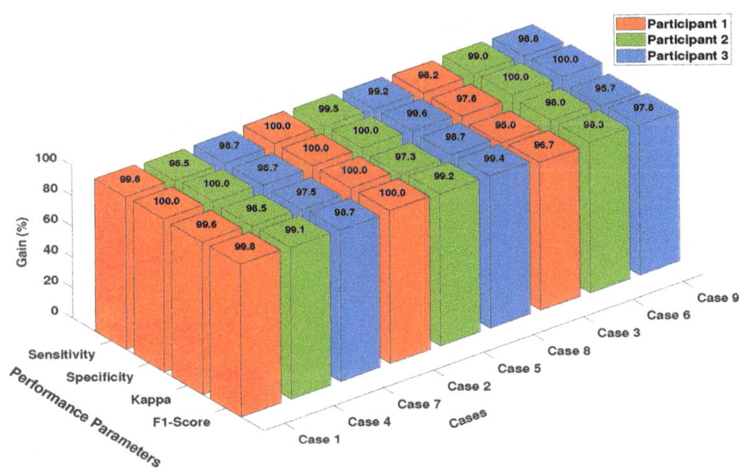

Figure 9. Performance parameters of FFNN classifier for Dataset V.

Table 6. CADMMI-SDI application.

Application Components	Description
Load EEG Data	Load Sample EEG data for a specified destination. The file type must be *.csv or *.xlsx
Test EEG Signal	Load test data from a specific folder. The file format should be *.csv or *.xlsx
Classifiers	Choose a classifier by drop-down selection.
Start	A key to initiate/start the process
Channel #	Input desired number of channels and press "Plot" to display. The channel number should be separated by a comma
Summary	Text section to demonstrate the specifics of the process underway
Signals	2D plot window to display EEG signals
Features Scatter Plot	2D plot window to display SDI feature corresponding to each channel

The demonstration of the GUI application can be seen in link https://www.youtube.com/watch?v=ugWbq4JUtuI. A copy of the GUI application is freely available and interested readers are suggested to write an email to corresponding author.

6.8. Computational Complexity of SDI Feature

Figure 11 shows the computational time for feature extraction, training and testing for all subjects and classifiers using the system specifications given in Section 5. First of all, Figure 11a presents the all trials feature extraction time for each subject of dataset IVa and dataset IVb. It can be seen that the highest feature extraction time of 1.36 s is taken by subject "al" followed by subject "aa" and dataset IVb with 1.06 s and 0.65 s, respectively. The average single-trial feature extraction time is calculated to be 0.85 milliseconds. Next, Figure 11b shows all trials training time for individual subjects and all classifiers. It is observed that CFNN classifier takes the highest training time for all subjects followed by FFNN classifier. The highest training time of 1.8 s, 1.75 s and 1.5 s was recorded for subjects "al", "aa", and dataset IVb, respectively, using CFNN classifier. The highest training time recorded for FFNN classifier is 1.2 s, 1.1 s and 1.08 s for dataset IVa, subject "al" and "aa", respectively. The average single-trial training time for FFNN classifier is calculated to be 1.27 milliseconds. Last, Figure 11c shows all trials testing time for individual subjects and all classifiers. As noted, SVM classifier takes the highest testing time of 70 milliseconds and 60 milliseconds for subjects "al", and "aa", respectively. The time taken by FFNN classifier is minimum in most cases and the average single-trial training time is recorded to be 0.01 milliseconds. By accumulating the single trials computational times for FFNN classifier, it comes out to be 2.13 milliseconds which is very nominal as compared to other complex signal decomposition methods and it shows that besides noise robustness and classification accuracy, SDI features are computationally less complex and efficient and hence it can be employed in the production of practical BCI systems.

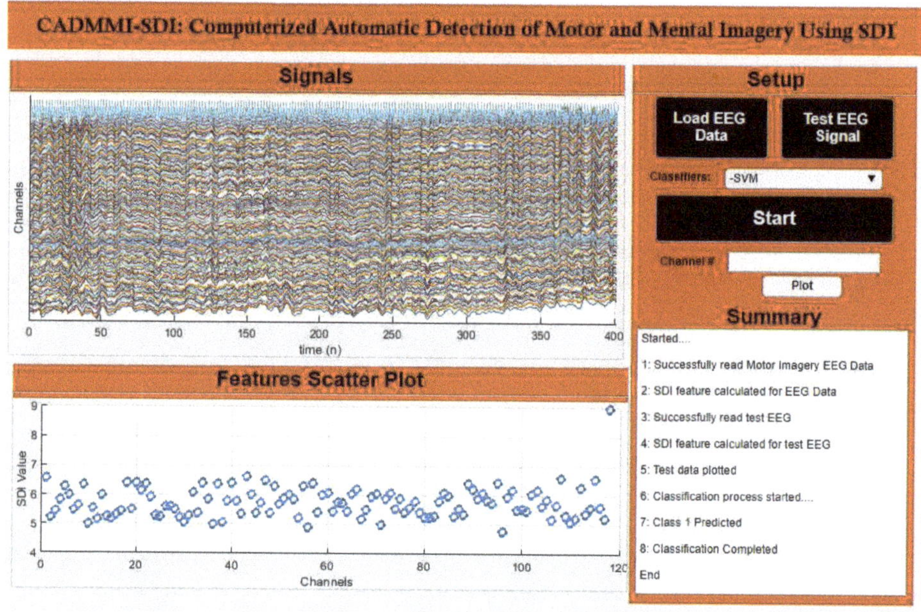

Figure 10. A display of CADMMI-SDI portraying all features and functionalities.

6.9. Performance Comparison with Other Literature

This section presents a comparative analysis of the proposed SDI framework with other recent state of art methods. Table 7 compares the classification accuracies for dataset IVa individual subjects and the best-case results are highlighted to make a fair comparison of other methods with the proposed approach. It can be seen from the table that subjects "aa" and "ay" attained 100% classification accuracy which is the highest among other methods. The results for subjects "al","av", and "aw" are above 90% and very close to the best results achieved by other methods. Comparing the results of SDI feature method with our previous studies [23,24], it is worth noting that our current method outperforms the complex signal decomposition and modes selection-based methods. It can be noted from Table 7 that our method achieved the highest average classification accuracy of 97.54% with minimal heterogeneity. Moreover, there is a 24.04% maximum gain in accuracy comparing to other state of the art methods and hence it suffices that SDI feature extraction is not only efficient and non-complex but also robust in estimating motor imagery EEG signals and this is validated by a fair comparison with other widely acclaimed studies.

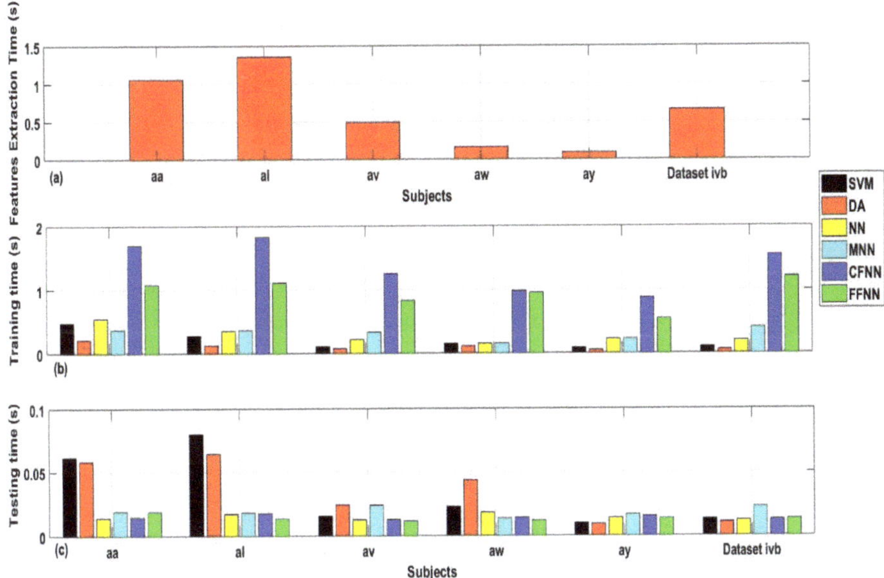

Figure 11. Bar plots representing time complexity: (**a**) Execution Time for SDI feature extraction method. (**b**) Training time. (**c**) Testing time.

Table 8 shows the comparative results for multiclass dataset V. The outcomes are presented in terms of average classification accuracies and the highest case results are highlighted to make the best combination stand out. It is worth noting that the proposed SDI method outperformed all other methods in terms of individual subject results. It can be seen that the SDI method attained an average classification outcome of 99.07%, 98.16%, and 98.37% for participant 1, participant 2, and participant 3, respectively, and these are highest as compared to other methods. In terms of overall average results, the proposed SDI framework scored the highest 98.53% accuracy with a standard deviation of 0.387 that shows the consistency of overall results. Last, it is inferred from the comparison that SDI feature extraction method gains a minimum of 15.26% average classification accuracy, which is a significant improvement and it shows that the proposed method is not only useful for binary class motor imagery datasets but equally significant for multiclass mental imagery dataset as well.

Table 7. Performance comparison of motor imagery EEG signals in terms of classification accuracy (%) with other literature.

Methods By	Suggested Methods	Classification Accuracy(%)						
		"aa"	"al"	"av"	"aw"	"ay"	"Avg."	"Std."
Our Present Work	**"Successive decomposition index tested with feedforward neural network"/the proposed**	**100**	98.3	92.7	96.7	**100**	**97.5**	2.7
our previous work in 2019 [24]	"Multivariate empirical wavelet transform tested with least-square support vector machines"	95	95	95	**100**	**100**	97	2.7
our previous work in 2019 [23]	"Empirical wavelet transform tested with least-square support vector machines"/our last work	94.5	91.7	97.2	95.6	97	95.2	2.3
work by "Wu" et al. in 2008 [48]	"Iterative spatio-spectral patterns learning"	93.6	**100**	79.3	99.6	98.6	94.2	8.7
work by "Kevric" et al. in 2017 [22]	"Wavelet packet decomposition tested with K nearest neighbors"	96	92.3	88.9	95.4	91.4	92.8	2.9
work by "Siuly" et al. in 2011 [49]	"Clustering tested with least-square support vector machines"	92.6	84.9	90.8	86.5	86.7	88.3	3.2
work by "Song" et al. in 2007 [50]	"Common spatial pattern tested with support vector machines"	87.4	97.4	69.7	96.8	88.6	87.9	11.2
work by "Lu" et al. in 2010 [19]	"Regularized common spatial pattern tested with aggregation"	76.8	98.2	74.5	92.2	77	83.7	10.7
work by "Zhang" et al. in 2013 [51]	"Z-score tested with linear discriminant analysis"	77.7	**100**	68.4	99.6	59.9	81.1	18.1
work by "Lotte" et al. in 2010 [18]	"Regularized common spatial pattern with selected subjects". "Common spatial pattern with Tikhonov regularization". "Common spatial pattern with weighted Tikhonov regularization". "Spatially regularization common spatial pattern".	70.5 71.4 69.6 72.3	96.4 96.4 98.2 96.4	53.5 63.3 54.6 60.2	71.9 71.9 71.9 77.7	75.4 86.9 85.3 86.5	73.6 77.9 75.9 78.6	15.3 13.4 16.6 13.8
work by "Yong" et al. in 2008 [52]	"Sparse spatial filter optimization"	57.5	86.9	54.4	84.4	84.3	73.5	16

Besides classification results, it is important to compare the complexity of other methods with SDI feature extraction method. As mentioned earlier in this study, our method has no signal decomposition, complex multidomain features extraction, or features selection procedures involved, which makes it computationally simple and less time-consuming. The studies in [22–24] use signal decomposition techniques that involves resolution of a time signal into different modes, then extraction of complex features and lastly selection of highly uncorrelated features. Such systems might be useful for the research analysis but they are not feasible to be adopted for practical BCI systems. Similarly, the studies [18,19,53] employs common spatial pattern (CSP)-based methods, which is another complex method for the analysis of EEG signals. The crux of the matter is whether we consider robustness, efficiency and complexity, the proposed SDI method outperforms all state-of-the-art methods in every aspect and gives us a feasible solution to be considered for the development of practical BCI systems.

Table 8. Performance comparison of mental imagery EEG signals in terms of classification accuracy (%) with other literature.

Methods By	Suggested Methods	Classification Accuracy (%)				
		"P1"	"P2"	"P3"	"Avg."	"Std."
Our Present Work	**"Successive decomposition index tested with feedforward neural network"/The proposed**	88.9	**94.1**	**92.4**	**91.8**	**2.6**
work by "Siuly" et al. in 2017 [54]	"Principal component analysis employed with random forest"	**91.8**	75.2	82.8	83.3	8.3
research by "Lin" et al. in 2009 [55]	"Modified partical swarm optimization employed with neural networks"	78.3	75.2	56.5	69.9	11.81
work by "Siuly" et al. in 2011 [49]	"Clustering employed with least-square support vector machines"	68.2	64.8	52.1	61.7	8.5
experiments by "Sun" et al. in 2008 [56]	"Selection of electrodes with the help of Ensemble method"	68.7	56.4	44.8	56.7	11.9
work by "Sun" et al. in 2007 [57]	"Ensemble Methods"	70.6	48.9	40.9	53.4	15.4
work by "Sun" et al. in 2009 [53]	"Automated common spatial method"	67.7	68.1	59.6	65.1	4.82

6.10. Future Recommendations

In the present study, we utilized data with class labels, however, semisupervised learning or transductive learning methods are attracting attention these days. In future, researchers are encouraged to implement these methods for MI classification and information for these methods can be found in [58,59]. It is also worth mentioning that here, in the present study, we focused on at most three classes and presented the results in Table 5. However, for more number of classes readers should focus on more innovative strategies such as available in [60].

7. Conclusions

This study exploits the successive decomposition index (SDI) for the feature estimation of motor and mental imagery tasks. Three publicly available datasets namely dataset IVa, dataset IVb and dataset V from BCI competition III were utilized to attest the effectiveness of proposed method. Initially, the data was denoised with MSPCA and distributed into individual trials. Then, the SDI algorithm was used to calculate the feature corresponding to each trial and build a feature matrix for individual class instances. For the analysis purpose, a statistical test was performed that comprised mean, median, standard deviation, and Kruskal–Wallis nonparametric test for individual trials and it confirmed the efficacy of SDI as a potential feature. Moreover, a single evaluation metric named polygon area metric is employed to avoid looking into long tables. To validate the performance of the said method corresponding to the number of channels, four different channel selection criteria were tested and it confirmed that the 118-channel scheme has the leading results among other combinations. Furthermore, the classifier parameters were varied and a comparison between denoised and noisy data was performed to certify its effect on the classification performance of SDI feature. We also carried out a test for multiclass dataset V, and it was concluded that the proposed method is equally significant for the binary class as well as multiclass data. In the end, a computerized automated system CADMMI-SDI was developed for the practical realization of the proposed method. A comprehensive comparison of this study is made with other state of the art methods and it confirmed that the proposed method is robust, efficient, less complex and it can be utilized for the development of practical BCI systems.

Author Contributions: M.T.S., X.Y., and M.Z.A. modeled the problem and wrote the manuscript. X.Y. thoroughly checked the mathematical modeling and English. M.T.S. and M.Z.A. solved the problem using MATLAB software. M.T.S. and M.Z.A. contributed to the results and discussion. Writing—review and editing, M.T.S., X.Y., M.Z.A., and Z.Y. All authors finalized the manuscript after its internal evaluation. All authors have read and agreed to the published version of the manuscript.

Funding: This work was supported in part by the Fundamental Research Funds for the Central Universities (G2018KY0308), the Chinese Postdoctoral Science Foundation (2018M641013), and Postdoctoral Science Foundation of Shaanxi Province (Grant No. 2018BSHYDZZ05).

Conflicts of Interest: The authors declare no conflict of interest.

References

1. Birbaumer, N.; Murguialday, A.R.; Cohen, L. Brain–computer interface in paralysis. *Curr. Opin. Neurol.* **2008**, *21*, 634–638. [CrossRef]
2. Jochumsen, M.; Knoche, H.; Kjaer, T.W.; Dinesen, B.; Kidmose, P. EEG Headset Evaluation for Detection of Single-Trial Movement Intention for Brain-Computer Interfaces. *Sensors* **2020**, *20*, 2804. [CrossRef]
3. Pfurtscheller, G.; Neuper, C.; Muller, G.; Obermaier, B.; Krausz, G.; Schlogl, A.; Scherer, R.; Graimann, B.; Keinrath, C.; Skliris, D.; et al. Graz-BCI: State of the art and clinical applications. *IEEE Trans.* **2003**, *11*, 1–4. [CrossRef] [PubMed]
4. Sadiq, M.T.; Shabbir, N.; Kulesza, W.J. Spectral subtraction for speech enhancement in modulation domain. *Ijcsi* **2013**, *10*, 282.
5. Chaudhary, U.; Birbaumer, N.; Curado, M. Brain-machine interface (BMI) in paralysis. *Ann. Phys. Rehabil. Med.* **2015**, *58*, 9–13. [CrossRef] [PubMed]
6. Chaudhary, U.; Birbaumer, N.; Ramos-Murguialday, A. Brain–computer interfaces for communication and rehabilitation. *Nat. Rev. Neurosci.* **2016**, *12*, 513. [CrossRef] [PubMed]
7. Chaudhary, U.; Mrachacz-Kersting, N.; Birbaumer, N. Neuropsychological and neurophysiological aspects of brain-computer-interface (BCI) control in paralysis. *J. Physiol. Paris* **2020**, JP278775. [CrossRef]
8. Cincotti, F.; Mattia, D.; Aloise, F.; Bufalari, S.; Schalk, G.; Oriolo, G.; Cherubini, A.; Marciani, M.G.; Babiloni, F. Non-invasive brain–computer interface system: Towards its application as assistive technology. *Brain Res. Bull.* **2008**, *75*, 796–803. [CrossRef]
9. Siuly, S.; Li, Y. Improving the separability of motor imagery EEG signals using a cross correlation-based least square support vector machine for brain–computer interface. *IEEE Trans.* **2012**, *20*, 526–538. [CrossRef]
10. Rehman, A.U.; Naqvi, R.A.; Rehman, A.; Paul, A.; Sadiq, M.T.; Hussain, D. A Trustworthy SIoT Aware Mechanism as an Enabler for Citizen Services in Smart Cities. *Electronics* **2020**, *9*, 918. [CrossRef]
11. Wu, X.; Zhou, B.; Lv, Z.; Zhang, C. To explore the potentials of independent component analysis in brain-computer interface of motor imagery. *IEEE J. Biomed Health* **2019**, *24*, 775–787. [CrossRef] [PubMed]
12. Jiang, X.; Bian, G.B.; Tian, Z. Removal of artifacts from EEG signals: A review. *Sensors* **2019**, *19*, 987. [CrossRef] [PubMed]
13. Alickovic, E.; Subasi, A. Effect of multiscale PCA de-noising in ECG beat classification for diagnosis of cardiovascular diseases. *Circ. Syst. Signal Process.* **2015**, *34*, 513–533. [CrossRef]
14. Gokgoz, E.; Subasi, A. Effect of multiscale PCA de-noising on EMG signal classification for diagnosis of neuromuscular disorders. *J. Med. Syst.* **2014**, *38*, 31. [CrossRef]
15. Lotte, F.; Bougrain, L.; Cichocki, A.; Clerc, M.; Congedo, M.; Rakotomamonjy, A.; Yger, F. A review of classification algorithms for EEG-based brain–computer interfaces: A 10 year update. *J. Neural Eng.* **2018**, *15*, 031005. [CrossRef]
16. Polat, K.; Güneş, S. Classification of epileptiform EEG using a hybrid system based on decision tree classifier and fast Fourier transform. *Appl. Math. Comput.* **2007**, *187*, 1017–1026. [CrossRef]
17. Gupta, A.; Agrawal, R.K.; Kirar, J.S.; Andreu-Perez, J.; Ding, W.P.; Lin, C.T.; Prasad, M. On the Utility of Power Spectral Techniques With Feature Selection Techniques for Effective Mental Task Classification in Noninvasive BCI. *IEEE Trans. Syst. Man Cybern. Syst.* **2019**, *99*, 1–13. [CrossRef]

18. Lotte, F.; Guan, C. Regularizing common spatial patterns to improve BCI designs: Unified theory and new algorithms. *IEEE. Trans. Biomed. Eng.* **2010**, *58*, 355–362. [CrossRef]
19. Lu, H.; Eng, H.L.; Guan, C.; Plataniotis, K.N.; Venetsanopoulos, A.N. Regularized common spatial pattern with aggregation for EEG classification in small-sample setting. *IEEE. Trans. Biomed. Eng.* **2010**, *57*, 2936–2946.
20. Rodríguez-Bermúdez, G.; García-Laencina, P.J. Automatic and adaptive classification of electroencephalographic signals for brain computer interfaces. *J. Med. Syst.* **2012**, *36*, 51–63. [CrossRef]
21. Pfurtscheller, G.; Neuper, C.; Schlogl, A.; Lugger, K. Separability of EEG signals recorded during right and left motor imagery using adaptive autoregressive parameters. *IEEE Trans. Neural Syst. Rehabilit. Eng.* **1998**, *6*, 316–325. [CrossRef] [PubMed]
22. Kevric, J.; Subasi, A. Comparison of signal decomposition methods in classification of EEG signals for motor-imagery BCI system. *Biomed. Signal Process.* **2017**, *31*, 398–406. [CrossRef]
23. Sadiq, M.T.; Yu, X.; Yuan, Z.; Fan, Z.; Rehman, A.U.; Li, G.; Xiao, G. Motor imagery EEG signals classification based on mode amplitude and frequency components using empirical wavelet transform. *IEEE Access* **2019**, *7*, 127678–127692. [CrossRef]
24. Sadiq, M.T.; Yu, X.; Yuan, Z.; Zeming, F.; Rehman, A.U.; Ullah, I.; Li, G.; Xiao, G. Motor Imagery EEG Signals Decoding by Multivariate Empirical Wavelet Transform-Based Framework for Robust Brain–Computer Interfaces. *IEEE Access* **2019**, *7*, 171431–171451. [CrossRef]
25. Chatterjee, R.; Bandyopadhyay, T.; Sanyal, D.K.; Guha, D. Comparative analysis of feature extraction techniques in motor imagery EEG signal classification. In Proceedings of the First International Conference on Smart System, Innovations and Computing, Jaipur, India, 15–16 April 2017; Springer: Jaipur, India, 2017.
26. Wang, L.; Xu, G.; Wang, J.; Yang, S.; Yan, W. Application of Hilbert-Huang transform for the study of motor imagery tasks. In Proceedings of the 30th Annual International Conference of the IEEE Engineering in Medicine and Biology Society, Vancouver, BC, Canada, 20–24 August 2008.
27. Chaudhary, P.; Agrawal, R. Non-dyadic wavelet decomposition for sensory-motor imagery EEG classification. *Brain Comput. Interfaces* **2020**, *7*, 1–11. [CrossRef]
28. Xu, J.; Zheng, H.; Wang, J.; Li, D.; Fang, X. Recognition of EEG signal motor imagery intention based on deep multi-view feature learning. *Sensors* **2020**, *20*, 3496. [CrossRef]
29. Chen, K.; Ai, Q.; Chen, J.; Zhou, S.; Zhao, Y. NAO Robot Walking Control System Based on Motor Imagery. *J. Phys. Conf. Ser.* **2020**, *1453*, 12123. [CrossRef]
30. Raghu, S.; Sriraam, N.; Rao, S.V.; Hegde, A.S.; Kubben, P.L. Automated detection of epileptic seizures using successive decomposition index and support vector machine classifier in long-term EEG. *Neural Comput. Appl.* **2019**, *32*, 8965–8984. [CrossRef]
31. Teixeira, C.; Direito, B.; Feldwisch-Drentrup, H.; Valderrama, M.; Costa, R.; Alvarado-Rojas, C.; Nikolopoulos, S.; Le Van Quyen, M.; Timmer, J.; Schelter, B.; et al. EPILAB: A software package for studies on the prediction of epileptic seizures. *J. Neurosci. Methods* **2011**, *200*, 257–271. [CrossRef]
32. Delorme, A.; Makeig, S. EEGLAB: An open source toolbox for analysis of single-trial EEG dynamics including independent component analysis. *J. Neurosci. Methods* **2004**, *134*, 9–21. [CrossRef]
33. Oostenveld, R.; Fries, P.; Maris, E.; Schoffelen, J.M. FieldTrip: Open source software for advanced analysis of MEG, EEG, and invasive electrophysiological data. *Comput. Intell. Neurosci.* **2011**, *2011*, 156869. [CrossRef] [PubMed]
34. Alickovic, E.; Subasi, A. Ensemble SVM method for automatic sleep stage classification. *IEEE Trans. Instrum. Meas.* **2018**, *67*, 1258–1265. [CrossRef]
35. Taran, S.; Bajaj, V. Motor imagery tasks-based EEG signals classification using tunable-Q wavelet transform. *Neural. Comput. Appl.* **2019**, *31*, 6925–6932. [CrossRef]
36. Chaudhary, S.; Taran, S.; Bajaj, V.; Siuly, S. A flexible analytic wavelet transform based approach for motor-imagery tasks classification in BCI applications. *Comput. Methods Programs Biomed.* **2020**, *187*, 105325. [CrossRef]
37. Chatterjee, R.; Bandyopadhyay, T. EEG based Motor Imagery Classification using SVM and MLP. In Proceedings of the 2nd international conference on Computational Intelligence and Networks (CINE), Bhubaneswar, India, 11 January 2016.

38. Miao, C.; Ma, Y.; Wu, W.; Zhang, Q.; Gao, Y. Pattern Classification of Motor Imagery EEG-NIRS Based on SVM with Artificial Fish Swarm Algorithm. In Proceedings of the 5th International Conference on Information, Cybernetics, and Computational Social Systems (ICCSS), Hangzhou, China, 16–19 August 2018.
39. Kumar, S.; Sharma, R.; Sharma, A.; Tsunoda, T. Decimation filter with common spatial pattern and fishers Discriminant analysis for motor imagery classification. In Proceedings of the 2016 international joint conference on neural networks (IJCNN), Vancouver, BC, Canada, 24–29 July 2016.
40. Fu, R.; Tian, Y.; Bao, T. Recognition method of single trial motor imagery electroencephalogram signal based on sparse common spatial pattern and Fisher discriminant analysis. *J. Biomed. Eng.* **2019**, *36*, 911–915.
41. Maksimenko, V.A.; Kurkin, S.A.; Pitsik, E.N.; Musatov, V.Y.; Runnova, A.E.; Efremova, T.Y.; Hramov, A.E.; Pisarchik, A.N. Artificial neural network classification of motor-related eeg: An increase in classification accuracy by reducing signal complexity. *Complexity* **2018**, *2018*, 9385947. [CrossRef]
42. Gonzalez, C.D.V.; Azuela, J.H.S.; Espino, E.R.; Ponce, V.H.P. Classification of motor imagery EEG signals with CSP filtering through neural networks models. In Proceedings of the Mexican International Conference on Artificial Intelligence, Guadalajara, Mexico, 22–27 October 2018; Springer: Guadalajara, Mexico, 2018.
43. Jana, G.C.; Swetapadma, A.; Pattnaik, P.K. Enhancing the performance of motor imagery classification to design a robust brain computer interface using feed forward back-propagation neural network. *Ain Shams Eng. J.* **2018**, *9*, 2871–2878. [CrossRef]
44. Akhter, M.P.; Jiangbin, Z.; Naqvi, I.R.; Abdelmajeed, M.; Mehmood, A.; Sadiq, M.T. Document-level text classification using single-layer multisize filters convolutional neural network. *IEEE Access* **2020**, *8*, 42689–42707. [CrossRef]
45. Akhter, M.P.; Jiangbin, Z.; Naqvi, I.R.; Abdelmajeed, M.; Sadiq, M.T. Automatic Detection of Offensive Language for Urdu and Roman Urdu. *IEEE Access* **2020**, *8*, 91213–91226. [CrossRef]
46. Aydemir, O. A New Performance Evaluation Metric for Classifiers: Polygon Area Metric. *J. Classif.* **2020**, 1–11. [CrossRef]
47. Siuly; Li, Y.; Wen, P. Comparisons between motor area EEG and all-channels EEG for two algorithms in motor imagery task classification. *Biomed. Eng. Appl. Basis C* **2014**, *26*, 1450040. [CrossRef]
48. Wu, W.; Gao, X.; Hong, B.; Gao, S. Classifying single-trial EEG during motor imagery by iterative spatio-spectral patterns learning (ISSPL). *IEEE. Trans. Biomed. Eng.* **2008**, *55*, 1733–1743. [CrossRef] [PubMed]
49. Li, Y.; Wen, P.P. Clustering technique-based least square support vector machine for EEG signal classification. *Comput. Methods Programs Biomed.* **2011**, *104*, 358–372.
50. Song, L.; Epps, J. Classifying EEG for brain-computer interface: Learning optimal filters for dynamical system features. *Comput. Intell. Neurosci.* **2007**, *2007*, 57180. [CrossRef]
51. Zhang, R.; Xu, P.; Guo, L.; Zhang, Y.; Li, P.; Yao, D. Z-score linear discriminant analysis for EEG based brain-computer interfaces. *PLoS ONE* **2013**, *8*, e74433. [CrossRef]
52. Yong, X.; Ward, R.K.; Birch, G.E. Sparse spatial filter optimization for EEG channel reduction in brain-computer interface. In Proceedings of the 2008 IEEE International Conference on Acoustics, Speech and Signal Processing, Las Vegas, NV, USA, 31 March–4 April 2008.
53. Sun, S.; Zhang, C. Adaptive feature extraction for EEG signal classification. *Med. Biol. Eng. Comput.* **2006**, *44*, 931–935. [CrossRef]
54. Siuly, S.; Zarei, R.; Wang, H.; Zhang, Y. A new data mining scheme for analysis of big brain signal data. In Proceedings of the Australasian Database Conference, Brisbane, Australia, 25–28 September 2017.
55. Lin, C.J.; Hsieh, M.H. Classification of mental task from EEG data using neural networks based on particle swarm optimization. *Neurocomputing* **2009**, *72*, 1121–1130. [CrossRef]
56. Sun, S.; Zhang, C.; Lu, Y. The random electrode selection ensemble for EEG signal classification. *Pattern Recognit.* **2008**, *41*, 1663–1675. [CrossRef]
57. Sun, S.; Zhang, C.; Zhang, D. An experimental evaluation of ensemble methods for EEG signal classification. *Pattern Recognit. Lett.* **2007**, *28*, 2157–2163. [CrossRef]
58. Taewijit, S.; Theeramunkong, T.; Ikeda, M. Distant supervision with transductive learning for adverse drug reaction identification from electronic medical records. *J. Healthc. Eng.* **2017**, *2017*, 7575280. [CrossRef]

59. Wan, S.; Mak, M.W.; Kung, S.Y. Transductive learning for multi-label protein subchloroplast localization prediction. *TCBB* **2016**, *14*, 212–224. [CrossRef] [PubMed]
60. Melvin, I.; Ie, E.; Kuang, R.; Weston, J.; Noble, W.S.; Leslie, C. SVM-Fold: A tool for discriminative multi-class protein fold and superfamily recognition. *BMC Bioinform.* **2007**, *8*, S2. [CrossRef] [PubMed]

Article

CNN and LSTM-Based Emotion Charting Using Physiological Signals

Muhammad Najam Dar [1,*], Muhammad Usman Akram [1], Sajid Gul Khawaja [1] and Amit N. Pujari [2,3]

1 Department of Computer and Software Engineering, College of Electrical and Mechanical Engineering, National University of Sciences and Technology, Islamabad 44000, Pakistan; usman.akram@ceme.nust.edu.pk (M.U.A.); sajid.gul@ceme.nust.edu.pk (S.G.K.)

2 School of Engineering and Technology, University of Hertfordshire, Hatfield AL10 9AB, England, UK; amit.pujari@ieee.org

3 School of Engineering, University of Aberdeen, Aberdeen AB24 3UE, Scotland, UK

* Correspondence: najam.dar77@ce.ceme.edu.pk

Received: 14 June 2020; Accepted: 4 August 2020; Published: 14 August 2020

Abstract: Novel trends in affective computing are based on reliable sources of physiological signals such as Electroencephalogram (EEG), Electrocardiogram (ECG), and Galvanic Skin Response (GSR). The use of these signals provides challenges of performance improvement within a broader set of emotion classes in a less constrained real-world environment. To overcome these challenges, we propose a computational framework of 2D Convolutional Neural Network (CNN) architecture for the arrangement of 14 channels of EEG, and a combination of Long Short-Term Memory (LSTM) and 1D-CNN architecture for ECG and GSR. Our approach is subject-independent and incorporates two publicly available datasets of DREAMER and AMIGOS with low-cost, wearable sensors to extract physiological signals suitable for real-world environments. The results outperform state-of-the-art approaches for classification into four classes, namely High Valence—High Arousal, High Valence—Low Arousal, Low Valence—High Arousal, and Low Valence—Low Arousal. Emotion elicitation average accuracy of 98.73% is achieved with ECG right-channel modality, 76.65% with EEG modality, and 63.67% with GSR modality for AMIGOS. The overall highest accuracy of 99.0% for the AMIGOS dataset and 90.8% for the DREAMER dataset is achieved with multi-modal fusion. A strong correlation between spectral- and hidden-layer feature analysis with classification performance suggests the efficacy of the proposed method for significant feature extraction and higher emotion elicitation performance to a broader context for less constrained environments.

Keywords: convolutional neural network (CNN); long short-term memory (LSTM); emotion recognition; EEG; ECG; GSR; deep neural network; physiological signals

---

1. Introduction

Recent trends in the field of affective computing have shifted towards a more reliable source of physiological signals [1–4] such as Electroencephalogram (EEG), Electrocardiogram (ECG), and Galvanic Skin Response (GSR) due to their significance in human–computer interaction (HCI). Emotions can be distinctively expressed as a non-verbal form of everyday social interaction. These non-verbal cues are generally reflected through facial expressions and tone of voice. However involuntary physiological responses (such as EEG and ECG) to the emotional stimuli are more reliable compared to voluntary response (such as sound or facial expressions), because the involuntary response cannot be masked intentionally [4] (a sad person may smile, which may also be the indication of depression). Similarly, external factors such as lighting conditions, accessories like glasses or a hat, and surrounding audio noise may affect the performance of these external voluntary modalities of expression, limiting their use in computing applications [5]. Biosensors help to monitor and collect physiological signals from heart (ECG), brain (EEG), or skin (GSR), and proves to be the most significant for the detection of stress levels and emotions [6,7]. Applications of physiological signal-based emotion recognition encompass psychological health-care monitoring for hospitalized patients [8], real-time stress-level detection of drivers, emotion-inspired multimedia applications [9], various bio-inspired human–machine interfaces, and health-care applications [10].

The problem of emotion elicitation is generally approached in the literature by measuring valence and arousal as represented in the emotion circumplex model [11]. A simple representation of the circumplex model is shown in Figure 1, which exhibits the positions of six basic emotions (happiness, anger, disgust, fear, sadness, and surprise) on the scales of valence and arousal. Valence expresses the positivity or negativity of the emotion; arousal expresses the degree of excitement covered by the emotion stimuli. Most of the affective computing literature based on physiological signals classifies valence (high/low) and arousal (high/low) as separate binary classification tasks [12,13]. However, little other research has approached the problem in the broader context of the classification of emotions into four categories [14,15] of High Valence—High Arousal (HVHA), High Valence—Low Arousal (HVLA), Low Valence—High Arousal (LVHA) and Low Valence—Low Arousal (LVLA).

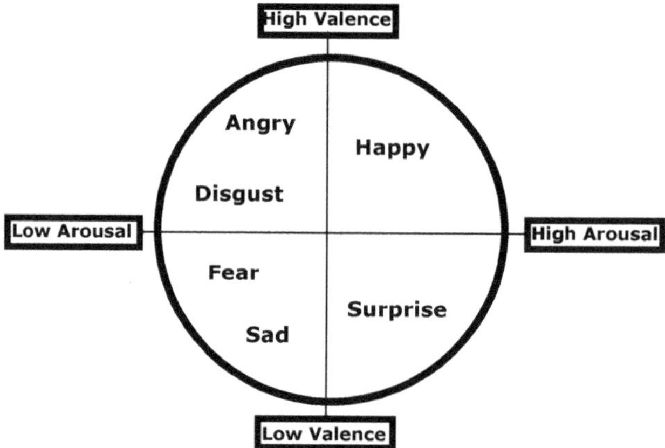

Figure 1. Valence arousal model for emotion elicitation.

The process of emotion elicitation in the literature generally revolves around conventional feature extraction from physiological signals. For EEG-based classification, the conventional statistical features, wavelet features, and Empirical Mode Decomposition (EMD)-based features were applied to test the efficacy of these algorithms for emotion elicitation. For instance, Support Vector Machine (SVM) was tested with the efficacy of EMD in combination with the genetic algorithm [16]. Stable statistical features such as band power, coherence, and entropy were proposed [17] for the classification of four emotions, namely happiness, fear, pleasure, and anger. Similar conventional approaches for ECG-based emotion recognition were also applied in the literature. In [18], time-domain ECG features after computation of P-QRS-T wave detection were extracted and selected through hybrid Particle Swarm Optimization (PSO) with a Fisher classifier for emotion recognition. However, these approaches used their private datasets for emotion classification. The more standard approaches incorporated publicly available datasets of DEAP for EEG and the MAHNOB-HCI dataset for ECG. For instance, relative wavelet energy with SVM was used in [19] for the DEAP dataset, with the recognition rate of 61.8%. SVM classification was proposed [20] for the wavelet-based extracted EEG features from the DEAP dataset to achieve 65.13% and 65.33% accuracy for binary classification of valence and arousal, respectively. One of the approaches [2] for ECG signals from the MAHNOB-HCI dataset was based on neighborhood component analysis for binary classification of valence and arousal. In [21], heart rate variability-based statistical features were incorporated with SVM for the recognition of five emotions using the MAHNOB-HCI dataset.

Novel trends in the computation of emotion elicitation exploit Deep Neural Networks (DNN) for physiological signals with improvement in recognition rates. One of the earliest attempts was made by [22], using Wavelet Transform and Back-Propagation Neural Networks (BPNN)-based emotion recognition from Electromyogram (EMG) signals to recognize four emotions of joy, sadness, anger, and pleasure. Another early attempt proposed [23] Radial Basis Function (RBF) Neural Network for an emotion elicitation task. However, both approaches used private datasets for the computation of emotions; the publicly available dataset of DEAP is also incorporated using DNN architectures. Long Short-Term Memory (LSTM) was deployed [24] for EEG-based emotion elicitation and reported recognition rates of 72.06% and 74.12% for binary classification of valence and arousal using DEAP dataset. A more recent study [25] proposed Graph-regularized Extreme Learning Machine (GELM) for the classification of HVHA, HVLA, LVHA, and LVLA. They incorporated spectral features from EEG with the GELM classifier for SEED [26] and DEAP datasets, respectively. In [27], LSTM networks were incorporated in 5*sec* segments of EEG signal from the DEAP dataset to report recognition rates of 85.45% and 85.65% for binary classification of valence and arousal, respectively. Similarly, Ref. [28] claimed the significance of deep CNN architecture for EEG-based emotion recognition using the DEAP dataset. The authors reported a recognition rate of 73.36% and 81.4% for binary classification of arousal and valence, respectively. Recognition results obtained from conventional and DNN-based approaches as described in [20,24,27] point towards the significance of DNN-based algorithms trends as compared to conventional approaches.

Various modalities such as face videos, eye movement, ECG, EEG, and GSR, etc. were incorporated in the literature for emotion elicitation. A multi-modal approach combines any of these modalities for better recognition performance. DNN-based multi-modal approach in the literature generally works in two dimensions—one of the dimension involves the joint representation of multi-modality with feature-level fusion, and the other dimension involves the coordinated representation of multi-modality with decision-level fusion [29]. Feature-level fusion concatenates sets of extracted features from various modalities before the classification learning stage, while decision-level fusion merges the classification decisions of various classifiers or deep networks for specific modalities [30]. In [31], feature-level fusion emotion recognition is applied using physiological signals. This study incorporated the decision-level fusion scheme of majority voting. In [32], the authors proposed multi-layer LSTM for the multi-modal fusion of EEG signals with videos for binary classification of valence and arousal respectively for MAHNOB-HCI dataset. In [33], eye movement and EEG-based decision-level multi-modal fusion was applied by analyzing confusion matrices. The significance of decision-level fusion lies in the accessibility of modality comparison in terms of its overall contribution to the recognition results [29]. In our study, we incorporate decision-level fusion based on majority voting due to the simplicity of the procedure for the combination of modalities.

Based on the previous research examination, there are the following few things that need to be addressed for the contribution of existing knowledge. ECG is less explored in the literature as compared to EEG, despite its higher significance in emotion elicitation. There is a need to build a high-performance non-invasive emotion recognition process using low-cost wearable sensors for broader classes of emotions. Connectivity among spectral features and the features extracted by deep neural networks need to be examined to understand and build more robust neural network architectures specific to the spectral properties of modality. EEG, ECG, and GSR are continuous-time signals with high memory content, which could be exploited by LSTM for better results. LSTM has the property of selectively remembering patterns for a long duration of time useful in feature extraction from physiological signals. Due to the irregular structure of EEG [13], 14 channels need to be arranged in the 2D image to exploit the usefulness of a 2D Convolutional Neural Network (CNN). Therefore, the proposed 1D-CNN + LSTM architecture is used to exploit their usefulness for time-series data of ECG and GSR as well as 2D-CNN architecture for image representation of 14-channel EEG. We selected recently published, publicly available datasets of AMIGOS [34] and DREAMER [35] for the emotion recognition in less constrained scenarios with the additional ensemble of modalities at the decision level.

The rest of the paper is organized as follows. Section 2 examines the state-of-the-art literature in detail. It also exposes research gaps identified as the implication of previous research and our contribution to fill those gaps in Sections 2.1 and 2.2. The details of the datasets used in this study and the details of proposed methodology are discussed in Section 3. Section 4 provides results on both datasets, and detailed discussion and spectral analysis of results are provided in Section 5. The discussion section also includes the detailed analysis and representation of deep features from hidden layers of proposed DNN architecture. Conclusions drawn from this research are finally presented in Section 6.

2. Related Work

To establish a comparison of our computational framework with related works, most relevant studies based on AMIGOS and DREAMER datasets are described in this section. A recent study [36] proposed a fusion of statistical features extracted from EEG, ECG, and GSR from the AMIGOS dataset. They reported a recognition rate of 67% and 68.8% for valence and arousal respectively using an SVM classifier. Another recent GSR-based framework [37] using AMIGOS dataset proposed temporal and spectral features with SVM (RBF kernel) to report recognition performance of 83.9% and 65% for valence and arousal, respectively. For the AMIGOS dataset, the significance of DNN can be explained by two similar studies, where one of the studies [34] reported 55.1% and 54.4% F1 scores for valence and arousal, respectively, using Gaussian Naive Bayes, while another study [38] reported 71% and 81% accuracy for valence and arousal, respectively, using convolutional neural networks.

In a recent study [39], the authors proposed self-supervised learning instead of self-assessment labels for the AMIGOS dataset using a convolutional neural network. They reported a recognition rate of 84% and 85.8% for valence and arousal, respectively. LSTM-RNN is recently proposed [40], with the use of attention-based mechanism for AMIGOS dataset, and reported recognition rates of 79.4% and 83.3% for binary classification of valence and arousal, respectively. In [14], 3D-CNN and 1D-CNN models are compared and report the accuracy of 99.7% for four classes of emotion. The significance of results for four classes of emotion can be understood using the recent two studies. A study [41] reported an accuracy of 66.67% and 71.54% for binary classification of valence and arousal, respectively, and reported 38.28% classification accuracy for four classes (HVHA, HVLA, LVHA, and LVLA) of emotions using the same proposed methodology. Another similar recent study [15] reported accuracy of 83.02% and 82.74% for binary classification of valence and arousal respectively, while reported 58.57% classification accuracy for four classes (HVHA, HVLA, LVHA, and LVLA) of emotions.

Currently using the DREAMER dataset, Graph Convolutional Broad Network (GCB-net) is suggested [13] to announce an accuracy of 86.99% and 89.32% for binary classification of valence and arousal respectively. In [12], using DREAMER, the accuracy of 86.23% and 84.54% are reported for classification of valence and arousal, respectively. In [35], a multi-modal fusion of EEG and ECG of the DREAMER dataset was incorporated with 61.84% and 63.32% recognition rates for valence and arousal, respectively.

To evaluate studies using both the DREAMER and AMIGOS dataset, Ref. [15] reported an accuracy of 79.95% for binary classification of both valence and arousal using the DREAMER dataset. They also reported accuracy of 83.94% and 82.76% for binary classification of valence and arousal respectively for the multi-modal fusion of physiological signals in the AMIGOS dataset. The most recent study [42] based on the AMIGOS dataset proposed the Bayesian network to achieve accuracy of 90% for binary classification of the high and low level of valence. They also reported accuracy of 86% for binary classification of the high and low level of valence for the DREAMER dataset.

2.1. Implications

It is interesting to state a few implications here from an extensive literature review based on a physiological signal-based emotion recognition. First, the deep-learning approaches especially convolutional and LSTM-based neural networks performed much better in terms of recognition performance of emotions as compared to conventional feature extraction and simple classifier-based approaches [9]. Second, EEG is extensively used for the emotion elicitation process; however, the separate significance of ECG and GSR modalities in terms of long duration sequence data are much less explored for this purpose especially using DNN-based approaches [43]. Third, most of the studies in the literature [44,45] with the most promising results are based on the evaluation of binary classification of valence (high valence, low valence) and arousal (high arousal, low arousal) separately. However, in the case of four classes of emotions (HVHA, HVLA, LVHA, and LVLA) the reported recognition rates decrease to a much larger extent [17]. Fourth, due to the highly subjective and dynamic nature of emotion, few pieces of research have proposed subject-dependent approaches with better recognition performance; however, subject-dependent approaches besides higher recognition rates require large training data from each subject and are not capable of being applied to unseen subjects [46]. Therefore, studies reported for subject-independent approaches [47,48] are more reliable for emotion elicitation tasks from a broader perspective as compared to the results obtained using subject-dependent approaches [49,50]. Fifth, recognition performance of emotion is more reliable and comparable when reported for publicly available datasets acquired using portable and wearable sensors [34,35] for their use in real-world environments as compared to invasive and restricted lab environments.

2.2. Our Contribution

We proposed the combination of LSTM and CNN architecture to improve the recognition performance for four classes (HVHA, HVLA, LVHA, and LVLA) of emotion with multi-modal fusion while exploiting the significance of various modalities such as ECG, EEG, and GSR. Our approach is subject-independent and recognition rates are reported against two of the publicly available datasets of AMIGOS and DREAMER. Both datasets acquired signals using wearable, low-cost sensors to represent the significance and reliability of proposed methodology in real-world environments.

3. Material and Methods

This section will elaborate the significance and details of datasets used in this study. After elaboration of materials used, proposed methodology will be discussed in detail.

3.1. Datasets

Two of the publicly available datasets named DREAMER and AMIGOS were incorporated for the evaluation of proposed deep neural network architectures. Both of these datasets are newly published yet extensively used in recent literature for physiological signal-based emotion elicitation, primarily because of their non-invasive nature of the data acquisition process using low-cost off-the-shelf devices. In both of the datasets, a 14-channel Emotiv Epoc [51] wireless headset for EEG and two-channel Shimmer ECG sensor [52] was used as compared to quite an invasive headset of Biosemi active two [53] used in DEAP [54] and MAHNOB-HCI [55] datasets. Biosemi active two is much more accurate and precise, and leverages more channels for EEG collection; however, invasive data acquisition procedure is required using this device. The DREAMER and AMIGOS dataset helps to improve emotion classification performance for the non-invasive environment with low-cost sensors. DREAMER data is provided with raw ECG and EEG signals, therefore dedicated basic pre-processing steps were applied in Section 3.3.1 to raw DREAMER dataset to be consistent with the basic pre-processed signals available for AMIGOS dataset. The concise details of both datasets are given below.

3.1.1. DREAMER

This dataset used audio-visual stimuli for affect elicitation from 23 subjects (14 males and 9 females). Each subject was exposed to 18 different trials of variable length from 65 s–393 s duration with the addition of a 61 s baseline signal (where there was no stimulus provided as a neutral state) for each trial. Each subject was asked to label each trial with the valence and arousal values for the scale of 1–5 using Self-Assessment Manikins (SAM). 14 channels of EEG with *AF*3, *F*7, *F*3, *FC*5, *T*7, *P*7, *O*1, *O*2, *P*8, *T*8, *FC*6, *F*4, *F*8, *AF*4 channels were recorded using Emotiv Epoc portable sensor with the sampling frequency of 128 Hz. ECG with two channels was recorded using the Shimmer sensor with a sampling rate of 256 Hz. 57 s of baseline signals (segmenting out the first two and last two seconds) and the last 60 s of each stimulus trial signals is incorporated for this study.

3.1.2. AMIGOS

The AMIGOS dataset also acquires 14-channel EEG (same channel position as described for DREAMER) and two ECG channels using similar sensors and sampling frequency used for the DREAMER dataset. However, the AMIGOS dataset used audio-visual stimuli of 16 different short-term trails with a variable duration from 51 s–150 s and the addition of a 5 s baseline (without stimulus) for each trial. The signals were self-annotated (using SAM) for each trial with the valence and arousal values on the scale of 1–9. Basic pre-processed data from AMIGOS is used for the complete length of trials and 5 s of baseline for 33 subjects with valid data out of total 40 subjects with the exemption of 7 subjects (with ID number 33, 24, 23, 22, 21, 12, 9) with invalid data. DREAMER published a dataset of 23 out of 25 persons with valid data only and exempted 2 persons with invalid data, therefore all the 23 person data from the DREAMER dataset is used. AMIGOS dataset in comparison to DREAMER additionally provided with GSR data, which is also incorporated in this study for comparison.

3.2. Methodology

The three main components of the proposed algorithm are pre-processing, classification, and multi-modal fusion. First, all physiological signals are required to be pre-processed before their use in neural network architecture [56]. EEG is quite different from other peripheral signals in terms of the number of channels, frequency, and amplitude. Therefore, the sequence of steps applied to the EEG is generally represented in parallel to the sequence of steps applied for other modalities before multi-modal fusion. Figure 2 represents a general block diagram of all the steps applied in this study. The following subsections represent a detailed description of these steps applied in the proposed methodology.

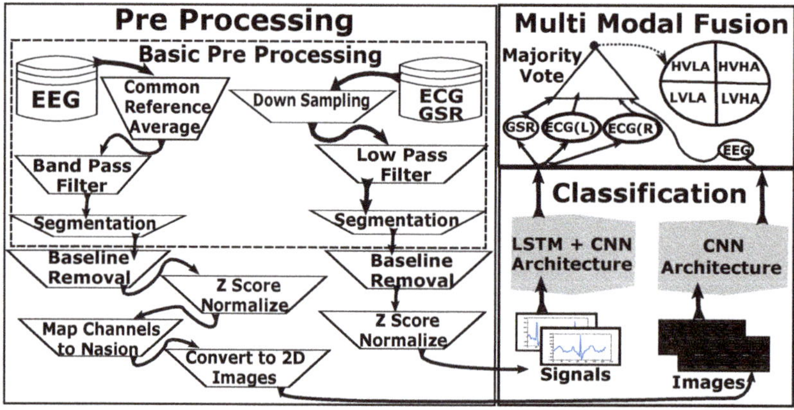

Figure 2. Block Diagram of complete methodology.

3.3. Pre-Processing

Pre-processing steps are divided into two categories, namely basic pre-processing and specialized pre-processing. The basic pre-processing steps are common to the generic applications of physiological signals. The basic pre-processing steps were followed by the steps used in basic pre-processing of AMIGOS dataset [34] to standardize physiological signals from both datasets. These steps involve signal filtering and noise removal, and are common for both GSR and ECG signals. However, EEG signals required different pre-processing steps as compared to ECG and GSR due to different inherent properties of EEG [57]. The second category with specialized pre-processing consists of steps required specific to the emotion elicitation algorithm. These steps include baseline removal and Z-score normalization, inspired by its significance in many studies such as [46,58–61].

3.3.1. Basic EEG Pre-Processing

The Basic EEG pre-processing framework consists of referencing common average, band-pass filtering, and segmentation. First, raw EEG signals are required to be re-referenced to remove channel biases introduced by online reference [62]. These channel biases affect the amplitude of EEG channels based on their spatial locality from online reference. The common average was computed by extracting the average EEG signal of all the 14 EEG channels as shown in Equation (1), where c represents the channel number from 1 to $N = 14$ and l is the length of EEG channel (64 s with the total of 8192 samples) each for DREAMER dataset.

$$meanEEG_l = \frac{1}{N}\left(\sum_{c=1}^{N} EEG_{c,l}\right) \tag{1}$$

$$acrEEG_{c,l} = EEG_{c,l} - meanEEG_l \tag{2}$$

Now each of the 14 EEG channels with length l are subtracted from the computed common average as described in Equation (2) to obtain EEG signals with averaged to a common reference. The results of basic EEG pre-processing are depicted in Figure 3. Figure 3a represents raw EEG signals of all the 14 channels and Figure 3b represents all channels of EEG signals after averaged to a common reference.

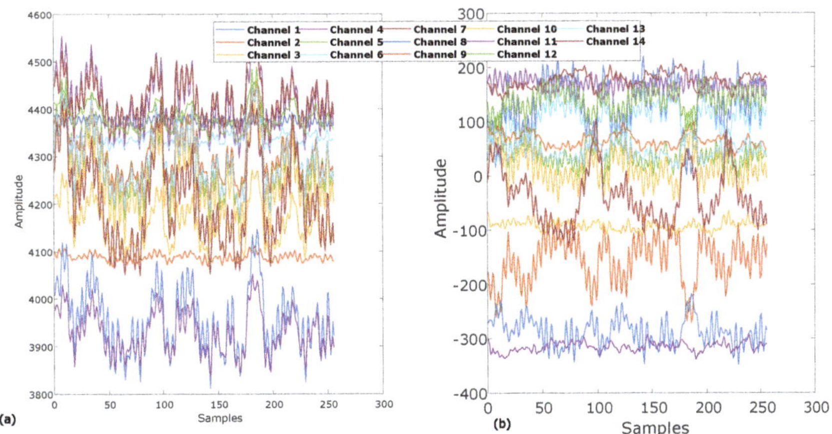

Figure 3. (**a**) Raw EEG signals of 14 channels. (**b**) EEG Signals of 14 channels averaged to common reference.

The second step of EEG pre-processing is band-pass filtering in which each channel of common referenced EEG signal is individually passed through band-pass filtering for noise removal. To compare against AMIGOS results, a passband frequency range of 4 Hz to 45 Hz is applied with transition band steepness of 0.85 and a stopband attenuation of 60 dB selected as a parameter. Figure 4 represents the original signal as an individual channel of EEG along with filtered signal after artifact removal using band-pass filtering. The effect of the band-pass filter is also illustrated as the power spectrum against frequency in Figure 4, where artifacts of below 4 Hz and around 50 Hz are removed from each EEG channel. After band-pass filtering, each channel of length 64 s is segmented by removing a 2 s signal from the start and 2 s of signal from the end to counter the effect of the filter at the edges. These three basic pre-processing steps result in EEG signals of length 60 s each (total of 7680 samples) for the DREAMER dataset. Therefore, the total size of DREAMER data for EEG after basic pre-processing is 5796×7680 (23 subjects, 18 trials, and 14 EEG channels).

3.3.2. Basic ECG and GSR Pre-Processing

Peripheral signals of raw DREAMER data also undergo basic pre-processing steps to be consistent with the AMIGOS dataset. Therefore, both channels of ECG signals are individually down-sampled to 128 Hz from 256 Hz. After down-sampling, a low-pass filter of 60 Hz was applied to remove the high-frequency noise components from ECG data. After low-pass filtering, each channel of length 64 s is segmented by removing 2 s from start and 2 s of signal from the end to counter the effect of the filter at the edges. These three basic pre-processing steps result in ECG signals of length 60 s each (total of 7680 samples) for the DREAMER dataset. Therefore, the total size of DREAMER data for ECG after basic pre-processing is 828×7680 (23 subjects, 18 trials, and 2 ECG channels).

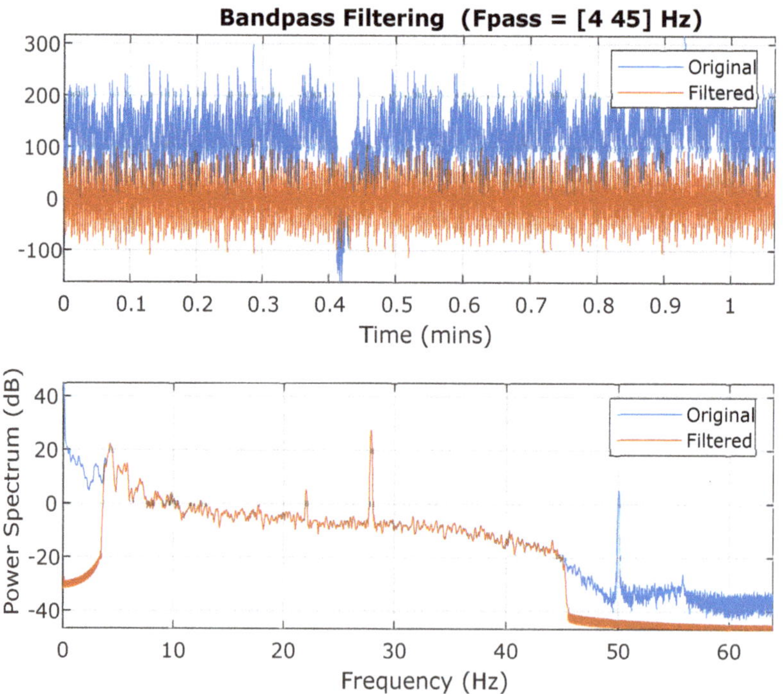

Figure 4. Band-pass filtering 4–45 Hz of individual EEG channel with power spectrum.

3.3.3. Baseline Removal

Both DREAMER and AMIGOS datasets have baseline signals, where there is no stimulus provided to the subjects. AMIGOS has 5 s and DREAMER has 57 s of baseline signal recorded with no emotional activity. It is, therefore, useful to remove this neutral baseline activity from all the EEG, ECG, and GSR signals as a specific pre-processing step for emotion elicitation. For this purpose, baseline signals were divided into 1 s segments and then the mean of these segments is computed as the mean baseline activity of each signal as shown in Equation (3).

$$meanBL = \frac{1}{S}\left(\sum_{s=1}^{S} BL_s\right) \tag{3}$$

EEG, ECG, and GSR signals data of 60*sec* emotional activity is also divided into 60 segments of 1 s each. After getting the mean of a specific channel for EEG, ECG, and GSR, each segment of emotional activity is subtracted from their corresponding mean segment of baseline activity to remove the neutral emotional effect as shown in Equation (4).

$$blrSig_s = Sig_s - meanBL \tag{4}$$

These steps are performed to remove baseline from all the 14 EEG channels, 2 ECG channels, and from GSR to enhance and emphasize the emotional effect of corresponding stimuli. For the DREAMER dataset, the baseline was computed from the mean of all 57 segments of baseline activity. Figure 5a illustrates a segment of the ECG signal (Left Channel) of 1 s of emotional activity after basic pre-processing steps applied. Figure 5b represents five baseline segments of ECG with neutral activity and the computed mean of these five segments, while Figure 5c shows the output of baseline removal after subtracting segment of emotional activity represented in Figure 5a from the mean signal of five baseline segments represented in Figure 5b.

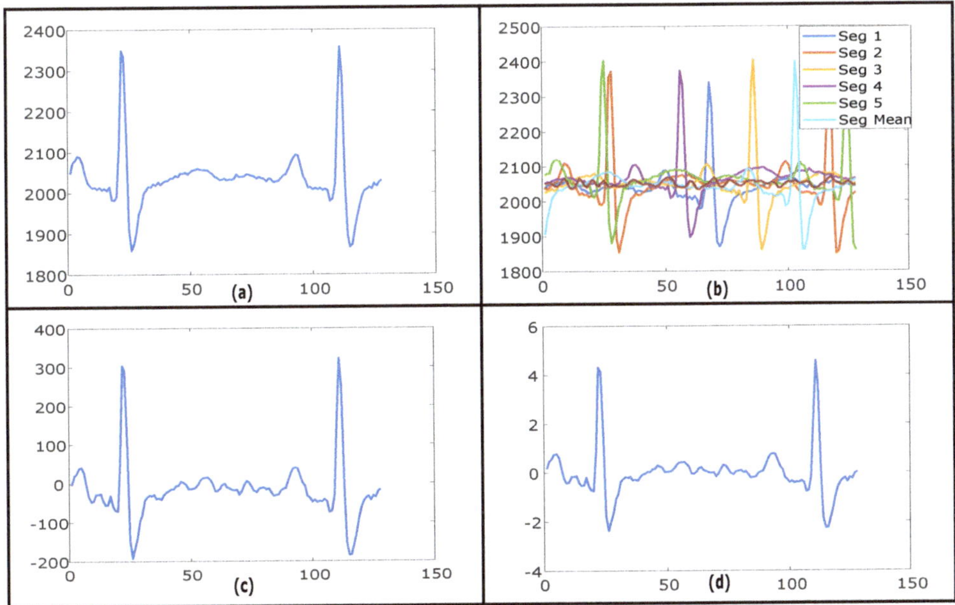

Figure 5. (**a**) ECG signal after basic pre-processing. (**b**) Five segments of ECG baseline activity with computed mean signal. (**c**) ECG signal (left channel) 1 s segment of emotional activity after removing mean of baseline activity. (**d**) After Z-score normalization.

3.3.4. Z-Score Normalization

Each of EEG, ECG, and GSR segments after baseline removal are normalized using Z-score normalization. Figure 5d illustrates the Z-score normalization of the already baseline removed ECG signal. This step is performed to prepare signals to use as a feature for neural network architecture after conversion to a common scale with unity standard deviation and zero mean. Both ECG channels and GSR are now prepared to use as a sequence input to deep neural network architecture for classification. Total size of ECG after pre-processing for DREAMER dataset is 828 × 60 × 128 (18 trials, 2 channels and 23 persons, 60 segments of 128 samples each) and for AMIGOS dataset is 1056 × 86.125 × 128 (16 trials, 2 channels and 33 persons, 86.125 average segments of 128 samples each). AMIGOS dataset has GSR with a total size of 528 × 86.125 × 128 after pre-processing.

3.3.5. Preparation of EEG-Based 2D Images

Total size of EEG data after Z-score normalization for DREAMER dataset is 5796 × 60 × 128 (18 trials, 14 channels and 23 persons) and for AMIGOS dataset is 7392 × 86.125 × 128 (16 trials, 14 channels and 33 persons). Each example of EEG containing 14 channels of 128 samples each is mapped to Nasion 10–20 system [63] with their corresponding positioning of channels to a 1D topological vector of size 81. Each corresponding sample of 14 channels is mapped to 1D topological vector using the matching sequence of set of channels of $AF3$, $F7$, $F3$, $FC5$, $T7$, $P7$, $O1$, $O2$, $P8$, $T8$, $FC6$, $F4$, $F8$, $AF4$ to the set of indices of 4, 13, 19, 21, 29, 31, 37, 39, 47, 49, 55, 57, 67, 76 respectively while keeping all other indices of 1D topological vector as zero. After mapping, a feature matrix of 81×128 dimension is obtained as single EEG example of specific trial of a specific person. This feature matrix is now converted to a 2D image (PNG format) for further processing. After the preparation of 2D images, the DREAMER dataset contains 24,840 images and the AMIGOS dataset contains 45,474 images of 81 × 128 size each. These images are now ready to be used as input to the image input layer of deep neural network architecture.

3.4. Proposed DNN Classification Architectures

ECG and GSR modalities can be used as sequence input to DNN architecture, while 14 channels EEG after converted to 81 × 128 size of images can be used as image input to DNN architecture. Therefore two different DNN architectures were developed for classification through the neural network, one for EEG that would be useful for classification of images and another architecture for ECG and GSR, which would hypothetically be useful for classification of signals or time-series sequence data. Architecture for EEG primarily builds upon 2D convolutional layers, while the architecture for ECG and GSR primarily builds upon the combination of LSTM and 1D convolutional layers, as LSTM is expected to fully exploit the potential of sequence or time-series data [64]. The details of both approaches and the design of their deep network architectures are given below.

3.4.1. DNN Architecture Design for EEG-Based Images

The detailed design of DNN architecture used for EEG image data is shown in Figure 6. This architecture accepts image data of size 81 × 128 in the image input layer. Then there are three sets of 2D convolutional layers, the first convolutional layer contains 8 filters of size 3 × 3, the second convolutional layer contains 16 filters of size 3 × 3 while third convolutional layer contains 32 filters of size 3 × 3 with padding of same values at the border. Each of these convolutional layers was followed by three layers of batch normalization, Rectified Linear Unit (ReLU) and 2D max-pooling layer. Each of the 2D max-pooling layers contains a pool size of 2 × 2 and a stride of 2 × 2 as well along with zero paddings. The third max-pooling layer is connected to a fully connected (FC) layer of size 4 which is then attached to the output layer after passing through the SoftMax layer. The findings from [65] suggest that for 2D-CNN architecture, wider datasets which have more number of classes require more FC layers as compared to deeper datasets which have fewer classes and more samples per class, and require fewer FC layers. Therefore, we used single FC layer in our 2D-CNN architecture.

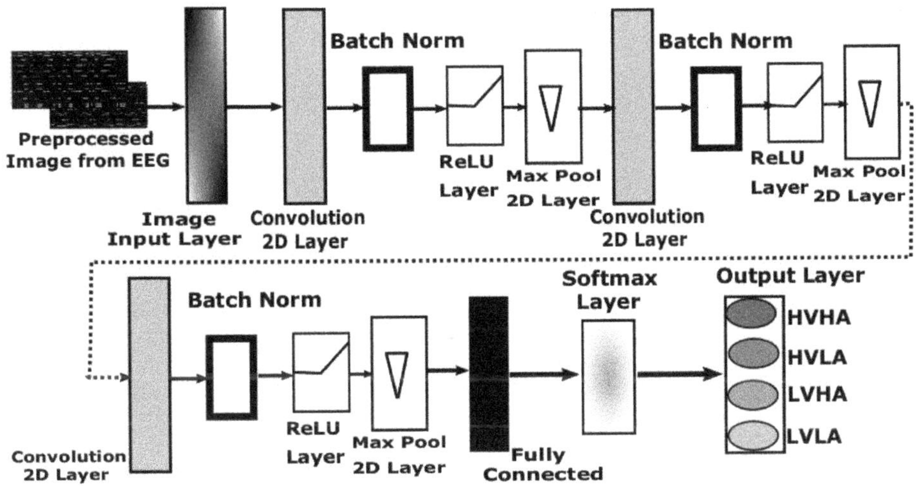

Figure 6. Design of deep neural network (2D-CNN) for EEG-based image data.

The input image to this network is computed using a segment of 1 s of 14 channels of EEG with 128 samples. Therefore, the trained network requires only one second of EEG signals to classify into four basic emotions. This robust neural network has total learnable parameters of 80, 1168, 2320, and 43,012 for first, second, third convolution layers, and fully connected layer, respectively. The detailed parameters internal to the neural network architecture are represented in Table 1.

Table 1. Details of 2D-CNN Architecture.

Serial	Layer Type	Activations	Weights/Offset	Bias/Scale	Learnables
1	Image Input	81 × 128 × 1	-	-	0
2	Convolution	81 × 128 × 1	3 × 3 × 1 × 8	1 × 1 × 8	80
3	Batch Normalization	81 × 128 × 1	1 × 1 × 8	1 × 1 × 8	16
4	ReLU	81 × 128 × 1	-	-	0
5	Ma × Pooling	41 × 64 × 8	-	-	0
6	Convolution	41 × 64 × 16	3 × 3 × 8 × 16	1 × 1 × 16	1168
7	Batch Normalization	41 × 64 × 16	1 × 1 × 16	1 × 1 × 16	32
8	ReLU	41 × 64 × 16	-	-	0
9	Ma × Pooling	21 × 32 × 16	-	-	0
10	Convolution	21 × 32 × 16	3 × 3 × 16 × 16	1 × 1 × 16	2320
11	Batch Normalization	21 × 32 × 16	1 × 1 × 16	1 × 1 × 16	32
12	ReLU	21 × 32 × 16	-	-	0
13	Fully Connected	1 × 1 × 4	4 × 10,752	4 × 1	43,012
14	Softma×	1 × 1 × 4	-	-	0
15	Classification Output	-	-	-	0

3.4.2. DNN Architecture for ECG and GSR

Figure 7 present 1D-CNN architecture proposed for ECG and GSR modalities. One second signal from either ECG or GSR data is captured as a sequence. This sequence is then passed through two one-dimensional convolutional layers, each followed by ReLU activation and max-pooling layers to extract temporal features directly from time-series data. Extracted features are then flattened for the LSTM layer. This LSTM layer learns the order dependence between extracted temporal features, suitable for the classification of time-series data. Now three dense layers provide the learning of prediction probabilities from extracted features for four classes of emotion.

The detailed design of DNN architecture for ECG and GSR-based sequence data is illustrated in Figure 7. This architecture accepts sequence inputs of size 1×128 in its sequence input layer. After the sequence folding layer, two sets of 1D convolutional layers were added. First convolutional layer contains 16 filters of size 3×1, while second convolutional layer contains 32 filters of size 3×1. Each of these convolutional layers is followed by the ReLU layer and 1D max-pooling layer of size 2×1 with a stride of one. After max pooling, sequences are unfolded based on mini-batch size, and flatten layer was applied to get a feature vector. This feature vector is now passed through the LSTM layer with 128 hidden units, state activation function of $tanh$ and gate activation function of $sigmoid$ is used. LSTM layer is then followed by a series of three fully connected layers of size 256 for FC1, 128 for FC2, and a size of 4 for FC3. Each fully connected layer is followed by a dropout layer of 0.5, discarding 50% of random features to avoid over-training of sequence data from LSTM. The last FC layer after dropout is connected to the classification output layer through the SoftMax layer for the classification of HVHA, HVLA, LVHA, and LVLA classes of emotion.

Total learnable parameters of 160, 4640 and 2,163,200 exists for first convolution, second convolution and LSTM layer, respectively. Recurrent weights to train for LSTM are 512×128 with the 512×1 bias for T4096 vector size of input from flatten layer. The detailed parameters of activations, weights, bias, and learnables are provided in Table 2. Intermediate hidden-layer results are also represented in Section 5 for comparison of features using this deep neural network architecture.

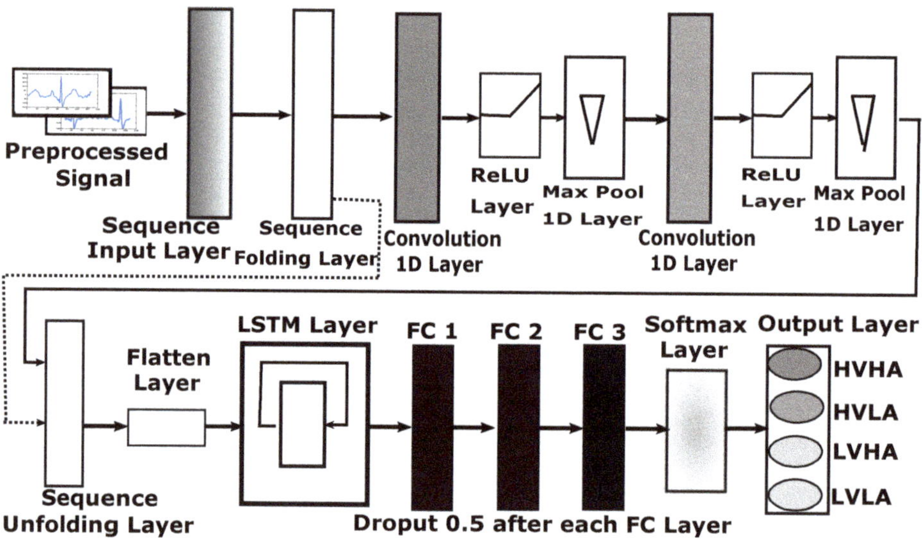

Figure 7. Design of deep neural network (1D-CNN + LSTM) for ECG and GSR-based sequence data.

3.5. Multi-Modal Fusion

To establish and compare the efficacy of the various combination of modalities, majority voting is applied. As every modality have their unique properties to depict the emotional state of a person, multi-modal fusion using majority voting at the decision level is used. This fusion helps to contribute comprehension of the effect of the various combination of modalities between GSR, ECG channels, and EEG.

Table 2. Details of 1D-CNN + LSTM Architecture.

Serial	Layer Type	Activations	Weights	Bias	Learnables
1	Sequence Input	128 × 1 × 1	-	-	0
2	Sequence Folding	128 × 1 × 1	-	-	0
3	Convolution	128 × 1 × 16	3 × 1 × 1 × 16	1 × 1 × 16	64
4	ReLU	128 × 1 × 16	-	-	0
5	Ma × Pooling	128 × 1 × 16	-	-	0
6	Convolution	128 × 1 × 32	3 × 1 × 16 × 32	1 × 1 × 32	1568
7	ReLU	128 × 1 × 32	-	-	0
8	Ma × Pooling	128 × 1 × 32	-	-	0
9	Sequence Unfolding	128 × 1 × 32	-	-	0
10	Flatten	4096	-	-	0
11	LSTM	128	Input: 512 × 4096, Recurrent: 512 × 128	512 × 1	2,163,200
12	Fully Connected	256	256 × 128	256 × 1	33,024
13	Dropout	256	-	-	0
14	Fully Connected	128	128 × 256	128 × 1	32,896
15	Dropout	128	-	0	
16	Fully Connected	4	4 × 128	4 × 1	516
17	Dropout	4	-	-	0
18	Softma×	4	-	-	0
19	Classification Output	-	-	-	0

4. Results

Pre-processed data for all modalities are now randomly split into 70% training data and 30% test data. After pre-processing, each modality contains 24,840 instances of data (81 × 128 size of images for EEG and 1 × 128 size of signals for ECG left, ECG right and GSR) for DREAMER dataset and 45,474 instances each for AMIGOS dataset. Two experiments were performed for each of AMIGOS and DREAMER datasets by randomly split into training and testing for the computation of results in both experiments. For the EEG approach, 13,642 images in the case of AMIGOS and 7452 images in the case of DREAMER were randomly selected as test data, while the remaining randomly selected images of these datasets were used as a training set of images. The same number of training and test samples respectively were randomly selected for ECG and GSR approaches as well.

For the computation of all the presented results, training parameters of deep neural network for both approaches of 2D-CNN and LSTM + 1D-CNN are consistent. Minimum batch size of 240, an initial learning rate of 0.001 with ADAM optimizer, and a gradient squared decay factor of 0.99 were used as training parameters. Core-i5 machine was used for the training of neural networks and testing of performance measures. Separate representation of results of both datasets is presented below.

4.1. AMIGOS Results

Table 3 illustrates the results computed for AMIGOS dataset. 2D-CNN-based approach was used to train 31,832 images randomly selected from the dataset. A similar number of instances from ECG left channel, ECG right channel, and GSR was incorporated to individually train LSTM + 1D-CNN-based neural network. ECG right channel among this second approach depicts the highest accuracy of 98.73%, while GSR with the lowest average accuracy of 63.67%. EEG signal using a 2D-CNN-based approach depicts quite low performance 74.65% as compared to ECG, but much better than the existing approaches in the literature for AMIGOS.

Table 3. Summary of results for AMIGOS dataset. Highlighted row represents best-performing modality.

Modality	1st Random Split	2nd Random Split	Average Accuracy
ECG (Left)	96.98%	96.93%	96.96%
ECG (Right)	**98.81%**	**98.65%**	**98.73%**
GSR	63.77%	63.56%	63.67%
EEG	76.39%	72.91%	74.65%

Table 3 depicts the better performance of the ECG right channel as compared to ECG left channel. Similarly, both of ECG channels performed better as compared to EEG modality, while EEG modality performs better than GSR. We used 2D-CNN architecture for EEG modality, while the same 1D-CNN architecture is used for both ECG channels and GSR modality. The relatively low performance of EEG is because of different architecture used for its 14 channel combinations, while the relatively low performance of GSR is because of the nature of its modality. GSR generally depicts less contribution in the evaluation of emotion elicitation as compared to ECG [66], which is enriched with more critical information regarding the emotional state of a person. The fusion of ECG left channel, ECG right channel, and EEG channel improve overall highest classification accuracy of 99%. The results for all combinations of the modalities are separately represented in Table 4.

Table 4. AMIGOS results for majority voting (fusion of modalities). Highlighted row represents the best-performing combination of modalities.

Modality	1st Random Split	2nd Random Split	Average Accuracy
ECGL + ECGR + GSR + EEG	97.5%	97.0%	97.25%
ECGL + ECGR	98.2%	97.7%	97.95%
ECGL + GSR	79.8%	81.6%	80.70%
ECGL + EEG	85.4%	78.9%	82.15%
ECGR + GSR	80.5%	82.5%	81.50%
ECGR + EEG	86.1%	79.5%	82.80%
GSR + EEG	68.7%	65.8%	67.25%
ECGL + ECGR + GSR	98.5%	98.2%	98.35%
ECGL + ECGR + EEG	**99.0%**	**98.6%**	**98.8%**
ECGL + GSR + EEG	91.5%	90.0%	90.75%
ECGR + GSR + EEG	92.2%	90.5%	91.35%

4.2. DREAMER Results

Table 5 illustrates the summary of results computed for two random splits for ECG channel 1, channel 2, and EEG. 2D-CNN network was trained with 17,388 randomly selected images and a similar number of samples from ECG channel 1 and ECG channel 2 were used to individually train the LSTM + 1D-CNN-based neural network. ECG channel 2 results in the highest average accuracy among these modalities, while EEG with another approach has low but comparable significant results for the classification of four classes in the literature. Table 6 represents the results of majority voting using combinations of channel 1 of ECG, channel of ECG, and EEG.

Table 5. Summary results for DREAMER dataset. Highlighted row represents best-performing modality.

Modality	1st Random Split	2nd Random Split	Average Accuracy
ECG (Channel 1)	90.46%	88.31%	89.39%
ECG (Channel 2)	**90.03%**	**90.96%**	**90.50%**
EEG	47.44%	49.64%	48.54%

Table 6. DREAMER results for majority voting. Highlighted row represents best-performing combination of modalities.

Modality	1st Random Split	2nd Random Split	Average Accuracy
ECGL + ECGR	89.5%	89%	89.25%
ECGL + EEG	63.9%	61.6%	62.75%
ECGR + EEG	63.9%	62.2%	63.05%
ECGL + ECGR + EEG	**90.8%**	**90.3%**	**90.55%**

The detailed results of the DREAMER dataset can be evaluated using confusion matrices for individual modalities as well as the fusion of modalities. Therefore, Tables 7 and 8 represents confusion matrices of ECG channel 1 and channel 2 respectively. Both channels of ECG depict promising results as compared to EEG, while Table 9 represents detailed results of EEG modality. Similarly, the detailed results of AMIGOS dataset in terms of confusion matrices are presented in Table 10 for ECG right-channel modality, Table 11 for GSR modality, and Table 12 for EEG modality respectively.

In both datasets, the recognition rate of ECG is better than EEG because of the difference in architecture. The combination of ECG and EEG yields better performance for both datasets because these two are the highest achieving modalities individually. Secondly, architecture for ECG and EEG is different, therefore EEG with relatively lower individual accuracy performed well on samples converted into images for 2D-CNN architecture. The architecture used for both modalities of ECG and GSR is the same, therefore the small accuracy value of individual GSR modality is due to less informative nature of GSR signals as compared to ECG signals, which results in deterioration in overall recognition rate when combined with ECG and EEG. In the DREAMER dataset, GSR signals are not available, therefore, GSR is not combined for the representation of that dataset results.

Table 7. Confusion matrix of ECG channel 1 of DREAMER dataset.

Output Class \ Target Class	HVHA	HVLA	LVHA	LVLA	Total
HVHA	2932 39.3%	69 0.9%	143 1.9%	60 0.8%	91.5%
HVLA	88 1.2%	1266 17%	69 0.9%	29 0.4%	87.2%
LVHA	121 1.6%	48 0.6%	2010 27%	47 0.6%	90.3%
LVLA	27 0.4%	3 0.0%	10 0.1%	530 7.1%	93.0%
Total	92.6%	91.3%	90.1%	79.6%	**90.4%**

Table 8. Confusion matrix of ECG channel 2 of DREAMER dataset.

Output Class \ Target Class	HVHA	HVLA	LVHA	LVLA	Total
HVHA	2926 39.3%	84 1.1%	138 1.9%	61 0.8%	91.2%
HVLA	78 1.0%	1221 16.4%	47 0.6%	39 0.5%	88.2%
LVHA	155 2.1%	76 1.0%	2037 27.3%	45 0.6%	88.1%
LVLA	9 0.1%	5 0.1%	10 0.1%	521 7.0%	95.6%
Total	92.4%	88.1%	91.3%	78.2%	**90.0%**

Table 9. Confusion matrix of EEG of DREAMER dataset.

Output Class \ Target Class	HVHA	HVLA	LVHA	LVLA	Total
HVHA	2142 28.7%	656 8.8%	1037 13.9%	256 3.4%	52.4%
HVLA	315 4.2%	397 5.3%	273 3.7%	82 1.1%	37.2%
LVHA	490 6.6%	223 3.0%	784 10.5%	116 1.6%	48.6%
LVLA	221 3.0%	110 1.5%	138 1.9%	212 2.8%	31.1%
Total	67.6%	28.6%	35.1%	31.8%	**47.4%**

Table 10. Confusion matrix of ECG right modality for AMIGOS dataset.

Output Class \ Target Class	HVHA	HVLA	LVHA	LVLA	Total
HVHA	3027 22.2%	14 0.1%	13 0.1%	8 0.1%	98.9%
HVLA	13 0.1%	3139 23.0%	10 0.1%	9 0.1%	99.0%
LVHA	14 0.1%	19 0.1%	3664 26.9%	22 0.2%	98.5%
LVLA	20 0.1%	10 0.1%	11 0.1%	3649 26.7%	98.9%
Total	98.5%	98.6%	99.1%	98.9%	**98.8%**

Table 11. Confusion matrix of GSR modality for AMIGOS dataset.

Output Class \ Target Class	HVHA	HVLA	LVHA	LVLA	Total
HVHA	1950 14.3%	332 2.4%	313 2.3%	444 3.3%	64.2%
HVLA	386 2.8%	1958 14.4%	380 2.8%	495 3.6%	60.8%
LVHA	420 3.1%	530 3.9%	2656 19.5%	617 4.5%	62.9%
LVLA	318 2.3%	362 2.7%	349 2.6%	2132 15.6%	67.4%
Total	63.4%	61.5%	71.8%	57.8%	**63.7%**

Table 12. Confusion matrix of EEG modality for AMIGOS dataset.

Output Class \ Target Class	HVHA	HVLA	LVHA	LVLA	Total
HVHA	2356 17.3%	282 2.1%	321 2.4%	287 2.1%	72.6%
HVLA	293 2.1%	2429 17.8%	350 2.6%	299 2.2%	72.1%
LVHA	231 1.7%	276 2.0%	2819 20.7%	285 2.1%	78.1%
LVLA	194 1.4%	195 1.4%	208 1.5%	2817 20.6%	82.5%
Total	76.6%	76.3%	76.2%	76.4%	**76.4%**

5. Discussion

After results analysis, it is observed that AMIGOS performed better as compared to DREAMER dataset with the same methodology. The second observation from Tables 9 and 10 exhibits that the response of all the four classes of HVHA, HVLA, LVHA, and LVLA is consistent for a specific modality, except a significantly better response of LVHA in case of GSR modality as shown in Table 11. The third observation is that the response of these four classes is also consistent for the specific modality of the DREAMER dataset as shown in Tables 7–9, but based on the contribution of each class in the overall dataset. This could also point towards the observation of more imbalance class instances for the DREAMER dataset as compared to the AMIGOS dataset.

Based on these above-mentioned observations, one possible reason for AMIGOS outperforming the DREAMER dataset is its more balanced distribution of classes as compared to DREAMER. Another reason for the better performance of AMIGOS as compared to DREAMER is due to the nature of the self-assessment acquisition process. Self-assessment for the AMIGOS dataset was obtained on a scale of 1–9 for arousal and valence separately. However, for the DREAMER dataset, self-assessment from subjects was acquired on the scale of 1–5 for both valence and arousal. The scale of 1–5 not only exhibits half the freedom of choice on an intensity scale of emotion but also restricts the imbalance created by avoiding the midpoint between 1–5 scale as participants can only provide integer data for the intensity of arousal and valence. However, AMIGOS gives participants the liberty to self-assess in a floating-point number for the scale of 1–9, hence better categorization of emotion can be made which implied better performance of the algorithm on this dataset comparatively. One more possible reason for variation in results is because the total number of instances in AMIGOS for the specific trial is variable and based on the actual length of the trial; however, in the case of DREAMER, only the last 60 segments (1 s each) of each trial were incorporated (because a few trials are much larger in length to be significant for specific emotion). As AMIGOS explores the full-length potential of each trial, this could also explain the variation between the performance of AMIGOS and the DREAMER dataset.

It is also interesting to investigate the spectral signature of ECG signals and compared it with GSR signals to elaborate on the significance of ECG results as compared to GSR results for the same deep neural network architecture. Best and worst-performing ECG instances from test data of AMIGOS, based on their prediction probabilities were selected for four classes individually are presented in Figure 8a,b respectively. Signals of all classes except LVHA are baseline shifted with high noise content hiding the actual shape of an ECG signal results in its worst performance for the use of emotion elicitation. The power spectrum comparison for four classes of best and worst of these signals is presented in Figure 8c,d, respectively. Figure 8c depicts higher inter-class spectral variability of best-performing signals as compared to lower inter-class spectral variability of worst signals except for LVHA in Figure 8d.

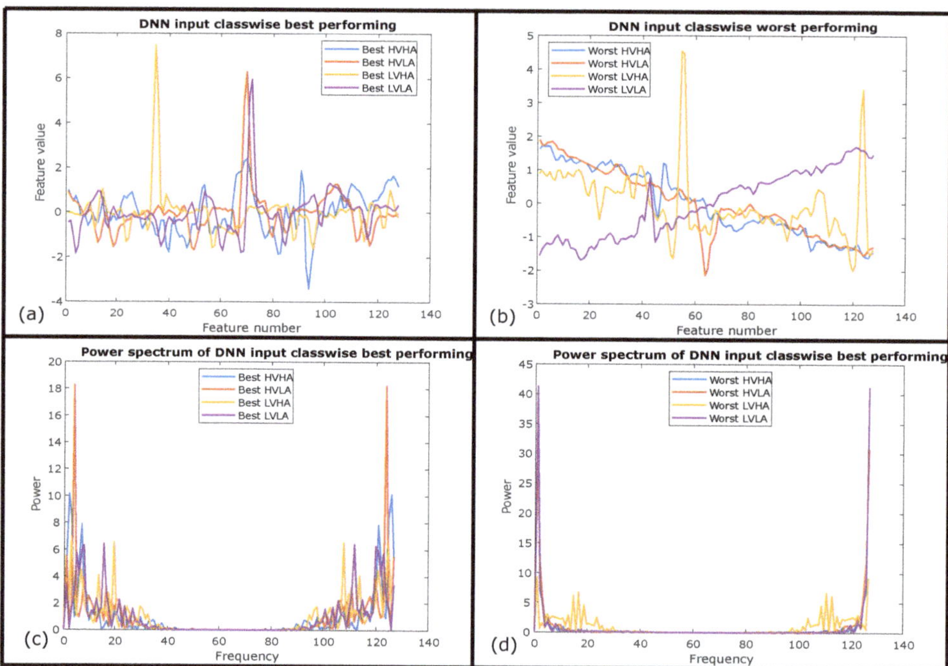

Figure 8. (**a**) Best-performing class-wise ECG samples. (**b**) Worst-performing class-wise ECG samples. (**c**) Spectrum of best-performing class-wise ECG samples. (**d**) Spectrum of worst-performing class-wise ECG samples.

A strong connection between the performance of proposed deep neural network architecture and spectral inter-class variability can also be proven through spectral analysis of comparatively less significant modality of GSR. Therefore, Figure 9a,c exhibits higher inter-class temporal and spectral variability of best-performing GSR signals as compared to lower inter-class temporal and spectral variability of worst-performing GSR signals as shown in Figure 9b,d respectively with the exception of LVHA class. This exception of LVHA class depicts higher recognition performance for LVHA which is also evident from confusion matrices of AMIGOS dataset presented in Tables 10 and 11 for both ECG and GSR respectively. Another observation can be drawn with the higher inter-class variability of the ECG signal in Figure 8c as compared to the GSR signal in Figure 8, representing the spectral significance for the emotion elicitation process as a performance comparison of these two modalities. Therefore, these temporal and spectral observations proved the significance of proposed CNN + LSTM-based architecture to extract those temporal-spectral features for emotion elicitation.

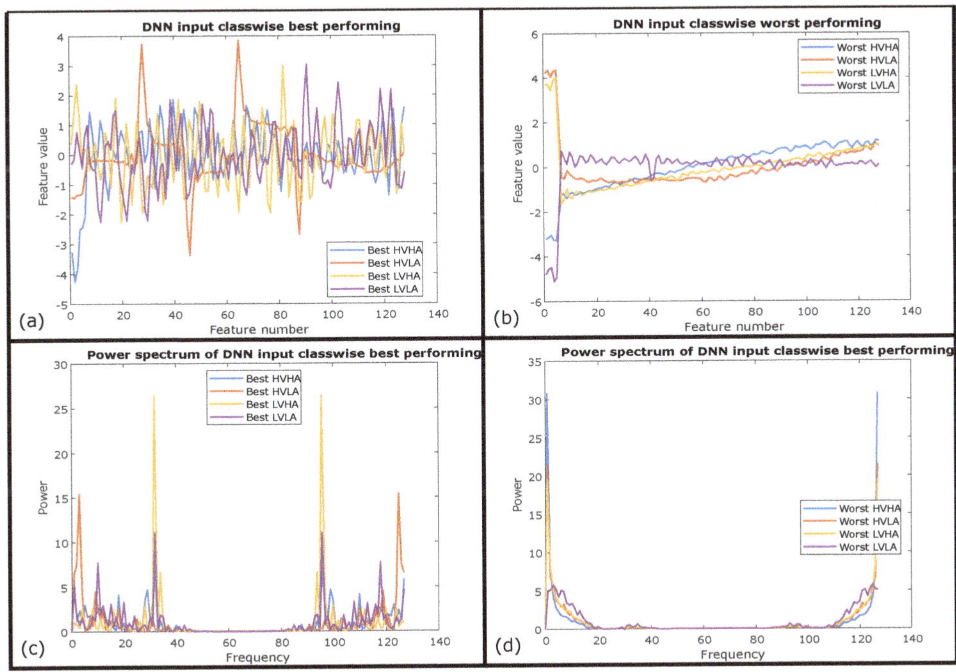

Figure 9. (**a**) Best-performing class-wise GSR samples. (**b**) Worst-performing class-wise GSR samples. (**c**) Spectrum of best-performing class-wise GSR samples. (**d**) Spectrum of worst-performing class-wise GSR samples.

Intermediate features from hidden layers of 1D-CNN + LSTM architecture are also interesting to investigate for best-performing features in Figure 10 and worst-performing features in Figure 11 respectively. These layers' wise deep features are the continuity of input signals of best-performing ECG instances and worst-performing ECG instances represented in Figure 8a,b respectively. Figure 10a,c represents extracted features after first and second convolution layers of 1D-CNN + LSTM architecture while Figure 10b,d represents extracted features after first and second hidden ReLU layers respectively. These hidden deep features from best ECG instances are about to evolve in next hidden layers, where the flatten layer converts 32 instances of activations of 128 samples each into 4096 size feature vector as shown in Figure 10e. This feature vector is passed through the LSTM layer with feature representation in Figure 10f and with feature representation of first, second, and third fully connected layers shown in Figure 10g–i respectively. Through all these hidden layers features evolve with the improvement in inter-class variability in the subsequent layers.

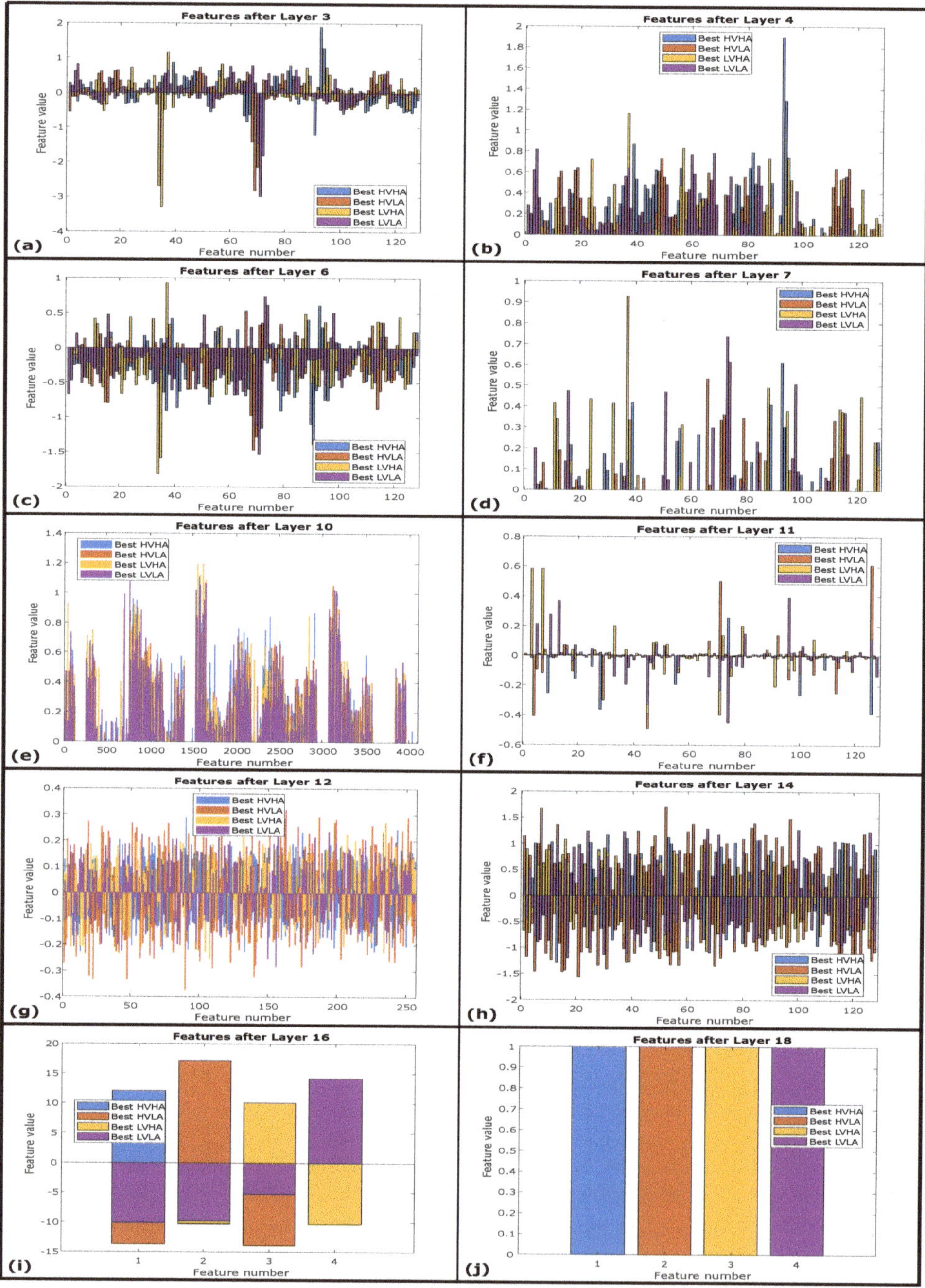

Figure 10. (**a**) Best class-wise features after first convolution layer. (**b**) Best class-wise features after first ReLU layer. (**c**) Best class-wise features after second convolution layer. (**d**) Best class-wise features after second ReLU layer. (**e**) Best class-wise features after flatten layer. (**f**) Best class-wise features after LSTM layer. (**g**) Best class-wise features after FC1 layer. (**h**) Best class-wise features after FC2 layer. (**i**) Best class-wise features after FC3 layer. (**j**) Best class-wise predictions after SoftMax layer.

Figure 11a–h also describe the hidden feature representations evolved through subsequent layers for worst ECG instances for the same 1D-CNN + LSTM architecture. It can be observed that the inter-class variability improves significantly after LSTM and for deep layers of architectures such as represented in Figure 11g,h. However, it is also evident that the overall inter-class variability enhancement for emotion classification is much smaller as compared to the improvement found for best-performing ECG instances. Signals performed worst primarily since these signals are inherited with strong input noise, which is unable to be removed through pre-processing steps performed before introducing the signals to DNN.

This investigation can be evaluated with the importance of a strong linkage between spectral inter-class variability and the performance of proposed deep neural network architecture. The better response of the LVHA class for both ECG and GSR modalities is strongly linked with its higher spectral inter-class variability for the AMIGOS dataset as highlighted in Tables 10 and 11. Therefore, besides the significance of LVHA class, it is wrongly classified as HVHA and other wrong decisions represented in Figure 11i,j. The best-performing signals are predicted with the highest prediction probabilities are represented in Figure 10i,j for FC3 and SoftMax layer respectively. It can also be concluded that the 1D-CNN + LSTM-based deep architecture is highly capable of extracting and distinguishing spectral features that support better performance of the architecture.

The main objective of this study is to improve recognition performance of four classes of emotions in a less constrained real-world environments. To evaluate the accomplishment of this objective, we need to explore the performance comparison of this study with previous literature using benchmark datasets. This comparison is made on the two publicly available datasets of AMIGOS and DREAMER. Physiological signals acquired for both of these recently published DREAMER and AMIGOS datasets use wearable low-cost sensors for EEG, ECG and GSR acquisition as compared to invasive acquisition of physiological signals in previously published DEAP and MAHNOB-HCI datasets. Therefore, comparison of emotion recognition of state of the art with proposed study for both datasets is presented in Table 13.

In Table 13, two classes of emotions represent either high/low levels of valence or arousal, while four classes represent HVHA, HVLA, LVHA, and LVLA. For instance, the study [36] with statistical features and SVM using AMIGOS dataset results in 68.8% and 67% of accuracy for arousal and valence respectively, while another study [37] improved valence results by using SVM-RBF. The recognition results from [12,13,35,39] and attention-based LSTM-RNN study [40] improved with deep-learning algorithms for either AMIGOS or DREAMER dataset for two classes of emotions. In a recent study using Bayesian DNN [42], only binary classification of high and low level of valence results are reported as 86% for DREAMER dataset and 90% for AMIGOS dataset. 3D-CNN architecture [14] yields good results for AMIGOS dataset with the cost of computational complexity and expensive training time and resources. In a more recent study [15], four classes of emotions with recognition performance of 55.56% and 58.57% is achieved using transfer learning of CNN-VGG16 model [67] for DREAMER and AMIGOS datasets respectively. The combination of 1D-CNN and LSTM used to develop a robust methodology that performed much better than the previous results for two of the latest publicly available datasets suitable for real-world emotion monitoring. Table 13 summarized the comparison between proposed methodology with state-of-the-art research using physiological-based emotion elicitation.

Therefore, our proposed architecture outperformed state-of-the-art approaches for the physiological signals acquired through less invasive methods. ECG proved to be a more significant modality as compared to EEG and GSR through the design of specialized deep neural networks based on the nature of the physiological signals. Spectral analysis with a performance comparison for both datasets appraises the adequacy of our methodology to inherently extract spectral features from these modalities as well.

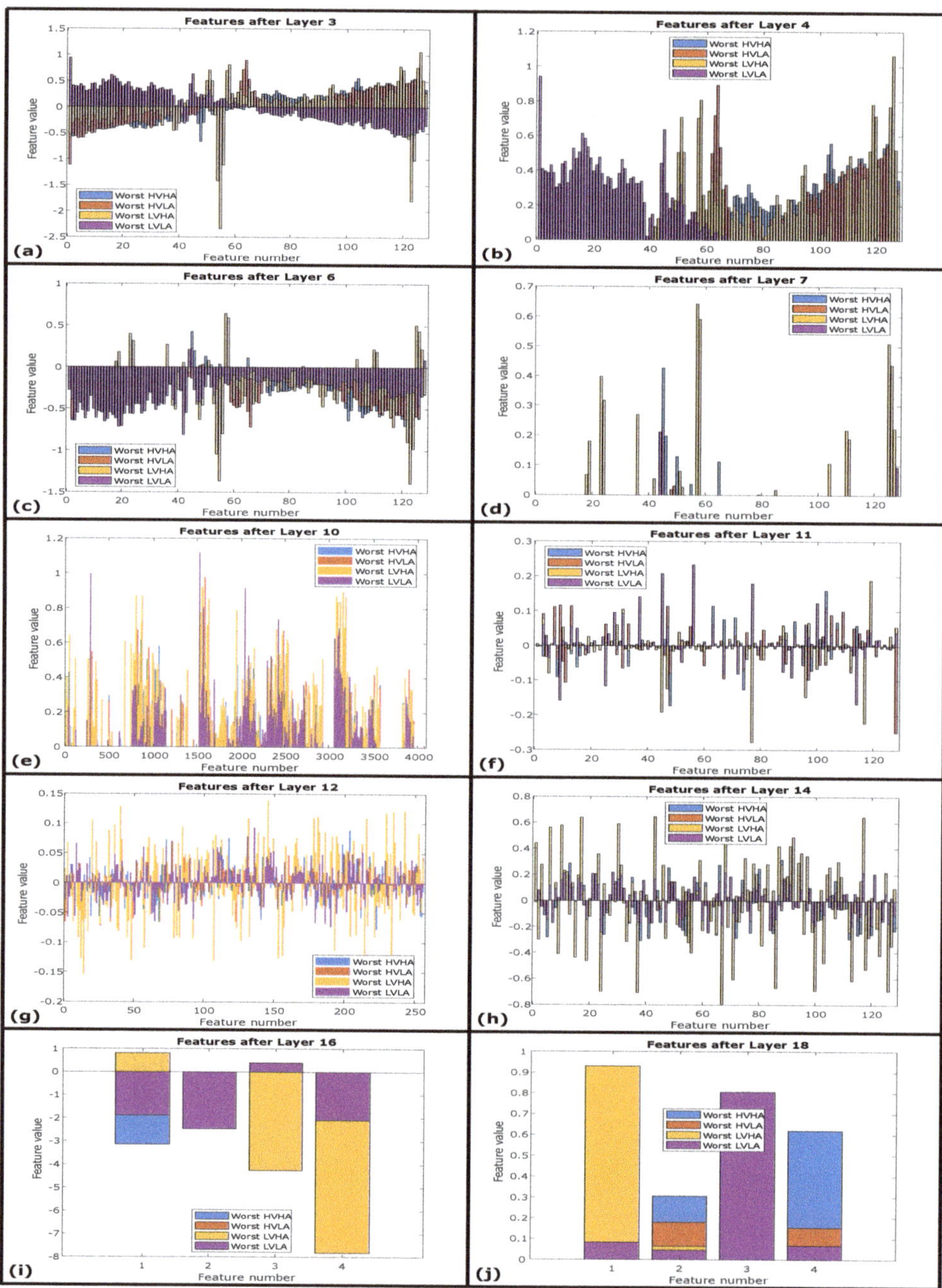

Figure 11. (**a**) Worst class-wise features after first convolution layer. (**b**) Worst class-wise features after first ReLU layer. (**c**) Worst class-wise features after second convolution layer. (**d**) Worst class-wise features after second ReLU layer. (**e**) Worst class-wise features after flatten layer. (**f**) Worst class-wise features after LSTM layer. (**g**) Worst class-wise features after FC1 layer. (**h**) Worst class-wise features after FC2 layer. (**i**) Worst class-wise features after FC3 layer. (**j**) Worst class-wise predictions after SoftMax layer.

Table 13. Comparison with state-of-the-art related work. V and A represents binary classification (High or Low) of Valence and Arousal respectively, while 4 classes are HVHA, HVLA, LVHA and LVLA.

Study	Methodology	Emotion Classes	Accuracy (DREAMER)	Accuracy (AMIGOS)
[36]	Statistical Features, SVM	2	-	67%(V), 68.8%(A)
[37]	SVM-RBF	2	-	83.9%(V), 65%(A)
[35]	Statistical Features, SVM	2	61.84%(V), 63.32%(A)	-
[12]	DGCNN	2	86.23%(V), 84.54%(A)	-
[13]	GCB-net	2	86.99%(V), 89.32%(A)	-
[39]	CNN	2	-	84%(V), 85.8%(A)
[40]	LSTM-RNN	2	-	79.4%(V), 83.3%(A)
[14]	3D-CNN	4	-	99.7%
[42]	Bayesian DNN	2	86%(V)	90%(V)
[15]	CNN-VGG16	4	55.56%	58.57%
Proposed Study	LSTM + CNN	4	90.8%	99%

6. Conclusions

Our proposed deep-learning architecture combines the usefulness of LSTM and CNN, which proved to be efficient for emotion recognition and outperforming previous approaches. Intermediate results of the deep neural network as a hidden feature representation helps us get an insight into features evolving through these layers. Furthermore, the significance of the proposed methodology lies in the higher performance for four classes of emotion elicitation based on subject-independent study, while the wireless acquisition of physiological signals is more suitable for the less constrained real-world environments.

The scope of this study is limited to the decision-level fusion of modalities using majority voting; however, feature-level fusion techniques with proposed architecture may yield significant results. Similarly, three-dimensional CNN architectures can be explored with the combination of LSTM as future investigation for performance improvement. Future dimensions will also assist EEG with more specialized deep neural networks for better performance as well for various other physiological signals such as skin temperature and respiration.

Author Contributions: Conceptualization, M.N.D. and M.U.A.; Data curation, M.N.D.; Formal analysis, M.N.D.; Investigation, S.G.K.; Methodology, M.N.D.; Resources, A.N.P.; Software, S.G.K.; Supervision, M.U.A.; Validation, A.N.P.; Writing—original draft, M.N.D.; Writing—review and editing, M.U.A. All authors have read and agreed to the published version of the manuscript.

Funding: The APC was partially funded by National University of Sciences and Technology (NUST), Islamabad, Pakistan.

Conflicts of Interest: The authors declare no conflict of interest.

References

1. Hwang, B.; You, J.; Vaessen, T.; Myin-Germeys, I.; Park, C.; Zhang, B.T. Deep ECGNet: An optimal deep learning framework for monitoring mental stress using ultra short-term ECG signals. *Telemed. e-Health* **2018**, *24*, 753–772. [CrossRef]
2. Ferdinando, H.; Seppänen, T.; Alasaarela, E. Enhancing Emotion Recognition from ECG Signals using Supervised Dimensionality Reduction. In Proceedings of the ICPRAM, Porto, Portugal, 24–26 February 2017; pp. 112–118.
3. Cheng, Z.; Shu, L.; Xie, J.; Chen, C.P. A novel ECG-based real-time detection method of negative emotions in wearable applications. In Proceedings of the 2017 International Conference on Security, Pattern Analysis, and Cybernetics (SPAC), Shenzhen, China, 15–17 December 2017; pp. 296–301.

4. Zhao, M.; Adib, F.; Katabi, D. Emotion recognition using wireless signals. In Proceedings of the 22nd Annual International Conference on Mobile Computing and Networking, New York, NY, USA, 3–7 October 2016; pp. 95–108.
5. Calvo, R.A.; D'Mello, S. Affect detection: An interdisciplinary review of models, methods, and their applications. *IEEE Trans. Affect. Comput.* **2010**, *1*, 18–37. [CrossRef]
6. Healey, J.A.; Picard, R.W. Detecting stress during real-world driving tasks using physiological sensors. *IEEE Trans. Intell. Transp. Syst.* **2005**, *6*, 156–166. [CrossRef]
7. Agrafioti, F.; Hatzinakos, D.; Anderson, A.K. ECG pattern analysis for emotion detection. *IEEE Trans. Affect. Comput.* **2011**, *3*, 102–115. [CrossRef]
8. Yuvaraj, R.; Murugappan, M. Hemispheric asymmetry non-linear analysis of EEG during emotional responses from idiopathic Parkinson's disease patients. *Cogn. Neurodyn.* **2016**, *10*, 225–234. [CrossRef]
9. Shu, L.; Xie, J.; Yang, M.; Li, Z.; Li, Z.; Liao, D.; Xu, X.; Yang, X. A review of emotion recognition using physiological signals. *Sensors* **2018**, *18*, 2074. [CrossRef] [PubMed]
10. Liu, Y.; Sourina, O.; Nguyen, M.K. Real-time EEG-based emotion recognition and its applications. In *Transactions on Computational Science XII*; Springer:Berlin, Heidelberg, 2011; pp. 256–277.
11. Russell, J.A. A circumplex model of affect. *J. Personal. Soc. Psychol.* **1980**, *39*, 1161. [CrossRef]
12. Song, T.; Zheng, W.; Song, P.; Cui, Z. EEG emotion recognition using dynamical graph convolutional neural networks. *IEEE Trans. Affect. Comput.* to be published.. [CrossRef]
13. Zhang, T.; Wang, X.; Xu, X.; Chen, C.P. GCB-Net: Graph convolutional broad network and its application in emotion recognition. *IEEE Trans. Affect. Comput.* to be published (Early Access). [CrossRef]
14. Zhao, Y.; Cao, X.; Lin, J.; Yu, D.; Cao, X. Multimodal Emotion Recognition Model using Physiological Signals. *arXiv* **2019**, arXiv:1911.12918.
15. Siddharth, S.; Jung, T.P.; Sejnowski, T.J. Utilizing deep learning towards multi-modal bio-sensing and vision-based affective computing. *arXiv* **2019**, arXiv:1905.07039.
16. Petrantonakis, P.C.; Hadjileontiadis, L.J. Emotion recognition from brain signals using hybrid adaptive filtering and higher order crossings analysis. *IEEE Trans. Affect. Comput.* **2010**, *1*, 81–97. [CrossRef]
17. Lan, Z.; Sourina, O.; Wang, L.; Liu, Y. Real-time EEG-based emotion monitoring using stable features. *Vis. Comput.* **2016**, *32*, 347–358. [CrossRef]
18. Xu, Y.; Liu, G.Y. A method of emotion recognition based on ECG signal. In Proceedings of the 2009 International Conference on Computational Intelligence and Natural Computing, Wuhan, China, 6–7 June 2009; Volume 1, pp. 202–205.
19. Candra, H.; Yuwono, M.; Handojoseno, A.; Chai, R.; Su, S.; Nguyen, H.T. Recognizing emotions from EEG subbands using wavelet analysis. In Proceedings of the 2015 37th annual international conference of the IEEE engineering in medicine and biology society (EMBC), Milan, Italy, 25–29 August 2015; pp. 6030–6033.
20. Candra, H.; Yuwono, M.; Chai, R.; Handojoseno, A.; Elamvazuthi, I.; Nguyen, H.T.; Su, S. Investigation of window size in classification of EEG-emotion signal with wavelet entropy and support vector machine. In Proceedings of the 2015 37th Annual international conference of the IEEE Engineering in Medicine and Biology Society (EMBC), Milan, Italy, 25–29 August 2015; pp. 7250–7253.
21. Guo, H.W.; Huang, Y.S.; Lin, C.H.; Chien, J.C.; Haraikawa, K.; Shieh, J.S. Heart rate variability signal features for emotion recognition by using principal component analysis and support vectors machine. In Proceedings of the 2016 IEEE 16th International Conference on Bioinformatics and Bioengineering (BIBE), Taichung, Taiwan, 31 October–2 November 2016; pp. 274–277.
22. Cheng, B.; Liu, G. Emotion recognition from surface EMG signal using wavelet transform and neural network. In Proceedings of the 2nd International Conference on Bioinformatics and Biomedical Engineering (ICBBE), Shanghai, China, 16–18 May 2008; pp. 1363–1366.
23. Xianhai, G. Study of emotion recognition based on electrocardiogram and RBF neural network. *Procedia Eng.* **2011**, *15*, 2408–2412. [CrossRef]
24. Li, X.; Song, D.; Zhang, P.; Yu, G.; Hou, Y.; Hu, B. Emotion recognition from multi-channel EEG data through convolutional recurrent neural network. In Proceedings of the 2016 IEEE International Conference on Bioinformatics and Biomedicine (BIBM), Shenzhen, China, 15–18 December 2016; pp. 352–359.
25. Zheng, W.L.; Zhu, J.Y.; Lu, B.L. Identifying stable patterns over time for emotion recognition from EEG. *arXiv* **2016**, arXiv:1601.02197.

26. Zheng, W.L.; Lu, B.L. A multimodal approach to estimating vigilance using EEG and forehead EOG. *J. Neural Eng.* **2017**, *14*, 026017. [CrossRef]
27. Alhagry, S.; Fahmy, A.A.; El-Khoribi, R.A. Emotion recognition based on EEG using LSTM recurrent neural network. *Emotion* **2017**, *8*, 355–358. [CrossRef]
28. Tripathi, S.; Acharya, S.; Sharma, R.D.; Mittal, S.; Bhattacharya, S. Using Deep and Convolutional Neural Networks for Accurate Emotion Classification on DEAP Dataset. In Proceedings of the Twenty-Ninth IAAI Conference, San Francisco, CA, USA, 6–9 February 2017.
29. Liu, W.; Qiu, J.L.; Zheng, W.L.; Lu, B.L. Multimodal Emotion Recognition Using Deep Canonical Correlation Analysis. *arXiv* **2019**, arXiv:1908.05349.
30. Planet, S.; Iriondo, I. Comparison between decision-level and feature-level fusion of acoustic and linguistic features for spontaneous emotion recognition. In Proceedings of the 7th Iberian Conference on Information Systems and Technologies (CISTI 2012), Madrid, Spain, 20–23 June 2012; pp. 1–6.
31. Chen, J.; Hu, B.; Xu, L.; Moore, P.; Su, Y. Feature-level fusion of multimodal physiological signals for emotion recognition. In Proceedings of the 2015 IEEE International Conference on Bioinformatics and Biomedicine (BIBM), Washington, DC, USA, 9–12 November 2015; pp. 395–399.
32. Liu, J.; Su, Y.; Liu, Y. Multi-modal emotion recognition with temporal-band attention based on LSTM-RNN. In *Pacific Rim Conference on Multimedia*; Springer: Cham, Switzerland, 2017; pp. 194–204.
33. Lu, Y.; Zheng, W.L.; Li, B.; Lu, B.L. Combining eye movements and EEG to enhance emotion recognition. In Proceedings of the Twenty-Fourth International Joint Conference on Artificial Intelligence, Buenos Aires, Argentina, 25–31 July 2015.
34. Correa, J.A.M.; Abadi, M.K.; Sebe, N.; Patras, I. Amigos: A dataset for affect, personality and mood research on individuals and groups. *arXiv* **2017**, arXiv:1702.02510.
35. Katsigiannis, S.; Ramzan, N. DREAMER: A database for emotion recognition through EEG and ECG signals from wireless low-cost off-the-shelf devices. *IEEE J. Biomed. Health Inform.* **2017**, *22*, 98–107. [CrossRef]
36. Yang, H.C.; Lee, C.C. An Attribute-invariant Variational Learning for Emotion Recognition Using Physiology. In Proceedings of the 2019 IEEE International Conference on Acoustics, Speech and Signal Processing (ICASSP), Brighton, UK, 12–17 May 2019; pp. 1184–1188.
37. Shukla, J.; Barreda-Angeles, M.; Oliver, J.; Nandi, G.; Puig, D. Feature Extraction and Selection for Emotion Recognition from Electrodermal Activity. *IEEE Trans. Affect. Comput.* to be published (Early Access).. [CrossRef]
38. Santamaria-Granados, L.; Munoz-Organero, M.; Ramirez-Gonzalez, G.; Abdulhay, E.; Arunkumar, N. Using deep convolutional neural network for emotion detection on a physiological signals dataset (AMIGOS). *IEEE Access* **2018**, *7*, 57–67. [CrossRef]
39. Sarkar, P.; Etemad, A. Self-supervised learning for ecg-based emotion recognition. In Proceedings of the 2020 IEEE International Conference on Acoustics, Speech and Signal Processing (ICASSP), Barcelona, Spain, 4–8 May 2020; pp. 3217–3221.
40. Li, C.; Bao, Z.; Li, L.; Zhao, Z. Exploring temporal representations by leveraging attention-based bidirectional LSTM-RNNs for multi-modal emotion recognition. *Inf. Process. Manag.* **2020**, *57*, 102185. [CrossRef]
41. Siddharth, S.; Jung, T.P.; Sejnowski, T.J. Multi-modal approach for affective computing. *arXiv* **2018**, arXiv:1804.09452.
42. Harper, R.; Southern, J. A bayesian deep learning framework for end-to-end prediction of emotion from heartbeat. *arXiv* **2019**, arXiv:1902.03043.
43. Min, S.; Lee, B.; Yoon, S. Deep learning in bioinformatics. *Briefings Bioinform.* **2017**, *18*, 851–869. [CrossRef]
44. Khorrami, P.; Le Paine, T.; Brady, K.; Dagli, C.; Huang, T.S. How deep neural networks can improve emotion recognition on video data. In Proceedings of the 2016 IEEE International Conference on Image Processing (ICIP), Phoenix, AZ, USA, 25–28 September 2016; pp. 619–623.
45. Tzirakis, P.; Trigeorgis, G.; Nicolaou, M.A.; Schuller, B.W.; Zafeiriou, S. End-to-end multimodal emotion recognition using deep neural networks. *IEEE J. Sel. Top. Signal Process.* **2017**, *11*, 1301–1309. [CrossRef]
46. Arevalillo-Herráez, M.; Cobos, M.; Roger, S.; García-Pineda, M. Combining Inter-Subject Modeling with a Subject-Based Data Transformation to Improve Affect Recognition from EEG Signals. *Sensors* **2019**, *19*, 2999. [CrossRef]
47. Liu, Y.; Sourina, O. Real-time subject-dependent EEG-based emotion recognition algorithm. In *Transactions on Computational Science XXIII*; Springer: Berlin/Heidelberg, Germany, 2014; pp. 199–223.

48. Liu, Y.; Sourina, O. EEG-based subject-dependent emotion recognition algorithm using fractal dimension. In Proceedings of the 2014 IEEE International Conference on Systems, Man, and Cybernetics (SMC), San Diego, CA, USA, 5–8 October 2014; pp. 3166–3171.
49. Pandey, P.; Seeja, K. Emotional state recognition with eeg signals using subject independent approach. In *Data Science and Big Data Analytics*; Springer: Singapore, 2019; pp. 117–124.
50. Bozhkov, L.; Georgieva, P.; Santos, I.; Pereira, A.; Silva, C. EEG-based subject independent affective computing models. *Procedia Comput. Sci.* **2015**, *53*, 375–382. [CrossRef]
51. Duvinage, M.; Castermans, T.; Petieau, M.; Hoellinger, T.; Cheron, G.; Dutoit, T. Performance of the Emotiv Epoc headset for P300-based applications. *Biomed. Eng. Online* **2013**, *12*, 56. [CrossRef]
52. Burns, A.; Greene, B.R.; McGrath, M.J.; O'Shea, T.J.; Kuris, B.; Ayer, S.M.; Stroiescu, F.; Cionca, V. SHIMMER™–A wireless sensor platform for noninvasive biomedical research. *IEEE Sensors J.* **2010**, *10*, 1527–1534. [CrossRef]
53. Ries, A.J.; Touryan, J.; Vettel, J.; McDowell, K.; Hairston, W.D. A comparison of electroencephalography signals acquired from conventional and mobile systems. *J. Neurosci. Neuroeng.* **2014**, *3*, 10–20. [CrossRef]
54. Koelstra, S.; Muhl, C.; Soleymani, M.; Lee, J.S.; Yazdani, A.; Ebrahimi, T.; Pun, T.; Nijholt, A.; Patras, I. Deap: A database for emotion analysis; using physiological signals. *IEEE Trans. Affect. Comput.* **2011**, *3*, 18–31. [CrossRef]
55. Soleymani, M.; Lichtenauer, J.; Pun, T.; Pantic, M. A multimodal database for affect recognition and implicit tagging. *IEEE Trans. Affect. Comput.* **2011**, *3*, 42–55. [CrossRef]
56. Singh, R.R.; Conjeti, S.; Banerjee, R. A comparative evaluation of neural network classifiers for stress level analysis of automotive drivers using physiological signals. *Biomed. Signal Process. Control* **2013**, *8*, 740–754. [CrossRef]
57. Subramanian, R.; Wache, J.; Abadi, M.K.; Vieriu, R.L.; Winkler, S.; Sebe, N. ASCERTAIN: Emotion and personality recognition using commercial sensors. *IEEE Trans. Affect. Comput.* **2016**, *9*, 147–160. [CrossRef]
58. Harper, R.; Southern, J. End-To-End Prediction of Emotion From Heartbeat Data Collected by a Consumer Fitness Tracker. In Proceedings of the 2019 8th International Conference on Affective Computing and Intelligent Interaction (ACII), Cambridge, UK, 3–6 September 2019; pp. 1–7.
59. Zhang, J.; Chen, M.; Zhao, S.; Hu, S.; Shi, Z.; Cao, Y. ReliefF-based EEG sensor selection methods for emotion recognition. *Sensors* **2016**, *16*, 1558. [CrossRef]
60. Zhuang, N.; Zeng, Y.; Tong, L.; Zhang, C.; Zhang, H.; Yan, B. Emotion recognition from EEG signals using multidimensional information in EMD domain. *BioMed Res. Int.* **2017**, *2017*. [CrossRef]
61. Cimtay, Y.; Ekmekcioglu, E. Investigating the use of pretrained convolutional neural network on cross-subject and cross-dataset EEG emotion recognition. *Sensors* **2020**, *20*, 2034. [CrossRef]
62. Ludwig, K.A.; Miriani, R.M.; Langhals, N.B.; Joseph, M.D.; Anderson, D.J.; Kipke, D.R. Using a common average reference to improve cortical neuron recordings from microelectrode arrays. *J. Neurophysiol.* **2009**, *101*, 1679–1689. [CrossRef]
63. Okamoto, M.; Dan, H.; Sakamoto, K.; Takeo, K.; Shimizu, K.; Kohno, S.; Oda, I.; Isobe, S.; Suzuki, T.; Kohyama, K.; et al. Three-dimensional probabilistic anatomical cranio-cerebral correlation via the international 10–20 system oriented for transcranial functional brain mapping. *Neuroimage* **2004**, *21*, 99–111. [CrossRef] [PubMed]
64. Karim, F.; Majumdar, S.; Darabi, H.; Chen, S. LSTM fully convolutional networks for time series classification. *IEEE Access* **2017**, *6*, 1662–1669. [CrossRef]
65. Basha, S.S.; Dubey, S.R.; Pulabaigari, V.; Mukherjee, S. Impact of fully connected layers on performance of convolutional neural networks for image classification. *Neurocomputing* **2020**, *378*, 112–119. [CrossRef]
66. Goshvarpour, A.; Abbasi, A.; Goshvarpour, A. An accurate emotion recognition system using ECG and GSR signals and matching pursuit method. *Biomed. J.* **2017**, *40*, 355–368. [CrossRef] [PubMed]
67. Girshick, R. Fast r-cnn. In Proceedings of the IEEE International Conference on Computer Vision, Santiago, Chile, 7–13 December 2015; pp. 1440–1448.

Article

Phonocardiogram Signal Processing for Automatic Diagnosis of Congenital Heart Disorders through Fusion of Temporal and Cepstral Features

Sumair Aziz [1,†], Muhammad Umar Khan [1,*,†], Majed Alhaisoni [2], Tallha Akram [3,*] and Muhammad Altaf [3]

1 Department of Electronics Engineering, University of Engineering and Technology Taxila, Taxila 47080, Pakistan; sumair.aziz@uettaxila.edu.pk
2 College of Computer Science and Engineering, University of Ha'il, Ha'il 55211, Saudi Arabia; majed.alhaisoni@gmail.com
3 Department of Electrical and Computer Engineering, COMSATS University Islamabad, Wah Campus, Wah Cantonment 47040, Pakistan; mohammadaltaf@gmail.com
* Correspondence: sa.umarkhan@gmail.com (M.U.K.); tallha@ciitwah.edu.pk (T.A.)
† These authors contributed equally to this work.

Received: 25 May 2020; Accepted: 1 July 2020; Published: 6 July 2020

Abstract: Congenital heart disease (CHD) is a heart disorder associated with the devastating indications that result in increased mortality, increased morbidity, increased healthcare expenditure, and decreased quality of life. Ventricular Septal Defects (VSDs) and Arterial Septal Defects (ASDs) are the most common types of CHD. CHDs can be controlled before reaching a serious phase with an early diagnosis. The phonocardiogram (PCG) or heart sound auscultation is a simple and non-invasive technique that may reveal obvious variations of different CHDs. Diagnosis based on heart sounds is difficult and requires a high level of medical training and skills due to human hearing limitations and the non-stationary nature of PCGs. An automated computer-aided system may boost the diagnostic objectivity and consistency of PCG signals in the detection of CHDs. The objective of this research was to assess the effects of various pattern recognition modalities for the design of an automated system that effectively differentiates normal, ASD, and VSD categories using short term PCG time series. The proposed model in this study adopts three-stage processing: pre-processing, feature extraction, and classification. Empirical mode decomposition (EMD) was used to denoise the raw PCG signals acquired from subjects. One-dimensional local ternary patterns (1D-LTPs) and Mel-frequency cepstral coefficients (MFCCs) were extracted from the denoised PCG signal for precise representation of data from different classes. In the final stage, the fused feature vector of 1D-LTPs and MFCCs was fed to the support vector machine (SVM) classifier using 10-fold cross-validation. The PCG signals were acquired from the subjects admitted to local hospitals and classified by applying various experiments. The proposed methodology achieves a mean accuracy of 95.24% in classifying ASD, VSD, and normal subjects. The proposed model can be put into practice and serve as a second opinion for cardiologists by providing more objective and faster interpretations of PCG signals.

Keywords: phonocardiogram; machine learning; empirical mode decomposition; feature extraction; mel-frequency cepstral coefficients; support vector machines; computer aided diagnosis; congenital heart disease; statistical analysis

---

1. Introduction

Congenital heart disease (CHD) is one the most common birth defects which affect the overall structure of the heart and vessels, found in not more than 1% of newborns [1]. CHD manifests itself at birth and symptoms may vary from mild asymptomatic cases to severe, life-threatening indications. With advances in treatment, there is an increasing population of adults surviving with congenital heart malformations. Globally, cardiovascular diseases (CVD) are the main cause of mortality. Many adult CHD survivors presenting an increased risk of CVD [2] may have long term health problems, which affect their quality of life. In Pakistan, CHD remains to be an important medical issue and the number of patients is increasing every day [3]. Among newborn children and youngsters, cardiac disorders are responsible for a large extent (30% to 50%) of mortality brought about by birth surrenders. The most common cardiac defects which represent about 85% of all congenital heart diseases are ventricular septal defects (VSDs; 34%), and atrial septal defects (ASDs), which contribute up to 13% [4]. Like any other medical issue, robust diagnosis methods are required for the timely diagnosis of the CHDs. Different non-obtrusive procedures are utilized in identifying heart defects. Using the electrocardiogram (ECG) is one of the most common paths for identifying heart issues; it is based on the electrical signals generated during the heart muscle contraction/relaxation. The ECG reveals the electrical activity of the heart and is mostly recorded by the placement of three electrodes for early diagnosis. It comprises five waves; i.e., P, Q, R, S, and T. These waves are prepared to make sense of different pathologies [5].

Another commonly used mechanism for diagnosis of heart disorder is through the analysis of the heart sound [6]. Easy access to digital stethoscopes allows medical staff to record and analyze heart sounds for diagnostic purposes. The phonocardiogram (PCG) records heart sounds and murmurs in the form of a plot and the machine by which these sounds are recorded is known as the phonocardiograph. It is one of the non-obtrusive systems, which records heart condition in audible form. Heart sounds are generated by the opening or closing of the heart valves. Blood flow through the valves' orifices or into the ventricular chambers also produces heart sounds. Recording of the PCG signal consists of four important heart sound constituents; namely, S1, S2, S3, and S4.

An atrial septal defect (ASD) [7] is a birth deformity of the heart in which there is a hole in the wall (septum) that isolates the upper chambers (atria) of the heart. A gap can fluctuate in size and requires a medical procedure. The reasons for CHD amongst most infants are obscure, but genetic factors are also important, as a few infants have heart defects as a result of changes in their genes or chromosomes [8]. A ventricular septal defect (VSD) is an opening in the heart, a typical heart imperfection that is present during childbirth (congenital).

Extensive research has been carried out for the detection and classification of congenital heart disorders using the PCG signal. The PCG signal classification approach was suggested using the nested set of classifiers; namely, random forest, cost-sensitive classifier, and LogitBoost (LB) [9]. A combination of time domain, statistical, and frequency domain features was used for effective classification. Cepstrum-analysis-based feature extraction was performed to classify normal and abnormal PCG signals through a support vector machine (SVM) classifier [10].

PCG signal classification was achieved through linear SVM and a combination of dynamic time wrapping (DTW) and Mel-frequency cepstral coefficient (MFCC) features in [11] to achieve 82.4% accuracy. The screening method of PCG signals using a modified Arash-band method and an SVM classifier has been used [12]. In [13], the PCG signal was first segmented into S1, systole, S2, and diastole through the hidden Markov model (HMM). Gammatone frequency cepstral coefficient (GFCC) features were extracted to perform classification using weighted SVM without segmentation and with segmented signals. The sensitivity of 90.3% and specificity of 89% were achieved through 10-fold cross-validation. Rubin et al. [14] proposed a method for classification of normal and abnormal PCG signals based on Mel-frequency cepstral coefficients (MFCCs) and a two-layer convolutional neural network (CNN). This method achieved an overall score of 83.99% with the PHY16 challenge database. Spectrogram features from PCG were used to train CNN and Adaboost classifiers [15]. A simple decision rule was implemented on outputs of both classifiers to generate final classification results with an overall reported accuracy of 89%. In another study [16], the authors used a Hamming filter for noise reduction in PCG signals. A four-layer 1D CNN for PCG signal classification was employed and the overall accuracy of this method was 79%. In a recent study [17], the CNN architecture was presented for heart sound classification. CNN was tested on different feature sets, such as Mel-Spectrogram, MFCC, and sub-band envelopes.

Zhihai Tu et al. performed filtration of heart sound signals using wavelet transform. Heart sound segmentation was performed using Hilbert transform [18,19], and cubic polynomial interpolation [20]. Samuel E Schmidt et al. presented an easy and cheap system for the identification of coronary artery disease (CAD) using acoustic features. A quadratic discriminant function was used to combine the different features. The accuracy to diagnose the CAD disease is 73% [21]. In another study [22], tunable Q-wavelet transformation [23–25] and signal second difference with the median filter were used for the detection of artifact in heart sound. In [26], the classification of heart sound was achieved through power MFCC features fused with fractal features. The nearest neighbor classifier was employed to perform classification. The overall accuracies achieved on three publicly available datasets were 92%, 81%, and 98%. In [27] heart sounds classification was performed through MFCC and linear predictive coding (LPC) features in conjunction with the Adaboost ensemble classifier. In [28], the authors used the least square support vector machine (LSSVM) with wavelet features for the detection of heart pathologies. VSD was diagnosed from the time-frequency feature matrix acquired from heart sounds [29]. The ellipse-based model achieved max accuracy of 97.6% on large VSD sounds. The authors used the auscultation jacket to detect heart abnormalities [30]. The system with a feed-forward neural network as the classifier achieved sensitivity and specificity of 84% and 86% respectively. In [31], normal and abnormal cardiac sounds were classified using ensemble EMD, auto-regressive models, and a neural network. The method showed sensitivity and specificity of 82% and 88% respectively. An efficient method for the detection of abnormal PCG signals was proposed [32] using MFCCs and SVM with a classification accuracy of 92.6%. Classification of CAD and non-CAD subjects from PCG and ECG [33] using a dual input neural network (DINN) achieved specificity, accuracy, and G-mean of 89.17%, 95.62%, and 93.69%, respectively. A combination of machine learning and a deep learning model [34] for identification of congestive heart failure (CHF) from audio PCG obtained an accuracy of 93.2%.

Classification of ASD and normal PCG signals collected from newborn subjects was performed using a combination of short-time Fourier transform (STFT) and MFCC and its derivatives features [35]. Accuracy of 93.2% was achieved through the KNN classifier. An approach based on discrete wavelet transform (DWT) and multilayer perceptron (MLP) for estimation of VSD were presented in [36]. Features such as power, standard deviation, skewness, kurtosis, and Shannon entropy were extracted from eight levels of detailed coefficients of DWT. In another similar study [37], a combination of wavelet and MFCC features was proposed to achieve 97% accuracy on normal and four abnormal classes of heart sounds. In [38], a comparative analysis of four features reduction methods for PCG signals is presented. Experiments were performed on normal patients, and those with three different classes of heart disorders; namely, ASD, VSD and AS. Double discriminant embedding (DDE), feature space discriminant analysis (FSDA), clustering-based feature extraction (CBEF), and feature extracting using attraction points (FEUAP) were used with a KNN classifier. Table 1 presents a comparative summary of existing literature in terms of feature extraction and classification methods and the number of classes used in the experimentation.

In the present research, a novel method for PCG signal analysis for the detection and classification of congenital heart diseases is presented. Classification of ASD and VSD based on PCG signals is targeted using empirical mode decomposition (EMD) and a fusion of MFCC and temporal features. Specifically, a new feature fusion-based approach for the classification of ASD and VSD using PCG signal analysis is proposed. The classification performances of MFCCs and temporal features 1D local texture patterns (1D-LTPs) were individually evaluated and followed by the evaluation over the proposed fused feature representation. The proposed method was shown to be accurate, reliable, and robust due to comprehensive PCG signal representation with reduced features.

Table 1. Comparison with existing literature.

Ref	Year	Dataset	Classes	Features	Classifier	Results
[9]	2016	Physionet Challenge 2016 [39]	Normal(2488), Abnormal(665)	Time-frequency, Wavelet and statistical	LogitBoost, Random Forest	Acc: 84.48%
[11]	2016	Physionet Challenge 2016 [39]	Normal(2575), Abnormal(665)	Dynamic time warping	SVM	Acc: 82.4%
[15]	2016	Physionet Challenge 2016	Normal(2575), Abnormal(665)	124 Time-frequency features	Adaboost, CNN	Acc: 89%
[12]	2016	Self-collected	Normal(132), Abnormal seven classes(131)	Arash-Band	SVM	Acc: 87.45%
[36]	2017	Self-collected	Small VSD(60), Large VSD(60)	Statistical, DWT features	Multilayer Perceptron (MLP)	Acc: 96.6%
[35]	2017	Self-collected	Normal, VSD	(STFT), MFCC	KNN	Acc: 93.2%
[13]	2018	PhysioNet Computing in Cardiology Challenge	Normal(2575), Abnormal(665)	GFCC	Weighted SVM	Sen: 90.3% Spec: 89%
[17]	2018	UoC-murmur database, PhysioNet-2016	Normal(336), CHD(130), Normal/Abnormal(2435)	Mel-Spectrogram, MFFC and sub-band envelopes	CNN	Acc: 81.5% Sen: 84.5%
[10]	2018	PhysioNet Computing in Cardiology Challenge-2016	Normal(50), Abnormal(50)	Cepstrum Analysis	SVM	Acc: 95%
[38]	2018	Self-collected	Normal(40), Abnormal(58)	CBFE, FEUAP, FSDA, DDE	KNN	Acc: 84.39%
[32]	2019	Self-collected	Normal(175), Abnormal(108)	MFCC, normalized average Shannon energy	SVM	Acc: 92.6%
This work	**2020**	**Self-collected**	**Normal(140), Abnormal(140)**	**MFCC + 1D-LTPs**	**SVM**	**Acc: 95.63%**
This work	**2020**	**Self-collected**	**Normal(140), ASD(85), VSD(55)**	**MFCC + 1D-LTPs**	**SVM**	**Acc: 95.24%**

The rest of this article is organized as follows. Section 2 describes details about the data acquisition and the proposed methodology. Section 3 presents results of detection and multiclass experiments. A comparative analysis of this work with previous studies is presented in Section 4. In Section 5, conclusions of this research and future directions are described.

2. Materials and Methods

2.1. Overview

A PCG signal acquired using a stethoscope was digitized through an analog-to-digital converter. Signal preprocessing was performed on the acquired signal to remove possible noise and distortions. A data-driven approach known as empirical mode decomposition (EMD) was applied to denoise the signal. After preprocessing, feature extraction was performed to capture the most significant and decisive information from different classes of PCG signals. MFCC and temporal features were extracted and fused to better represent the signal. Finally, the support vector machine classifier was employed to distinguish different classes of PCG data. A sketch of the proposed system is presented in Figure 1.

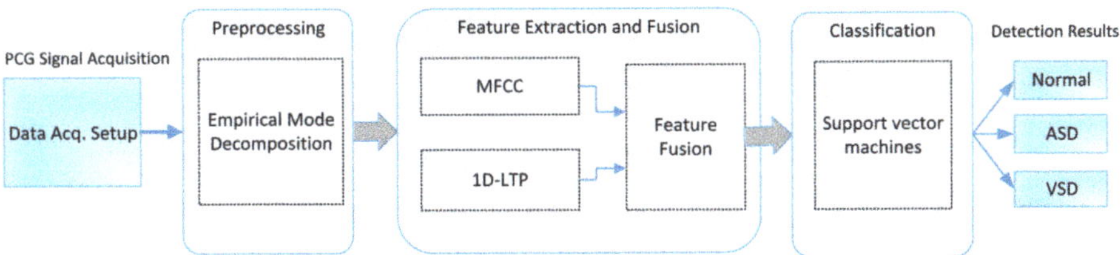

Figure 1. Sketch of the proposed cardiac disorder classification system.

2.2. Materials

One of the main challenges in studies related to the CHDs is the availability of respective PCG signals. There are several PCG signal datasets available [40,41], but they have following shortcomings.

1. The number of observations (signals) is limited.
2. Not recorded in a hospital environment.
3. Limited to two classes of data; namely, normal and abnormal.

Therefore, a new dataset of PCG signals was acquired that contains ASD, VSD, and normal data classes.

A self-built and low-cost data acquisition system (a microphone fitted in simple stethoscope) was utilized and connected with a computer for the acquisition of PCG signals in .wav format with 16-bit resolution and a sampling frequency of 44.1 kHz. PCG signal data were acquired by placing a stethoscope between the third and fourth left intercostal space. This site is best known for the detection of CHDs through auscultation.

PCG data were acquired from different patients admitted at Rawalpindi Institute of Cardiology, Rawalpindi, Pakistan; 85, 55, and 140 samples were collected from ASD, VSD, and normal subjects respectively. All recordings, each of five seconds, were taken in the hospital environment and under the supervision of an expert physician from the pulmonic, aortic, mitral, and tricuspid areas of the human heart. Labeling of the samples was done by an expert cardiologist who further validated through various tests of each participating subject. Table 2 provides a summary of the dataset according to each class, and examples of signals collected from normal, ASD, and VSD subjects are shown in Figure 2.

The reader may also be interested in the MATLAB codes of the newly developed feature extraction process [42]. However, it only provides experimental results on the PCG dataset comprised of the normal, ASD, and VSD classes.

Table 2. Description of PCG dataset.

Status	No. of Signals	No. of Subjects	Male	Female
Normal	140	28	17	11
ASD	85	17	12	5
VSD	55	11	7	4

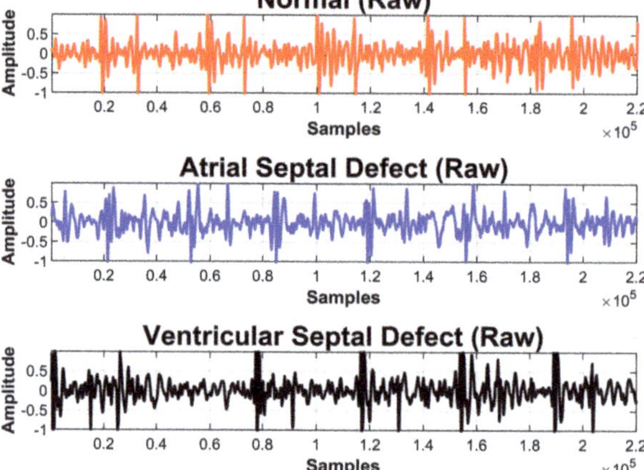

Figure 2. PCG signals collected from normal, arterial septal defect (ASD), and ventricular septal defect (VSD) subjects.

2.3. Preprocessing-Empirical Mode Decomposition

Acquired PCG signal gets corrupted due to embedded electronics, environmental noise, and other body organ artifacts. These noise elements suppress useful discriminative data associated with different classes of cardiac health and thus make the classification process more challenging. Signal denoising is a crucial preprocessing phase to obtain the unique region of interest for each data class, i.e., ASD, VSD, and normal. Empirical mode decomposition (EMD) [43–45] is a widely employed method in the domain of medical signal processing for denoising [46,47] and feature extraction [48,49]. EMD reduces the given data into a collection of subcomponents called intrinsic mode functions (IMFs). The process of IMF extraction is known as sifting. The original signal $q(t)$ can be expressed in terms of IMFs and residual signal $r(t)$ as follows:

$$q(t) = \sum_{k=1}^{N} h_k(t) + r(t) \tag{1}$$

where the number of extracted IMFs is represented by N and IMFs $h_k(t)$ are obtained from raw PCG signal $q(t)$ through an iterative process known as sifting. Major computing steps of the sifting process are listed below [50].

1. Calculate local minima and maxima from PCG signal $q(t)$.
2. Cubic spline interpolation is performed on local minima and maxima to form lower envelope $e_{min}(t)$ and upper envelope $e_{max}(t)$.
3. Calculate the mean of upper and lower envelopes as described by Equation (2).

$$a(t) = \frac{1}{2}\left(e_{\min}(t) + e_{\max}(t)\right) \tag{2}$$

4. Subtract $a(t)$ from the original signal $q(t)$ as:

$$y(t) = q(t) - a(t) \tag{3}$$

5. Repeat the steps (1)–(4) until the above mentioned two conditions of IMF are fulfilled.

Here, first, IMF is represented as $h_1(t) = y(t)$. Remaining IMFs from the residual signal are extracted as defined by Equation (4).

$$r_1(t) = q(t) - h_1(t) \tag{4}$$

To extract the remaining IMFs, $r_1(t)$ is now treated as a new signal and the sifting procedure is iteratively applied until a residual signal becomes monotonic functions. Figures 3–5 show IMFs extracted from PCG signals of normal, ASD, and VSD subjects. It was experimentally observed that the first and last two IMFs contain high-frequency noise and DC offset respectively. Therefore, they were subtracted from the remaining signal to acquire a good quality denoised signal represented by $x(t)$ as follows:

$$x(t) = \sum_{k=2}^{N-2} h_k(t) \tag{5}$$

Figure 6 illustrates the preprocessed signal $x(t)$ for normal, ASD, and VSD subjects.

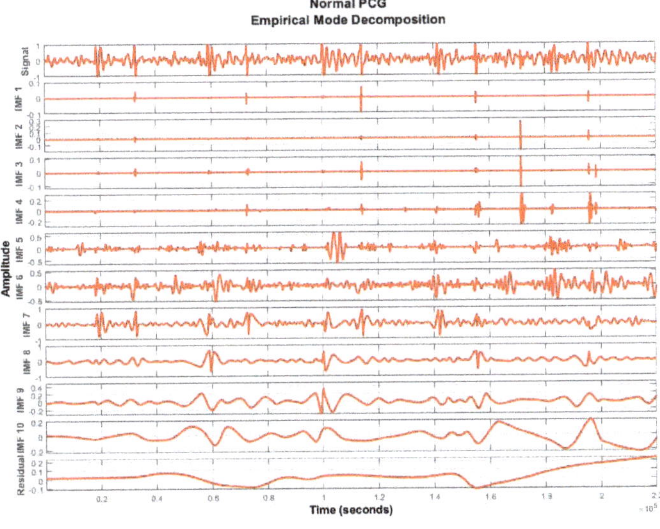

Figure 3. Intrinsic mode functions (IMFs) extracted from the PCG signal of a normal subject.

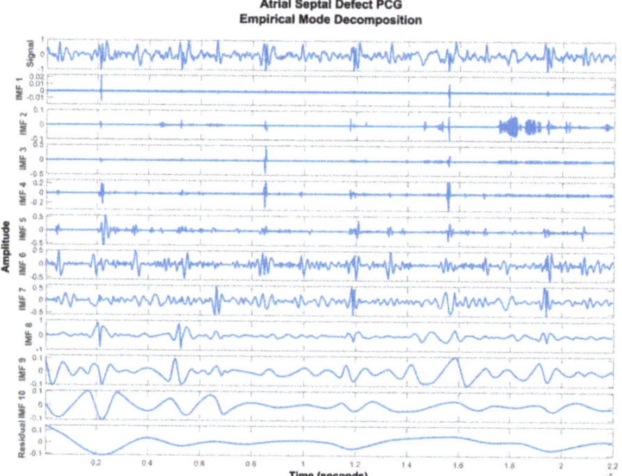

Figure 4. IMFs extracted from the PCG signal of an ASD subject.

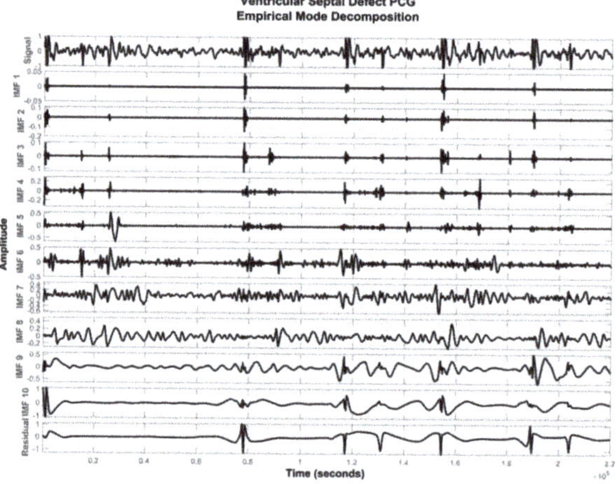

Figure 5. IMFs extracted from the PCG signal of a VSD subject.

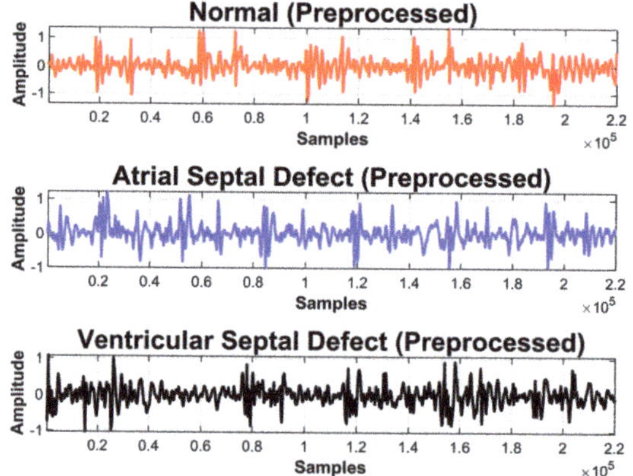

Figure 6. Preprocessed PCG signal of normal, ASD, and VSD subjects.

2.4. Feature Extraction

In this step, feature extraction was performed on the preprocessed PCG signal $x(t)$. Frequency-based features such as Mel-frequency cepstral coefficients (MFCCs) and temporal features 1D local texture patterns (1D-LTPs) were extracted. The final feature vector was constructed by fusion of these two feature sets to best represent the PCG signal data of different classes with minimum possible values.

2.4.1. 1D Local Ternary Patterns (1D-LTPs)

Local ternary patterns are an extended form of widely used temporal features known as local binary patterns [51] used extensively in the domain of computer vision [52–54]. One-dimensional local ternary patterns (1D-LTPs) are modified feature descriptors applied for signal processing applications [55–58]. Steps for extraction of 1D-LTP features are delineated in Figure 7.

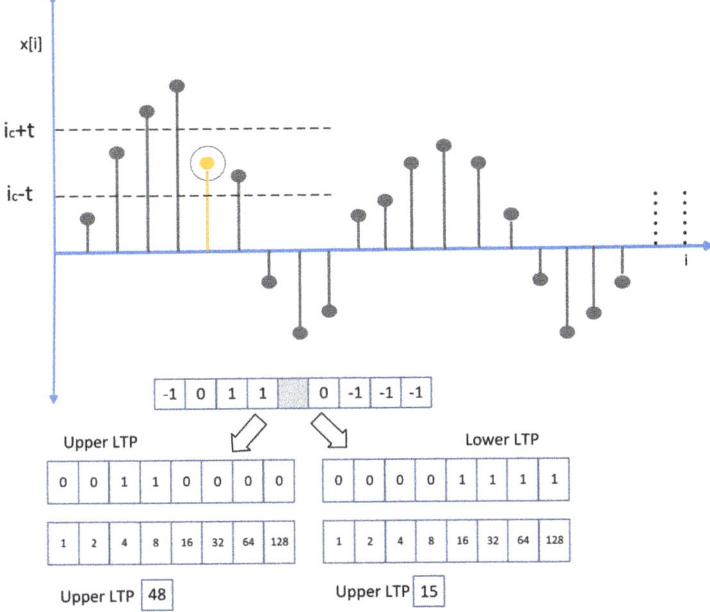

Figure 7. One-dimensional local ternary pattern (1D-LTP) feature extraction steps.

To extract 1D-LTP features from preprocessed signal $x(t)$, it is first divided into windows of size $W+1$. The center sample of each window is θ, the upper bound is $\theta+\phi$ and the lower bound is $\theta-\phi$. Each window of size $W+1$ is divided into left and right equal-sized frames around center sample $x[i]$.

$$F(x_i,\theta,\phi)) = \begin{cases} +1, & x_i-(\theta+\phi)\geq 0 \\ 0, & (\theta+\phi)<x_i<(\theta-\phi) \\ -1 & x_i-(\theta-\phi)\leq 0 \end{cases} \tag{6}$$

The $F(.)$ is the three-valued vector output having values +1, 0 and -1. $F(.)$ is split into upper and lower patterns using Equations (7) and (9).

$$LTP_{upper} = \sum_{p=1}^{8} S_u\left(F(p)\right).2^p \tag{7}$$

$$S_u = \begin{cases} 1, & \text{if } F(p) = 1 \\ 0, & \text{otherwise.} \end{cases} \tag{8}$$

$$LTP_{lower} = \sum_{p=1}^{8} S_l\left(F(p)\right).2^p \tag{9}$$

$$S_l = \begin{cases} 1, & \text{if } F(p) = -1 \\ 0, & \text{otherwise.} \end{cases} \tag{10}$$

LTP_{upper} is calculated by using Equation (8) and LTP_{lower} is computed from Equation (10). LTP_{upper} and LTP_{lower} were the resultant LTP feature vectors extracted from the PCG signal.

2.4.2. Mel Frequency Cepstral Coefficients (MFCC)

Mel-frequency cepstral coefficients (MFCCs), a well-known group of features for speech/speaker recognition systems, have recently gained importance as features for classifying heart sounds [26,32,59,60]. Mel frequencies are grounded in the nonlinear physiognomies of the human ear's sensitivity to different frequencies [61]. MEL frequency is related to linear frequency in Equation (11).

$$Mel(f) = 2595\log_{10}\left(1 + \frac{f}{700}\right) \tag{11}$$

The process of MFCCs' calculation is shown in Figure 8. The preprocessed PCG signal is pre-weighted to improve the signal to noise ratio. In a frame blocking stage, the segmented PCG signals are blocked into frames using a window length of 30 ms with a 20 ms window overlapping. For a sampling frequency of 44.1 kHz, a hamming window of length 1323 samples was chosen to avoid the parasitic spectral leakage. Fast Fourier transform (FFT) is applied to segmented PCG signals to transform each frame to its frequency domain version. The frequency-domain segmented PCG array is filtered by a group of band-pass Mel triangular filters and transformed into the Mel inverse spectrum domain. The logarithm of Mel spectrum coefficients from each Mel filter is used to compress the higher band of the PCG signal. In the final stage, the logarithmic Mel spectrum coefficients are transformed using the discrete cosine transform (DCT) illustrated in Equation (12).

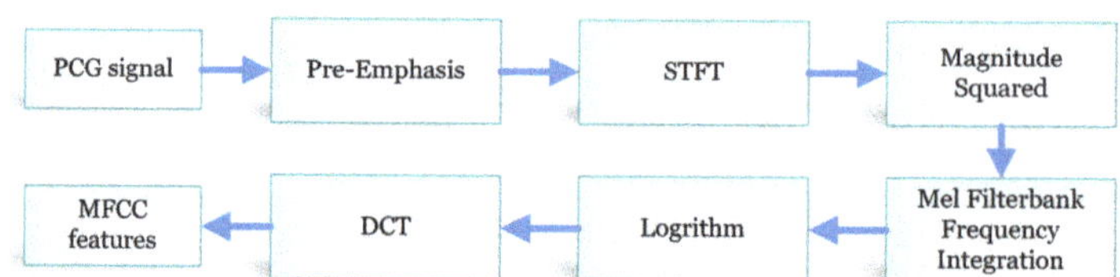

Figure 8. The process of mel-frequency cepstral coefficient (MFCC) feature extraction.

$$c[n] = \sum_{m=0}^{N-1} S[m] \cos\left(\frac{\pi n}{M}\left(m - \frac{1}{2}\right)\right), \quad n = 0, 1, 2..., M \tag{12}$$

where M is the total number of filter banks. For this study, 13 MFCCs were extracted from denoised heart sound.

2.5. Feature Fusion

MFCC and 1D-LTP features extracted in previous steps were fused to construct a joint feature vector having dimensions of 1×33. A combination of temporal and frequency features helps in extracting more discriminant information embedded in the PCG signal about heart disorders. Feature fusion is realized through a simple serial concatenation of MFCC and 1D-LTP features.

2.6. Classification—Support Vector Machines

The final feature vector from the PCG signal consists of a total of 33 features (20 LTPs + 13 MFCC). Features are extracted from each class (normal, ASD, VSD). The SVM classifier is a widely applied method of classification for biomedical signals [62–65] due to its excellent generalization capability. It obtains the optimal separating hyperplane for class separation by converting input features to higher dimensions through some nonlinear mapping [66]. The distance between patterns and the hyperplane is maximized using a maximum margin principle to get the best separation. Kernel functions, such as quadratic, cubic, and Gaussian ones, are used for mapping the data into higher dimensional space. Table 3 presents the parameters of classifiers used during training/testing. In this study, SVM was used in two different settings: (1) Binary SVM where input PCG features were labeled as "normal" and "abnormal." (2) Multiclass experiments where input PCG features were labeled as "normal" or according to the disease type; i.e., ASD or VSD.

Table 3. Parameters of selected classifiers.

Classifier	Kernel Function	Kernel Scale	Box Constraint Level	Multiclass Method	Standardize Data
SVM-L	Linear	Automatic	1	One-vs-one	True
SVM-Q	Quadratic	Automatic	1	One-vs-one	True
SVM-C	Cubic	Automatic	1	One-vs-one	True
SVM-G	Gaussian	44	1	One-vs-one	True

3. Results

In this study, an automated heart disease classification system using the PCG signal is proposed. Raw PCG signal was first preprocessed through EMD, followed by feature extraction through the fusion of MFCC and 1D-LTP features. 1D-LTPs extract the most discriminative information embedded in the PCG signal. Distribution of 1D-LTP features of different classes (normal/ASD/VSD) can be visualized from scatter plots shown (Figure 9). It can be observed that the intra-class difference between features is minimal, while the inter-class difference is maximal. This shows that the extracted features contain generous decisive information about different classes of PCG signals.

The performance of the proposed method was evaluated using standard statistical indices of accuracy, sensitivity (sen), and specificity (spec), which were calculated from the following four parameters

- True positive (TP): abnormal PCG signal correctly detected as abnormal.
- False negative (FN): PCG signal of an abnormal subject detected as normal.
- True negative (TN): normal PCG signal correctly detected as normal.
- False positive (FP): PCG signal of a normal subject detected as abnormal.

$$Accuracy = \frac{TP + TN}{(TP + TN + FP + FN)} \times 100 \tag{13}$$

$$Sen = \frac{TP}{(TP + FN)} \times 100 \tag{14}$$

$$Spec = \frac{TN}{(TN + FP)} \times 100 \tag{15}$$

In this study, the experiments were performed for two different problems.

1. Detection experiment (normal vs. abnormal): All feature vectors belonging to abnormal subjects (ASD, VSD) were labeled as abnormal.

2. Multiclass evaluation (normal vs. ASD vs. VSD): Feature data were labeled according to the disease type in the experiment.

Training and testing of classifiers were pursued through a 10-fold cross-validation method with each subset of features; i.e., MFFC, 1D-LTPs, and fusion of MFCC+1D-LTP. All simulations were performed in MATLAB 2018a on the core i5 computer. All results presented in this paper were averaged over 100 experiments.

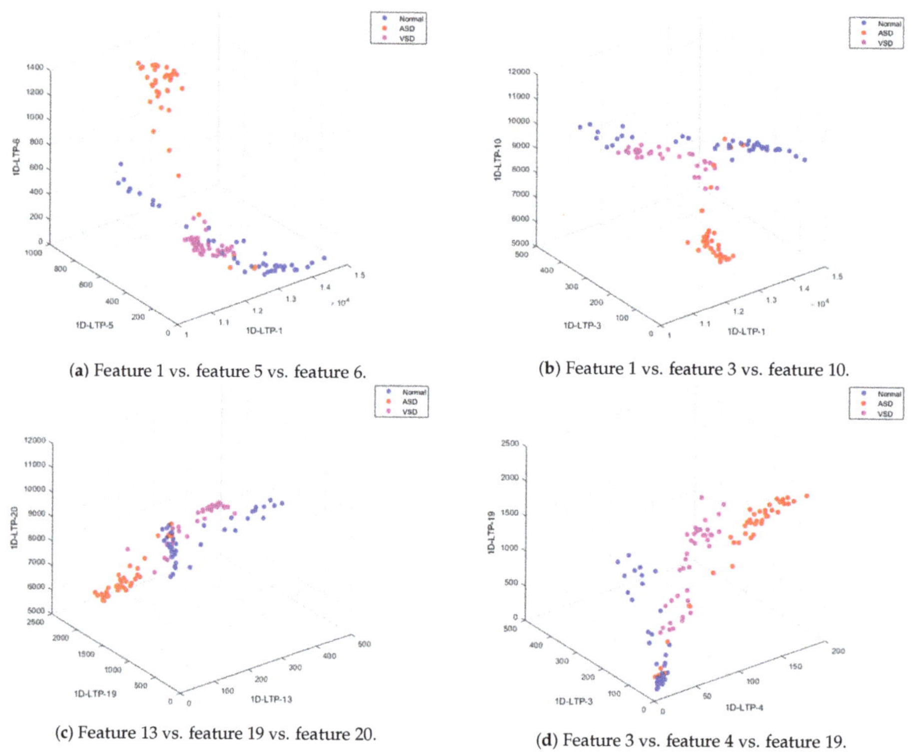

(**a**) Feature 1 vs. feature 5 vs. feature 6.

(**b**) Feature 1 vs. feature 3 vs. feature 10.

(**c**) Feature 13 vs. feature 19 vs. feature 20.

(**d**) Feature 3 vs. feature 4 vs. feature 19.

Figure 9. Scatter plots of 1D-LTP features.

3.1. Detection Experiment

The experiments for the detection of normal and abnormal subjects were performed on the self-collected dataset using a low-cost data acquisition setup. In detection experiments, the dataset was split into two classes; namely, normal and abnormal. All features vectors belonging to ASD and VSD patients were labeled as abnormal. An SVM classifier with different kernel functions, such as SVM-linear (SVM-L), SVM-quadratic (SVM-Q), SVM-cubic (SVM-C), and SVM-Gaussian (SVM-G), was employed to perform classification. The results of these experiments in terms of accuracy, sensitivity, specificity, positive predictive value (PPV), negative predictive value (NPV), and error rate are illustrated in Table 4. Results of applying individual feature sets (MFCC and 1D-LTP) on PCG signal data are also presented (Table 4). The highest results using only MFCC features were achieved through SVM-C (94.05%); 1D-LTP-only feature extraction achieved the highest accuracy of 94.05% with the SVM-Q classifier. The best results of 95.8% accuracy with SVM-C classifiers were acquired upon feature fusion of MFCCs and 1D-LTPs. Table 5 illustrates the confusion matrix showing individual class accuracy with SVM-C and a combination of MFCC and 1D-LTP features. It was evident from experimentation that the fusion of MFCC and 1D-LTP features provide a significant improvement in classification performance.

Table 4. Performance comparison of SVM on different feature sets for binary experiments. Bold font indicates the best result obtained against each feature set.

Feature Set	Classifier	Accuracy (%)	Sensitivity (%)	Specificity (%)	PPV (%)	NPV (%)	Error (%)
MFCC	SVM-L	89.88	76.19	94.44	82.05	92.25	10.12
	SVM-Q	89.29	80.95	92.06	77.27	93.55	10.71
	SVM-C	**92.26**	**88.1**	**93.65**	**82.22**	**95.93**	**7.74**
	SVM-G	75.6	7.14	98.41	60	76.07	24.4
1D-LTP	SVM-L	94.05	88.1	96.03	88.1	96.03	5.95
	SVM-Q	**94.05**	**83.33**	**97.62**	**92.11**	**94.62**	**5.95**
	SVM-C	91.07	76.19	96.03	86.49	92.37	8.93
	SVM-G	86.31	47.62	99.21	95.24	85.03	13.69
MFCC+1D-LTP	SVM-L	94.05	90.48	95.24	86.36	96.77	5.95
	SVM-Q	94.05	88.1	96.03	88.1	96.03	5.95
	SVM-C	**95.83**	**92.86**	**96.83**	**90.7**	**97.6**	**4.17**
	SVM-G	93.45	88.1	95.24	86.05	96	6.55

Table 5. Confusion matrix for detection (normal vs. abnormal) experiments.

	Predicted Class	
Actual Class	**Normal**	**Abnormal**
Normal	90%	10%
Abnormal	2%	98%

3.2. Multiclass Evaluation (Normal vs. ASD vs. VSD)

Multiclass experiments were performed to precisely identify the type of heart disorder. Features were labeled according to the disorder type; i.e., ASD, VSD, or normal. A multiclass SVM with different kernels was trained and tested using 10-fold cross-validation. The results of applying different multiclass SVM classifiers on individual feature sets (MFCC, 1D-LTP) and the fusions of both are illustrated in Table 6. The obtained results revealed that the SVM-C classifier achieved a peak accuracy of 88.69% with only MFCC features, while the same classifier provided 94.64% accuracy with 1D-LTP features. Performance results were further improved by the fusion of MFCC and 1D-LTP features with the SVM-C classifier; i.e., 95.24% accuracy. In Table 7, class-wise information of accuracy for ASD, VSD, and normal classes in the form of a confusion matrix with the SVM-C classifier are shown. The proposed feature fusion methodology effectively extracted the characteristic information from multiclass PCG signals.

Table 6. Performance comparison of SVM using different feature sets for multiclass experiments. Bold font indicates the best result obtained against each feature set.

Feature set	Classifier	Accuracy(%)	Sensitivity(%)	Specificity(%)	PPV(%)	NPV(%)	Error(%)
MFCC	SVM-L	83.93	92.86	85.71	68.42	97.3	16.07
	SVM-Q	86.9	90.48	90.48	76	96.61	13.1
	SVM-C	**88.69**	**90.48**	**94.44**	**84.44**	**96.75**	**11.31**
	SVM-G	83.33	97.62	81.75	64.06	99.04	16.67
1D-LTP	SVM-L	94.64	97.62	93.65	83.67	99.16	5.36
	SVM-Q	94.05	90.48	95.24	86.36	96.77	5.95
	SVM-C	**94.64**	**90.48**	**96.03**	**88.37**	**96.8**	**5.36**
	SVM-G	93.45	92.86	93.65	82.98	97.52	6.55
MFCC+1D-LTP	SVM-L	93.45	97.62	92.06	80.39	99.15	6.55
	SVM-Q	94.43	95.05	94.41	85.06	98.28	5.57
	SVM-C	**95.24**	**95.24**	**95.24**	**86.96**	**98.36**	**4.76**
	SVM-G	93.45	100	91.27	79.25	100	6.55

Table 7. Confusion matrix for multiclass experiments.

	Predicted Class		
Actual Class	**Normal**	**ASD**	**VSD**
Normal	90%	10%	0%
ASD	6%	94%	0%
VSD	0%	0%	100%

3.3. Statistical Significance

The primary objective behind performing this statistical analysis was to achieve a certain level of confidence in the proposed scheme. Analysis of variance (ANOVA) [67] was utilized to testify whether the results were statistically significant or not—simply by comparing the means of multiple distributions.

In this work, a proposed scenario (MFCC + 1D-LTP) was considered for two different classifiers (SVM-C, SVM-Q)—selected based on the improved performance compared to the rest. In using ANOVA, a series of tests were performed for the assumptions of normality and homogeneity of variance. A Shapiro–Wilk test [68] was performed for the former, and the Bartletts test [68] for the latter one—with the significance level α selected to be 0.01. The means of our approach were $\bar{x_1}$,$\bar{x_2}$, calculated from the overall accuracy of both classifiers. The null hypothesis H_0, given that $\bar{x_1} = \bar{x_2}$, while the alternative hypothesis H_a given that $\bar{x_1} \neq \bar{x_2}$. The p-value was computed and the null hypothesis was tested, H_0; if it was rejected, $p < \alpha$, then the Bonferroni posthoc test was applied.

For the proposed method (MFCC + 1D-LTP), and with selected classifiers (SVM-C and SVM-Q), the Shapiro–Wilk test generated p-value, $p_c = 0.6987$, and $p_q = 0.9352$. By following the Bartletts test, the associated chi-squared probabilities were: $p_c = 0.712$ and $p_q = 0.312$. The p-values of two different classifiers are significantly greater than α. Therefore, from the test results (normality and equality of variances), we failed to repudiate the null hypothesis H_0, and we are confident in claiming that the test data were normally distributed, and the variances were also homogeneous. The ANOVA test, including five different parameters (degrees of freedom (dfs), a sum of squared deviation (SS), mean squared error (MSE), F-statistics, and p-value) is shown in Table 8. The performance ranges of two selected classifiers based on the proposed method are shown in Figure 10.

Table 8. ANOVA test on two selected classifiers based on the proposed method.

Variance Source	SS	df	MSE	F-Statistics	p-Value
Between	1.8482	1	1.84815	0.63	0.4721
Within	11.7503	4	2.93758	-	-
Total	13.5985	5	-	-	-

The results were validated based on the Bonferroni post hoc test, Figure 11, which is the most common approach to be applied whenever there exists a chance of a significant difference between the means of multiple distributions. It was certified that the proposed method performed much better than conventional methods.

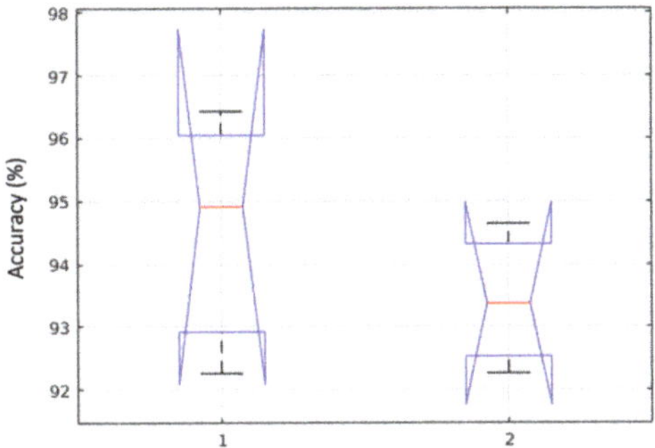

Figure 10. Box-plot of accuracy values for selected classifiers (1:SVM-C, 2:SVM-Q).

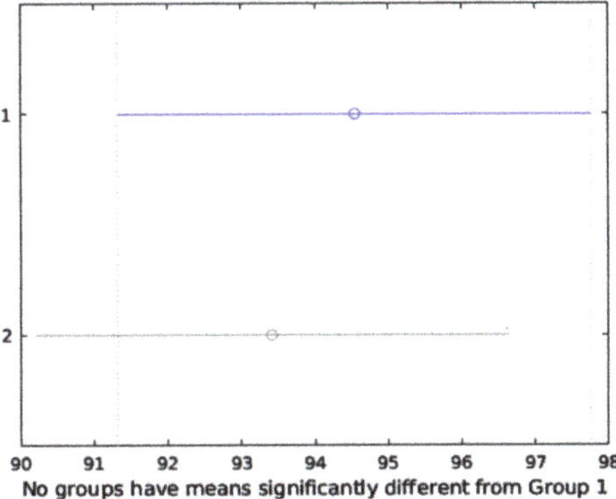

Figure 11. The means of both classifiers belong to a single group and are not significantly different.

4. Discussion

The proposed method of feature fusion with EMD-based signal denoising effectively extracted embedded information from PCG signals using the self-collected dataset of ASD and VSD cardiac disorders. The MFCC extracted frequency-domain features, while 1D-LTP features extracted temporal and texture information from the signal. Feature fusion of these two different types provided a powerful signal representation for different classes (normal, ASD, VSD) with a high degree of accuracy. Moreover, the proposed method classified normal and abnormal PCG data through SVM-C classifier with 95.83% accuracy, while 95.34% average accuracy was achieved on multiclass PCG data with the same classifier.

The numbers of classes, feature extraction techniques and classification methods of the proposed method were compared with the previously developed platforms (Table 1), which showed that several existing works [9–11,13,15,17] utilized the Physionet Challenge 2016 dataset [69] comprised of only two classes (healthy and unhealthy) while others used self-collected PCG signal data. MFCCs were widely employed by several studies [9,11,17,35], and acted as baseline features of choice. The SVM classifier is also widely adopted by existing works [10–13].

DWT and statistical features were used with a multilayer perceptron to achieve 96.6% accuracy on normal and ASD classes of PCG data [36]. In another work [38], a comparison of feature reduction methods was demonstrated. Experimental results are shown between normal and three different classes of heart diseases; i.e., ASD, VSD, and aortic stenosis. Feature reduction methods (DDE, FSDA, CBEF, EFUAP) were applied with K-nearest neighbor (KNN) classifier and 84.3% accuracy was achieved.

In contrast to the existing work, our research targeted the classification of multiple heart disorders (ASD, VSD) with the feature fusion approach of MFCC and new temporal feature descriptor 1D-LTP. The proposed method outperforms the existing approaches, as is evident from the presented results. To confirm the validity and robustness of our proposed method, confidence intervals against binary and multiclass experiments are also provided for the two best classifiers; i.e., SVM-C and SVM-Q. Figure 12a illustrates the confidence interval showing maximum, minimum, and average classification results of individual MFCC and 1D-LTP features and the feature fusion approach for binary experiments. Figure 12b presents a confidence interval of minimum, maximum, and average classification accuracy for multiclass experiments. From this comprehensive statistical analysis, it is quite straightforward to choose SVM-C as a standard classifier for this application.

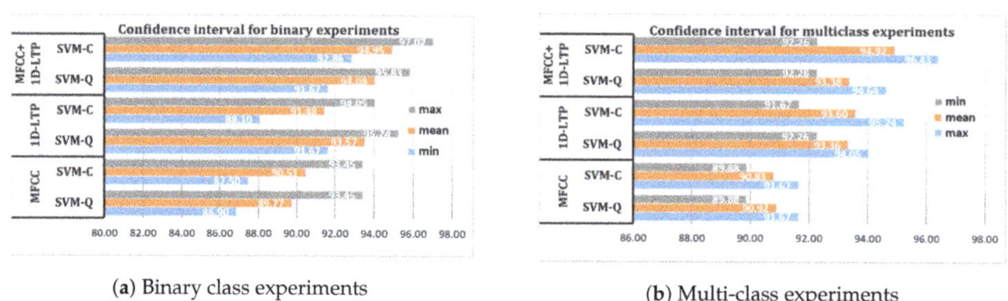

(**a**) Binary class experiments (**b**) Multi-class experiments

Figure 12. Confidence interval simulation results.

5. Conclusions

Preprocessing and classification of heart sounds is a challenging problem due to the addition of environmental noise. The addition of noise may hide the actual class information in the PCG signal. In this study, an effective classification framework was developed for the diagnosis of ASD, VSD, and normal subjects through PCG signal analysis. A feature fusion approach using novel 1D-LTP features along with strong MFCC features has shown to be an effective strategy exhibiting good discriminative properties of representing PCG signals. The proposed method was validated through different SVM kernels, and the best performance was achieved with SVM-C. The main findings of this research are the following:

- The proposed framework is non-invasive and reliable.
- The proposed scheme is independent of the morphological characteristics of the acquired PCG signal.
- This research introduces a new feature descriptor, i.e., 1D-LTP, that significantly improves the classification performance upon fusion with classical MFCCs.
- The proposed method is fully automated and works with all kinds of noisy PCG signals due to the adoption of a data-driven preprocessing approach; i.e., EMD.

This research has the following shortcomings:

- The dataset used is small in size.
- The selection of proper IMFs in EMD is not automated.

The proposed method for cardiac disorders can be enhanced by adding more data samples of PCG. In the future, we aim to apply feature reduction and fusion algorithms to further reduce the feature vector dimensions and increase system accuracy.

6. Compliance with Ethical Standards

6.1. Ethical Approval

All procedures performed in studies involving human participants were in accordance with the ethical standards of the institutional and/or national research committee and with the 1964 Helsinki declaration and its later amendments or comparable ethical standards.

6.2. Informed Consent

Informed consent was obtained from all individual participants included in the study.

Author Contributions: Conceptualization, S.A. and M.U.K.; methodology, S.A.; software, M.U.K.; validation, T.A. and M.A. (Majed Alhaisoni); formal analysis, T.A. and M.A.(Majed Alhaisoni); investigation, M.A. (Muhammad Altaf); writing—original draft preparation, S.A. and M.U.K.; writing—review and editing, M.A. (Muhammad Altaf); funding acquisition, M.A. (Majed Alhaisoni) All authors read and approved the final manuscript.

Funding: Research was funded by Deanship of Scientific Research at University of Ha'il.

Acknowledgments: Research was funded by Deanship of Scientific Research at University of Ha'il.

Conflicts of Interest: The authors declare no conflict of interest.

References

1. Reller, M.D.; Strickland, M.J.; Riehle-Colarusso, T.; Mahle, W.T.; Correa, A. Prevalence of congenital heart defects in metropolitan Atlanta, 1998–2005. *J. Pediatr.* **2008**, *153*, 807–813.
2. Wang, T.; Chen, L.; Yang, T.; Huang, P.; Wang, L.; Zhao, L.; Zhang, S.; Ye, Z.; Chen, L.; Zheng, Z.; et al. Congenital Heart Disease and Risk of Cardiovascular Disease: A Meta-Analysis of Cohort Studies. *J. Am. Heart Assoc.* **2019**, *8*, e012030.
3. Shafi, T.; Khan, M.; Atiq, M. Congenital heart disease and associated malformations in children with cleft lip and palate in Pakistan. *Br. J. Plast. Surg.* **2003**, *56*, 106–109.
4. Cuypers, J.J. *The Unnatural History of Congenital Heart Disease*; Optima Grafische Communicatie: Rotterdam, The Netherlands, 2015.
5. Venkatesan, C.; Karthigaikumar, P.; Paul, A.; Satheeskumaran, S.; Kumar, R.J.I.A. ECG Signal Preprocessing and SVM Classifier-Based Abnormality Detection in Remote Healthcare Applications. *IEEE Access* **2018**, *6*, 9767–9773.
6. Reed, T.R.; Reed, N.E.; Fritzson, P. Heart sound analysis for symptom detection and computer-aided diagnosis. *Simul. Model. Pract. Theory* **2004**, *12*, 129–146.
7. Ching, Y.H.; Ghosh, T.K.; Cross, S.J.; Packham, E.A.; Honeyman, L.; Loughna, S.; Robinson, T.E.; Dearlove, A.M.; Ribas, G.; Bonser, A.J.; et al. Mutation in myosin heavy chain 6 causes atrial septal defect. *Nat. Genet.* **2005**. *37*, 423–428.
8. Cuypers, J. The Unnatural History of Congenital Heart Disease: Outcome up to 40 Years after Surgical Repair in Childhood. Ph.D. Thesis, University Medical Center Rotterdam, Rotterdam, The Netherlands, 2015.
9. Homsi, M.N.; Medina, N.; Hernandez, M.; Quintero, N.; Perpiñan, G.; Quintana, A.; Warrick, P. Automatic heart sound recording classification using a nested set of ensemble algorithms. In Proceedings of the 2016 Computing in Cardiology Conference (CinC), Vancouver, BC, Canada, 11–14 September 2016; pp. 817–820.
10. Yadav, A.; Dutta, M.K.; Travieso, C.M.; Alonso, J.B. Automatic Classification of Normal and Abnormal PCG Recording Heart Sound Recording Using Fourier Transform. In Proceedings of the 2018 IEEE International Work Conference on Bioinspired Intelligence (IWOBI), San Carlos, Costa Rica, 18–20 July 2018; pp. 1–9.
11. Ortiz, J.J.G.; Phoo, C.P.; Wiens, J. Heart sound classification based on temporal alignment techniques. In Proceedings of the 2016 Computing in Cardiology Conference (CinC), Vancouver, BC, Canada, 11–14 September 2016; pp. 589–592.
12. Sepehri, A.A.; Kocharian, A.; Janani, A.; Gharehbaghi, A. An intelligent phonocardiography for automated screening of pediatric heart diseases. *J. Med. Syst.* **2016**, *40*, 16.

13. Alexander, B.; Nallathambi, G.; Selvaraj, N. Screening of Heart Sounds Using Hidden Markov and Gammatone Filterbank Models. In Proceedings of the 2018 17th IEEE International Conference on Machine Learning and Applications (ICMLA), Orlando, FL, USA, 17–20 December 2018; pp. 1460–1465.
14. Rubin, J.; Abreu, R.; Ganguli, A.; Nelaturi, S.; Matei, I.; Sricharan, K. Recognizing abnormal heart sounds using deep learning. *arXiv* **2017**, arXiv:1707.04642.
15. Potes, C.; Parvaneh, S.; Rahman, A.; Conroy, B. Ensemble of feature-based and deep learning-based classifiers for detection of abnormal heart sounds. In Proceedings of the 2016 Computing in Cardiology Conference (CinC), Vancouver, BC, Canada, 11–14 September 2016; pp. 621–624.
16. Ryu, H.; Park, J.; Shin, H. Classification of heart sound recordings using convolution neural network. In Proceedings of the 2016 Computing in Cardiology Conference (CinC), Vancouver, BC, Canada, 11–14 September 2016; pp. 1153–1156.
17. Bozkurt, B.; Germanakis, I.; Stylianou, Y. A study of time-frequency features for CNN-based automatic heart sound classification for pathology detection. *Comput. Biol. Med.* **2018**, *100*, 132–143.
18. Hung, T.; Chou, C.; Fang, W.; Li, A.H.; Chang, Y.; Hwang, B.; Shau, Y. Time-frequency analysis of heart sound signals based on Hilbert-Huang Transformation. In Proceedings of the 2012 IEEE 16th International Symposium on Consumer Electronics, Harrisburg, PA, USA, 4–6 June 2012; pp. 1–3. doi:10.1109/ISCE.2012.6241745.
19. Sharma, R.R.; Pachori, R.B. A new method for non-stationary signal analysis using eigenvalue decomposition of the Hankel matrix and Hilbert transform. In Proceedings of the 2017 4th International Conference on Signal Processing and Integrated Networks (SPIN), Noida, India, 2–3 February 2017; pp. 484–488. doi:10.1109/SPIN.2017.8049998.
20. Tu, Z.; Cao, G.; Li, Q.; Zhang, X.; Shi, J. Improved methods for detecting main components of heart sounds. In Proceedings of the 2010 Sixth International Conference on Natural Computation, Yantai, China, 10–12 August 2010; Volume 7, pp. 3585–3588.
21. Schmidt, S.E.; Holst-Hansen, C.; Hansen, J.; Toft, E.; Struijk, J.J. Acoustic Features for the Identification of Coronary Artery Disease. *IEEE Trans. Biomed. Eng.* **2015**, *62*, 2611–2619. doi:10.1109/TBME.2015.2432129.
22. Kumar, A.K.; Saha, G. Improved computerized cardiac auscultation by discarding artifact contaminated PCG signal sub-sequence. *Biomed. Signal Process. Control* **2018**, *41*, 48–62. doi:10.1016/j.bspc.2017.11.001.
23. Patidar, S.; Pachori, R.B. Constrained Tunable-Q Wavelet Transform based Analysis of Cardiac Sound Signals. *AASRI Procedia* **2013**, *4*, 57–63. doi:10.1016/j.aasri.2013.10.010.
24. Patidar, S.; Pachori, R.B.; Rajendra Acharya, U. Automated diagnosis of coronary artery disease using tunable-Q wavelet transform applied on heart rate signals. *Knowl.-Based Syst.* **2015**, *82*, 1–10. doi:10.1016/j.knosys.2015.02.011.
25. Patidar, S.; Pachori, R.B. Classification of cardiac sound signals using constrained tunable-Q wavelet transform. *Expert Syst. Appl.* **2014**, *41*, 7161–7170. doi:10.1016/j.eswa.2014.05.052.
26. Hamidi, M.; Ghassemian, H.; Imani, M. Classification of heart sound signal using curve fitting and fractal dimension. *Biomed. Signal Process. Control* **2018**, *39*, 351–359. doi:10.1016/j.bspc.2017.08.002.
27. Ali, S.; Adnan, S.; Nawaz, T.; Ullah, M.O.; Aziz, S. Human Heart Sounds Classification using Ensemble Methods. *Univ. Eng. Technol. Taxila Tech. J.* **2017**, *22*, 113.
28. Ari, S.; Hembram, K.; Saha, G. Detection of cardiac abnormality from PCG signal using LMS based least square SVM classifier. *Expert Syst. Appl.* **2010**, *37*, 8019–8026.
29. Sun, S.; Wang, H. Principal component analysis-based features generation combined with ellipse models-based classification criterion for a ventricular septal defect diagnosis system. *Austral. Phys. Eng. Sci. Med.* **2018**, *41*, 821–836.
30. Visagie, C.; Scheffer, C.; Lubbe, W.; Doubell, A.F. Autonomous detection of heart sound abnormalities using an auscultation jacket. *Austral. Phys. Eng. Sci. Med.* **2009**, *32*, 240–250.
31. Botha, J.; Scheffer, C.; Lubbe, W.; Doubell, A.F. Autonomous auscultation of the human heart employing a precordial electro-phonocardiogram and ensemble empirical mode decomposition. *Austral. Phys. Eng. Sci. Med.* **2010**, *33*, 171–183.
32. Ahmad, M.S.; Mir, J.; Ullah, M.O.; Shahid, M.L.U.R.; Syed, M.A. An efficient heart murmur recognition and cardiovascular disorders classification system. *Austral. Phys. Eng. Sci. Med.* **2019**, *42*, 733–743.

33. Li, H.; Wang, X.; Liu, C.; Wang, Y.; Li, P.; Tang, H.; Yao, L.; Zhang, H. Dual-Input Neural Network Integrating Feature Extraction and Deep Learning for Coronary Artery Disease Detection Using Electrocardiogram and Phonocardiogram. *IEEE Access* **2019**, *7*, 146457–146469.
34. Gjoreski, M.; Gradišek, A.; Budna, B.; Gams, M.; Poglajen, G. Machine Learning and End-to-End Deep Learning for the Detection of Chronic Heart Failure From Heart Sounds. *IEEE Access* **2020**, *8*, 20313–20324.
35. Ghaffari, M.; Ashourian, M.; İnce, E.A.; Demirel, H. Phonocardiography signal processing for automatic diagnosis of ventricular septal defect in newborns and children. In Proceedings of the 2017 9th International Conference on Computational Intelligence and Communication Networks (CICN), Girne, Cyprus, 16–17 September 2017; pp. 62–66.
36. Hassani, K.; Jafarian, K.; Doyle, D.J. Heart Sounds Features Usage for Classification of Ventricular Septal Defect Size in Children. In Proceedings of the 16th International Conference on Biomedical Engineering, Singapore, 7–10 December 2016; Springer: Singapore, 2016; pp. 28–31.
37. Son, G.Y.; Kwon, S. Classification of Heart Sound Signal Using Multiple Features. *Appl. Sci.* **2018**, *8*, 2344.
38. Imani, M.; Ghassemian, H. Phonocardiography Signal Classification by Applying Feature Space Transformations. In Proceedings of the 2018 9th International Symposium on Telecommunications (IST), Tehran, Iran, 17–19 December 2018; pp. 118–122.
39. Liu, C.; Springer, D.; Li, Q.; Moody, B.; Juan, R.A.; Chorro, F.J.; Castells, F.; Roig, J.M.; Silva, I.; Johnson, A.E. An open access database for the evaluation of heart sound algorithms. *Physiol. Meas.* **2016**, *37*, 2181.
40. Egeneral Medical Heart Murmur Database. 2020. Available online: http://www.egeneralmedical.com/listohearmur.html (accessed on: 23 October 2019).
41. Heart Sound. Murmur Library. University of Michigan. 2020. Available online: http://www.med.umich.edu/lrc/psb_open/html/repo/primer_heartsound/primer_heartsound.html (accessed on 23 October 2019).
42. Sumair, U. MATLAB Codes for Research. 2020. Available online: https://sites.google.com/view/seeduet/home/projects/pcg-project?authuser=0 (accessed on: 30 June 2020).
43. Pal, S.; Mitra, M. Empirical mode decomposition based ECG enhancement and QRS detection. *Comput. Biol. Med.* **2012**, *42*, 83–92.
44. Khan, M.U.; Aziz, S.; Bilal, M.; Aamir, M.B. Classification of EMG Signals for Assessment of Neuromuscular Disorder using Empirical Mode Decomposition and Logistic Regression. In Proceedings of the 2019 International Conference on Applied and Engineering Mathematics (ICAEM), Taxila, Pakistan, 27–29 August 2019; pp. 237–243.
45. Vargas-Lopez, O.; Amezquita-Sanchez, J.P.; De-Santiago-Perez, J.J.; Rivera-Guillen, J.R.; Valtierra-Rodriguez, M.; Toledano-Ayala, M.; Perez-Ramirez, C.A. A New Methodology Based on EMD and Nonlinear Measurements for Sudden Cardiac Death Detection. *Sensors* **2020**, *20*, 9.
46. Seera, M.; Lim, C.P.; Liew, W.S.; Lim, E.; Loo, C.K. Classification of electrocardiogram and auscultatory blood pressure signals using machine learning models. *Expert Syst. Appl.* **2015**, *42*, 3643–3652.
47. Khan, M.U.; Aziz, S.; Iqtidar, K.; Zainab, A.; Saud, A. Prediction of Acute Coronary Syndrome Using Pulse Plethysmograph. In Proceedings of the 2019 4th International Conference on Emerging Trends in Engineering, Sciences and Technology (ICEEST), Karachi, Pakistan, 10–11 December 2019; pp. 1–6.
48. Cheema, A.; Singh, M. An application of phonocardiography signals for psychological stress detection using non-linear entropy based features in empirical mode decomposition domain. *Appl. Soft Comput.* **2019**, *77*, 24–33.
49. Pachori, R.B.; Avinash, P.; Shashank, K.; Sharma, R.; Acharya, U.R. Application of empirical mode decomposition for analysis of normal and diabetic RR-interval signals. *Expert Syst. Appl.* **2015**, *42*, 4567–4581.
50. Huang, N.E.; Shen, Z.; Long, S.R.; Wu, M.C.; Shih, H.H.; Zheng, Q.; Yen, N.C.; Tung, C.C.; Liu, H.H. The empirical mode decomposition and the Hilbert spectrum for nonlinear and non-stationary time series analysis. *Proc. R. Soc. Lond. A Math. Phys. Eng. Sci.* **1998**, *454*, 903–995.
51. Ojala, T; Pietikainen, M; Maenpaa, T. Multiresolution gray-scale and rotation invariant texture classification with local binary patterns. *IEEE Trans. Pattern Anal. Mach. Intell.* **2002** *24*, 971–987.
52. Guo, Z.; Zhang, L.; Zhang, D.J. A completed modeling of local binary pattern operator for texture classification. *IEEE Trans. Image Process.* **2010**, *19*, 1657–1663.
53. Pietikäinen, M.; Hadid, A.; Zhao, G.; Ahonen, T. *Computer Vision Using Local Binary Patterns*; Springer Science & Business Media: London, United Kingdom, 2011

54. Nguyen, D.T.; Pham, T.D.; Baek, N.R.; Park, K.R. Combining deep and handcrafted image features for presentation attack detection in face recognition systems using visible-light camera sensors. *Sensors* **2018**, *18*, 699.
55. Adnan, S.M.; Irtaza, A.; Aziz, S.; Ullah, M.O.; Javed, A.; Mahmood, M.T. Fall detection through acoustic Local Ternary Patterns. *Appl. Acoust.* **2018**, *140*, 296–300.
56. Aziz, S.; Awais, M.; Akram, T.; Khan, U.; Alhussein, M.; Aurangzeb, K. Automatic Scene Recognition through Acoustic Classification for Behavioral Robotics. *Electronics* **2019**, *8*, 483.
57. Irtaza, A.; Adnan, S.M.; Aziz, S.; Javed, A.; Ullah, M.O.; Mahmood, M.T. A framework for fall detection of elderly people by analyzing environmental sounds through acoustic local ternary patterns. In Proceedings of the 2017 IEEE International Conference on Systems, Man, and Cybernetics (SMC), Banff, AB, Canada, 5–8 October 2017; pp. 1558–1563.
58. Kaya, Y.; Ertuğrul, Ö.F. A stable feature extraction method in classification epileptic EEG signals. *Austral. Phys. Eng. Sci. Med.* **2018**, *41*, 721–730.
59. Khan, M.U.; Aziz, S.; Sohail, M.; Shahid, A.A.; Samer, S. Automated Detection and Classification of Gastrointestinal Diseases using surface-EMG Signals. In Proceedings of the 2019 22nd International Multitopic Conference (INMIC), Islamabad, Pakistan, 29–30 November 2019; pp. 1–8.
60. Chowdhury, M.E.; Khandakar, A.; Alzoubi, K.; Mansoor, S.; M Tahir, A.; Reaz, M.B.I.; Al-Emadi, N. Real-Time Smart-Digital Stethoscope System for Heart Diseases Monitoring. *Sensors* **2019**, *19*, 2781.
61. Kamarulafizam, I.; Noor, A.M.; Harris, A.A.; Oemar, H.; Yusoff, K. Classification of heart sound based on multipoint auscultation system. In Proceedings of the 2013 8th International Workshop on Systems, Signal Processing and their Applications (WoSSPA), Algiers, Algeria, 12–15 May 2013; pp. 174–179.
62. Alturki, F.A.; AlSharabi, K.; Abdurraqeeb, A.M.; Aljalal, M. EEG Signal Analysis for Diagnosing Neurological Disorders Using Discrete Wavelet Transform and Intelligent Techniques. *Sensors* **2020**, *20*, 2505.
63. Dash, D.; Ferrari, P.; Dutta, S.; Wang, J. NeuroVAD: Real-Time Voice Activity Detection from Non-Invasive Neuromagnetic Signals. *Sensors* **2020**, *20*, 2248.
64. Aziz, S.; Khan, M.U.; Choudhry, Z.A.; Aymin, A.; Usman, A. ECG-based Biometric Authentication using Empirical Mode Decomposition and Support Vector Machines. In Proceedings of the 2019 IEEE 10th Annual Information Technology, Electronics and Mobile Communication Conference (IEMCON), Vancouver, BC, Canada, 17–19 October 2019; pp. 0906–0912.
65. Khan, M.U.; Aziz, S.; Ibraheem, S.; Butt, A.; Shahid, H. Characterization of Term and Preterm Deliveries using Electrohysterograms Signatures. In Proceedings of the 2019 IEEE 10th Annual Information Technology, Electronics and Mobile Communication Conference (IEMCON), Vancouver, BC, Canada, 17–19 October 2019; pp. 0899–0905.
66. Czabanski, R.; Horoba, K.; Wrobel, J.; Matonia, A.; Martinek, R.; Kupka, T.; Jezewski, M.; Kahankova, R.; Jezewski, J.; Leski, J.M. Detection of atrial fibrillation episodes in long-term heart rhythm signals using a support vector machine. *Sensors* **2020**, *20*, 765.
67. Akram, T.; Laurent, B.; Naqvi, S.R.; Alex, M.M.; Muhammad, N. A deep heterogeneous feature fusion approach for automatic land-use classification. *Inf. Sci.* **2018**, *467*, 199–218.
68. Hoffman, J.I. *Biostatistics for Medical and Biomedical Practitioners*; Academic Press: Cambridge, MA, USA, 2015.
69. Yazdani, S.; Azghani, M.R.; Sedaaghi, M.H. A new algorithm for ECG interference removal from single channel EMG recording. *Austral. Phys. Eng. Sci. Med.* **2017**, *40*, 575–584.

Article

EEG Headset Evaluation for Detection of Single-Trial Movement Intention for Brain-Computer Interfaces

Mads Jochumsen [1,*], Hendrik Knoche [2], Troels Wesenberg Kjaer [3], Birthe Dinesen [1] and Preben Kidmose [4]

1 Department of Health Science and Technology, Aalborg University, 9220 Aalborg, Denmark; bid@hst.aau.dk
2 Department of Architecture, Design and Media Technology, Aalborg University, 9000 Aalborg, Denmark; hk@create.aau.dk
3 Department of Neurology, Zealand University Hospital, Roskilde, Denmark. Department of Clinical Medicine, University of Copenhagen, 2200 Copenhagen, Denmark; twk@regionsjaelland.dk
4 Department of Engineering—Electrical and Computer Engineering, Aarhus University, 8200 Aarhus, Denmark; pki@eng.au.dk
* Correspondence: mj@hst.aau.dk

Received: 1 April 2020; Accepted: 13 May 2020; Published: 14 May 2020

Abstract: Brain–computer interfaces (BCIs) can be used in neurorehabilitation; however, the literature about transferring the technology to rehabilitation clinics is limited. A key component of a BCI is the headset, for which several options are available. The aim of this study was to test four commercially available headsets' ability to record and classify movement intentions (movement-related cortical potentials—MRCPs). Twelve healthy participants performed 100 movements, while continuous EEG was recorded from the headsets on two different days to establish the reliability of the measures: classification accuracies of single-trials, number of rejected epochs, and signal-to-noise ratio. MRCPs could be recorded with the headsets covering the motor cortex, and they obtained the best classification accuracies (73%–77%). The reliability was moderate to good for the best headset (a gel-based headset covering the motor cortex). The results demonstrate that, among the evaluated headsets, reliable recordings of MRCPs require channels located close to the motor cortex and potentially a gel-based headset.

Keywords: movement intention; brain–computer interface; movement-related cortical potential; neurorehabilitation

---

1. Introduction

Brain–computer interfaces (BCIs) have been proposed as a means for control of assistive devices and communication for patients with severe disabilities, such as spinal cord injury and amyotrophic lateral sclerosis (ALS) [1–3]. More recently, BCIs have been investigated for motor rehabilitation of patients with neural injuries such as stroke or spinal cord injury [4–8]. It is possible to induce neural plasticity, a proposed mechanism for motor learning and hence motor recovery [9,10], by pairing an intention to move (detected from the ongoing brain activity) with contingent somatosensory feedback from, e.g., electrical stimulation of nerves and muscles [11,12] or passive movements of the limbs through exoskeletons or rehabilitation robots [13]. BCI training has a positive effect on motor recovery after stroke (see [14] for a recent review). However, creating a stand-alone BCI training system for use in rehabilitation clinics or patients homes is difficult due to several factors. One of the factors is the setup of the EEG headset. It can be fairly time consuming to mount the EEG headset and ensure the cap is correctly placed, and a proper EEG signal quality is obtained. This is especially evident if the patients potentially should mount the EEG headset themselves [15]. These points are related to the usability of the headsets, which became a research topic when new and cheaper headsets and electrode types became available [15–18]. The usability evaluation consists of three factors: effectiveness, efficiency and satisfaction [19], and the former two factors are related to another impeding factor to create a stand-alone BCI system, the need for calibrating the BCI to ensure adequate detection performance. The non-stationarity of the electrical brain signals (electroencephalography (EEG) and electrocorticography), requires that the BCI system is calibrated to ensure adequate performance, which differs depending on the control signal and application. The BCI system performance is also affected by the signal quality, and different signal processing and pattern recognition techniques are used [20–22]. Several studies have investigated and compared different signal processing and pattern recognition techniques, and some studies have investigated the signal quality and BCI performance of different headsets or electrode types (dry vs. wet) [16,18,23–41]. The focus of these studies has primarily been on BCI control signals related to communication or control, such as P300 or steady-state visual evoked potentials (SSVEP). However, BCIs for inducing neural plasticity rely on control signals associated with movement preparation, such as movement-related cortical potentials (MRCPs) or event-related desynchronization (ERD). Hence, the ability to record MRCPs and ERD is crucial for any neurorehabilitation BCI, and this ability has been less explored in commercial headsets. A single study compared the signal quality of MRCP when recorded with two different amplifiers but with the same headset [42], and a couple of studies investigated if ERD patterns could be identified using dry and wet electrodes [27,28]. Thus, there is a need for an evaluation of different headset types that potentially could be used for BCI training in neurorehabilitation clinics. Various metrics have been used in the literature to evaluate the signal quality of the headsets or electrode types. For BCI applications, reasonable measures would be BCI performance-related metrics, such as classification accuracy or information transfer rate (especially for BCI applications within communication and control of external devices) [18,23,25,26,30,31,34,37]. The signal quality may be quantified in other ways as well. In a recent study, Oliveira et al [39] proposed a number of metrics to investigate the signal quality when comparing different headsets. These metrics include data or epoch rejection rate [16,39,40,43] and signal-to-noise ratio (SNR) or noise level [16,32,39,42,44]. This is also important for BCI applications, since the performances of pattern recognition techniques are affected by various types of noise that cannot be suppressed. Other important measures are related to the signal morphology, which can be quantified from grand averages across multiple trials (both event-related and evoked potentials) in which the amplitudes of the brain potentials are extracted, or scalp topographies in which amplitudes or spectral content are extracted from multiple channels over the scalp [32,36–39]. Lastly, it is important to evaluate the test–retest reliability of the measures [39].

In this study, the aim was to explore if MRCPs can be recorded with four different headset types. Two headsets covered the motor cortex, which is the traditional position to record MRCPs. The other two headsets were placed on the forehead and around the ear, which was preferred by stroke patients in a recent usability study [15]; however, it is not known if MRCPs can be recorded from these positions. We tested whether movement intentions (MRCPs) can be classified with respect to idle activity, which is the scenario that would be used in BCI training for stroke rehabilitation. Moreover, different signal quality measures are reported, as well as the test–retest reliability over two separate days.

2. Materials and Methods

2.1. Participants

Twelve healthy participants were included in this study (28 ± 3 years old, 2 females). All participants gave their written informed consent prior to the experiments. All procedures were approved by the local ethical committee of Region North Jutland, Denmark (N-20130081).

2.2. EEG Headsets

Four different types of headsets were included in this study. The four headsets are shown in Figure 1.

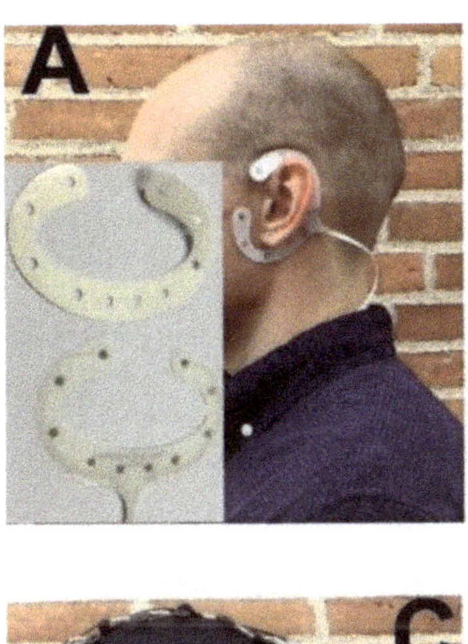

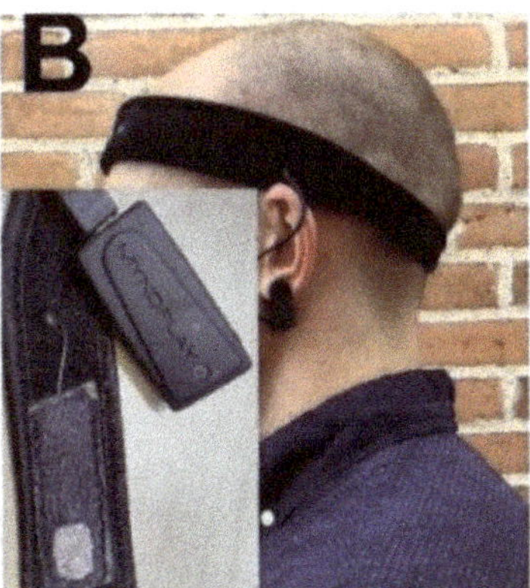

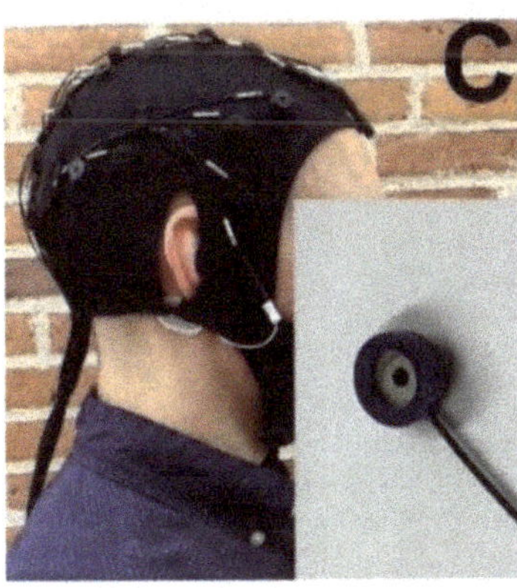

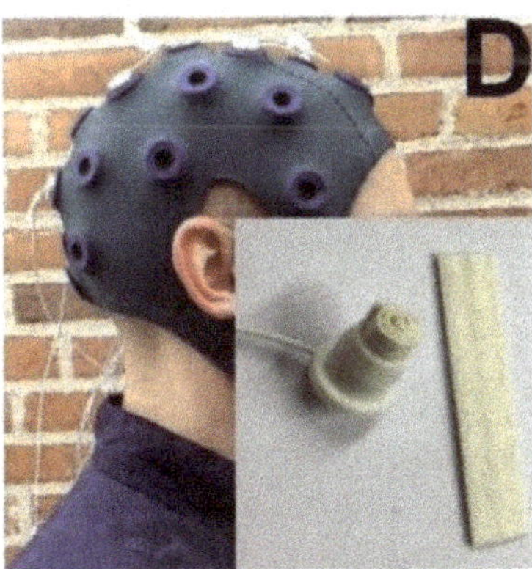

Figure 1. Overview of the four headsets and the electrode types. (**A**): cEEGrid from TMSi. The inset shows the sensor array and the double-sided adhesive tape. The channel in the upper part of the "C" when mounted is channel 1, and the channel numbers ascend clockwise on the electrode. (**B**): MyndBand from MyndPlay. The inset shows the dry electrode and the Bluetooth unit that transmits the data. (**C**): Quick-Cap from Compumedics. The inset shows the electrode type that is mounted in the cap. (**D**): Water-based headset from TMSi. The inset shows the electrode house and the felt insert.

2.2.1. cEEGrid: TMSi

The cEEGrid electrode (Figure 1A) contained 10 channels surrounding the ear [45]. In this study, it was placed around the left ear (close to T9 with respect to the International 10–20 System). The electrode consisted of a flex-PCB with screen-printed silver electrodes; the electrode was attached to the user with double-sided adhesive tape, which had to be fitted accurately to the recording sites. A small amount of electrode gel (ECI Electro-Gel™) was applied to each recording site. The channels were referenced to the channel at the lower end of the "C" (channel 10) and grounded to a moist wristband (left wrist). There was no measure of electrode impedance. The signals were sampled with 2000 Hz. The signals were recorded using a Mobita®amplifier from TMSi (Tucker, GA, United States).

2.2.2. MyndBand: MyndPlay

The MyndBand (Figure 1B) contained one dry electrode, which was placed on the forehead (close to F9 with respect to the International 10–20 System). The electrode was fixed with a neoprene headband. The signals were referenced to the left earlobe with an ear clip; there was no information about a ground electrode. A tool for measuring the impedance was available in the recording software;

the impedance could be in the following range: very poor–perfect. The impedance was perfect for all participants. However, it is not known what "perfect" corresponds to in terms of kΩ. The signals were sampled at 512 Hz.

2.2.3. Quick-Cap: Compumedics

The Quick-Cap (Figure 1C) electrodes covered the following positions with respect to the International 10–20 System: F3, Fz, F4, C3, Cz, C4, P3, Pz, and P4. The channels were referenced to Pz and grounded at AFz. The electrodes were filled with conductive gel (ECI Electro-Gel™) to establish contact between the electrodes and the scalp. A tool was available in the recording software to measure the impedance; the impedance of all channels was below 5 kΩ in all experimental sessions. The signals were sampled at 500 Hz. The signals were recorded using a Nuamp amplifier (EEG amplifiers, Nuamps Express, Neuroscan).

2.2.4. Water-Based Electrodes: TMSi

The water-based electrodes (Figure 1D) were placed in the following positions with respect to the International 10–20 System: F3, Fz, F4, C3, Cz, C4, P3, Pz, and P4. They were referenced to Pz, and they were grounded to a moist wristband (left wrist). The electrode consisted of a felt insert that had to be placed in an electrode house and soaked in water before placing it in the cap. There was no measure of electrode impedance. The signals were sampled at 2000 Hz. The signals were recorded using a Mobita®amplifier from TMSi.

2.3. *Experimental Procedure*

The participants participated in an experiment consisting of two experimental sessions, which were separated by at least 24 h. They sat in a comfortable chair and performed 100 cued ballistic palmar grasps of the right hand when continuous EEG was recorded. The hand was opened immediately after the grasp was performed; i.e, the contraction was not maintained. The participants were instructed to perform the movements as rapidly as possible and relax immediately after the grasp was performed. This movement type was performed for each of the four headsets, so, in total, 4×100 movements were performed in each of the two experiments. The order of the headsets was randomized on both recording days using Random.org. The participants were given two cues; one cue three seconds prior to the second cue, which indicated the onset of the task. The participants were instructed to sit as still as possible and avoid any eye movements and contractions of facial muscles. A trigger was used to mark the continuous EEG at the first cue; this was used to divide the continuous EEG into epochs. Each movement was separated by ten seconds. It took ~17 min to perform 100 movements. There was a break between the tests of the different headsets in which the subjects washed their hair.

2.4. *Data Analysis*

The EEG signals were analyzed in two different ways; 1) an analysis of the signal morphology, and 2) discrimination between movement intention and idle activity. For the signal quality analyses, the signals from Cz, or channel two for the cEEGrid, were used, since the pre-movement components of the MRCP can be recorded over the midline regardless of the site of movement [46]. For the classification of movement intentions and idle activity, all available channels were used. All analyses were performed in MATLAB 2019B (MathWorks).

2.4.1. Pre-Processing

The signals were band-pass-filtered from 0.05–10 Hz using a fourth-order zero-phaseshift Butterworth filter and downsampled to 500 Hz. The epochs were divided into "idle/noise" and "signal" epochs. Idle epochs were extracted from −5 to −3 seconds prior the movement onset, whereas the signal epochs were extracted from −1.5 to 0.5 s with respect to the task onset (0 s was the task onset). See Figure 2.

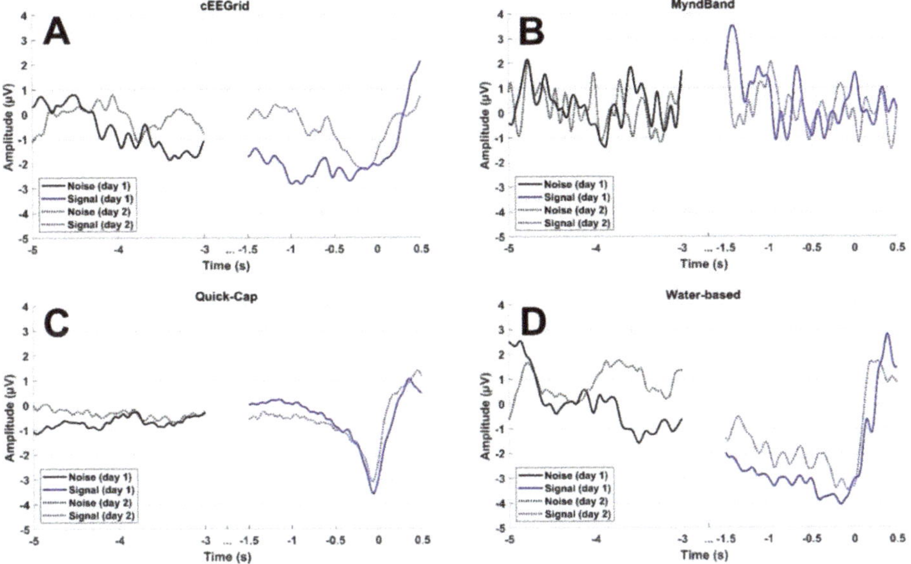

Figure 2. Grand average across participants for each headset and each day. "0 s" is the task onset. For the Quick-Cap and water-based electrodes, Cz was used, whereas channel 2 was used for the cEEGrid and the single electrode on the forehead for the MyndBand.

2.4.2. Signal-to-Noise Ratio, Epoch Rejection, and Peak Amplitudes

Initially, the signal and idle epochs that exceeded ±150 μV in peak–peak amplitude were rejected in the Cz channel (Quick-Cap and Water-based electrodes), channel 2 (cEEGrid, close to T9 according to the International 10–20 System) and in the single MyndBand channel (close to F9 according to the International 10–20 System). The analyses in this subsection were based on the specified individual channel for each headset. If more than 80% of the epochs were rejected, all data from that participant was removed from further analysis (see Table 1). The average of the signal and average of the idle epochs were calculated, and the root-mean-square (RMS) value was calculated of the two averages. The ratio between the signal and idle/noise RMS values were used as an estimate of the SNR. The average across the signal epochs was computed and the average amplitude was calculated from −0.2 to 0.2 s around the movement onset. The number of rejected epochs, SNR and averaged peak amplitudes were used as measures of signal quality.

Table 1. Results of the signal quality analyses for the four headsets. The signal-to-noise ratio (SNR), average amplitude around the movement onset, number of rejected epochs, and number of rejected participants are presented for both experimental sessions. The results are presented based on Cz (water-based electrodes and Quick-Cap) or the channel closest to that (MyndBand and cEEGrid).

	SNR		Amplitude (μV)		# Rejected Epochs		# Excluded Participants	
	Day 1	Day 2	Day 1	Day 2	Day 1	Day 2	Day 1	Day 2
cEEGrid	0.8/1.2/1.4	0.6/0.9/1.6	−4.9/−0.8/0.9	−3.6/−1.6/0.6	2/6/65	1/3/7	2	0
MyndBand	0.6/0.9/1.1	0.7/0.8/0.9	−0.7/0.3/0.9	−0.6/0.2/0.6	7/19/45	18/28/36	2	1
Quick-Cap	1.1/1.5/2.5	1.4/1.7/2.2	−3.4/−2.6/−0.9	−2.6/−1.0/0.0	0/1/2	0/0/2	0	0
Water-based	0.8/1.3/2.0	1.0/1.4/2.7	−7.4/−2.9/1.9	−4.3/−2.7/−0.3	2/2/6	2/3/9	0	0
	25% / Median / 75%							

2.4.3. Feature Extraction and Classification

Initially, the signal and noise epochs that exceeded ±150 μV in peak–peak amplitude in any of the recorded channels were excluded from further analysis. The feature extraction was performed for each channel. The feature extraction was based on previous studies, where MRCPs and ERD were detected from time domain analysis [47], frequency domain analysis [48], and template matching [49]. In the time domain, the mean amplitudes were extracted for 0.5-second windows without overlap and used as features, and the difference between the average amplitude in the first half and second half of the epoch. The power spectral density was estimated for the entire epoch with a 1-second Hamming window with 0.5-second overlap; the features were the power spectrum in 1-Hz bins from 6 to 30 Hz. For the template matching, the epochs were filtered from 0.05–10 Hz. An average of the signal epochs in each channel was obtained, and the cross-correlation was calculated between the template and the epochs at zero time lag. The classification was performed using a Random Forest classifier in a leave-one-out cross-validation scheme. The classifier was trained using 512 trees. The classification accuracy and number of rejected epochs are reported.

2.5. Test–Retest Reliability

The test–retest reliability was estimated using the intraclass correlation coefficient between the two recording days for participants that had a complete dataset using a 2-way mixed effect model with absolute agreement (IBM®, SPSS®). The intraclass correlation coefficient was calculated for SNR, average peak amplitudes and classification accuracies.

2.6. Statistics

Five two-way repeated measures analyses of variance (ANOVA) were performed with 'Headset' (four levels: cEEGrid, MyndBand, Quick-Cap, and Water-based) and 'Time' (two levels: Day 1, and Day 2) as factors on the following measures: 1) SNR, 2) amplitude, 3) number of rejected epochs (single-channel), 4) classification accuracy, and 5) number of rejected epochs (multiple channels). If the assumption of sphericity was violated, the Greenhouse–Geisser correction was applied. Significant tests were followed up with a posthoc test with Bonferroni correction. Significant test statistics were assumed when $p < 0.05$. The effect size was reported as well, using the partial eta squared value (η2).

3. Results

The results are summarized in Figures 3 and 4 and Tables 1–3.

3.1. Signal Quality

The results of the signal quality analyses are presented in Table 1 and Figures 2 and 3. The average amplitude around the movement onset was most prominent for the Quick-Cap and Water-based headset. It was expected to see a negative potential, such as the one shown in Figure 2C. The peak amplitudes recorded from the other two headsets were less prominent, and the clear MRCP morphology was absent. There was no interaction between headset and time ($F_{(3,21)} = 0.51$; $p = 0.68$; $\eta^2 = 0.07$), and no effect of time ($F_{(1,7)} = 0.38$; $p = 0.56$; $\eta^2 = 0.05$) and headset ($F_{(1.7,12.2)} = 1.95$; $p = 0.19$; $\eta^2 = 0.22$).

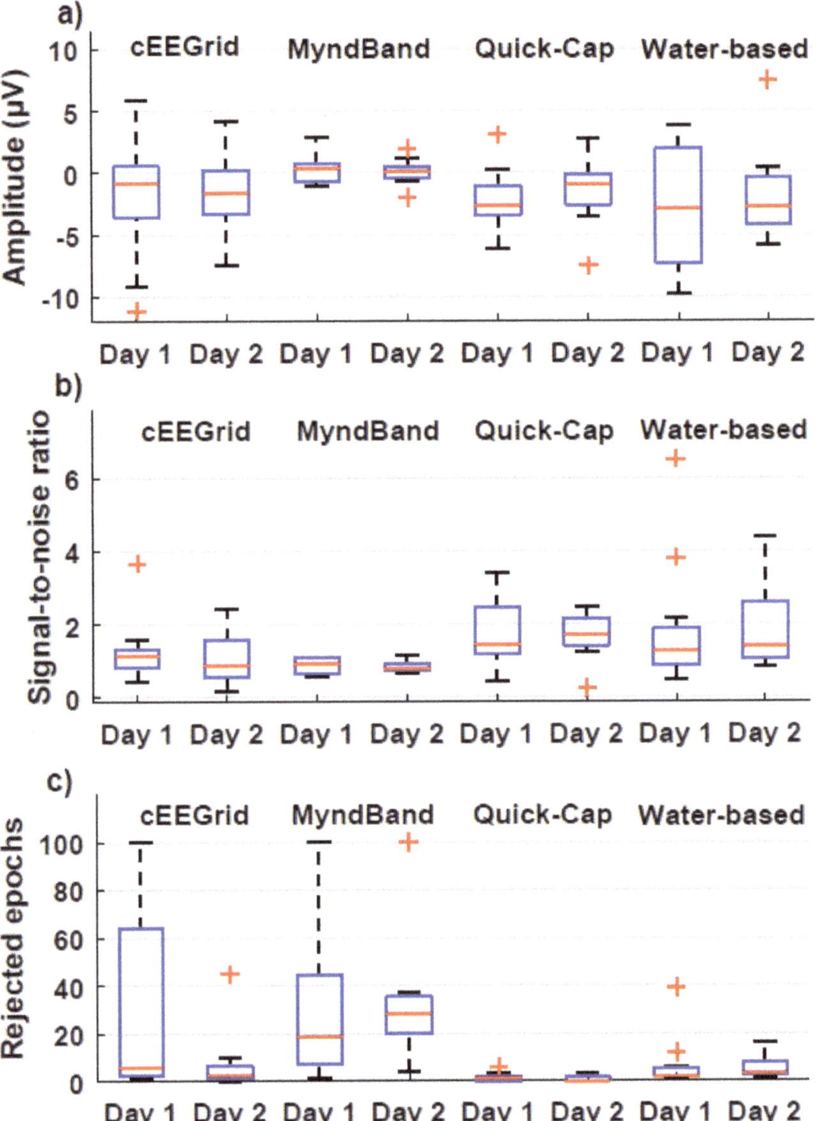

Figure 3. Boxplots with quartiles for the signal quality analyses. (**a**): Average amplitude around the movement onset; (**b**): Signal-to-noise ratio; (**c**): Number of rejected epochs for a single channel.

The SNR was highest for the Quick-Cap and the water-based headset, whereas the other two had similar SNRs. The statistical analysis showed no interaction between headset and time ($F_{(1.2,7.1)} = 0.03$; $p = 0.89$; $\eta^2 = 0.06$) and no effect of time ($F_{(1,6)} = 0.01$; $p = 0.91$; $\eta^2 = 0.02$), but the effect of headset was significant ($F_{(3,18)} = 6.67$; $p = 0.003$; $\eta^2 = 0.53$). The posthoc test revealed that the SNR associated with Quick-Cap was higher than the SNR obtained with the MyndBand.

The median number of rejected epochs was in the range of 0–6 rejected epochs for the Quick-Cap, water-based headset, and cEEGrid, whereas the median number of rejected epochs for the MyndBand was 19 and 28. This is also reflected in the number of participants that were rejected based on the criterion of 80% of the samples that should be within ±150 μV peak–peak amplitudes. There was no interaction between headset and time ($F_{(1.3,14.2)} = 1.38$; $p = 0.27$; $\eta^2 = 0.11$) and no effect of time ($F_{(1,11)} = 4.37$; $p = 0.06$; $\eta^2 = 0.28$), but the effect of headset ($F_{(1.9,20.6)} = 11.71$; $p < 0.001$; $\eta^2 = 0.52$) was significant. The posthoc test showed that more epochs were rejected for the MyndBand compared to the Quick-Cap and water-based headset, and more epochs were rejected for the water-based headset than the Quick-Cap.

3.2. Movement Intention vs. Idle Classification

The results are presented in Table 2 and Figure 4. The number of rejected epochs was high for the MyndBand, and the 75% quartiles for the water-based and cEEGrid headset. This is also reflected in the number of participants that were excluded from further analysis (more than 80% of the epochs were rejected). The accuracies were close to the significance threshold of random classification (chance level = 50%), calculated with 95% confidence limits (threshold for signficance = 60%) [50] for the cEEGrid and MyndBand, whereas accuracy was well above the threshold for significance for the Quick-Cap and Water-based headset. They had similar median classification accuracies, 74%–77% and 72%–73% for the Quick–Cap and water-based headset, respectively, on the two different days, but the classification accuracies were based on fewer samples and subjects for the water-based headset. The statistical analyses revealed no interaction between headset and time ($F_{(3,9)} = 0.25$; $p = 0.86$; $\eta^2 = 0.08$) and no effect of time ($F_{(1,3)} = 0.09$; $p = 0.79$; $\eta^2 = 0.03$) and headset ($F_{(3,9)} = 3.41$; $p = 0.07$; $\eta^2 = 0.53$). However, the sample size was limited. For the number of rejected epochs, there was no interaction between headset and time ($F_{(2.1,23.4)} = 2.36$; $p = 0.09$; $\eta^2 = 0.18$), and no effect of time ($F_{(1,11)} = 0.06$; $p = 0.81$; $\eta^2 = 0.006$), but the effect of headset was significant ($F_{(3,33)} = 10.25$; $p < 0.001$; $\eta^2 = 0.48$). The posthoc analysis showed that more epochs were rejected for the water-based headset and Myndband compared to the Quick-Cap, and that more epochs were rejected for the MyndBand compared to the cEEGrid.

Table 2. Results of the classification analyses for the four headsets. The classification accuracy and number of rejected epochs are presented for both experimental sessions. The results are based on all available channels; therefore, the number of rejected epochs differ from Table 1, which was based on a single channel.

	Classification Accuracy (%)		**# Rejected Epochs**		**# Excluded Participants**	
	Day 1	**Day 2**	**Day 1**	**Day 2**	**Day 1**	**Day 2**
cEEGrid	48/56/70	55/60/63	0/7/54	0/1/3	2	0
MyndBand	49/56/59	50/56/60	19/31/85	31/39/70	3	2
Quick-Cap	70/77/82	69/74/78	0/1/1	0/1/1	0	0
Water-based	64/73/78	65/72/75	0/3/88	0/63/100	3	4
	25% / Median / 75%					

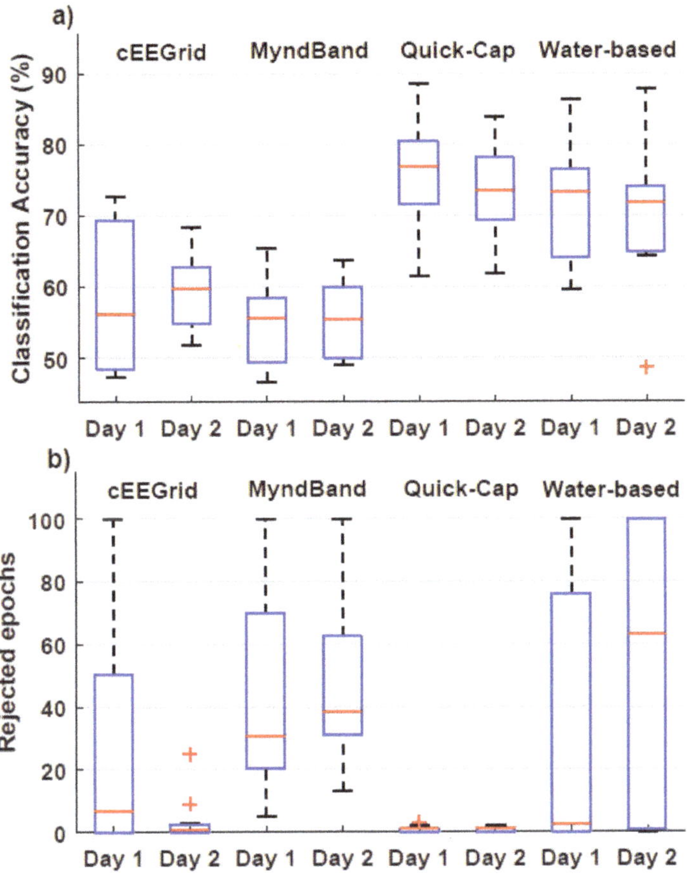

Figure 4. Boxplots with quartiles for the classification accuracies (**a**) and number of rejected epochs (**b**).

3.3. Test–Retest Reliability

The test–rerest reliability was assessed using the intraclass correlation coefficient, and the results are presented in Table 3. A moderate (intraclass correlation coefficient: 0.50–0.75) to good (intraclass correlation coefficient: 0.75–0.90) reliability was obtained for the Quick-Cap, whereas poor (intraclass correlation coefficient <0.50) to moderate reliability was obtained for the other headsets [51]. The negative intraclass correlation coefficients are likely due to bad estimates from a limited sample [52], where the mean-square error is larger than the mean square of the rows [51]. The negative values are obtained for the headsets where subjects were excluded (see Table 2).

Table 3. Test–retest reliability of the signal-to-noise ratio (SNR), average amplitude around the movement onset and the classification accuracies.

	Intraclass Correlation Coefficient (ICC)		
	ICC_SNR	**ICC_Amplitude**	**ICC_Classification Accuracy**
cEEGrid	−0.3	0.32	0.63
MyndBand	0.43	−0.21	0.33
Quick-Cap	0.78	0.83	0.59
Water-based	−0.29	0.06	−0.11

4. Discussion

In this study, the aim was to test four different types of headsets and electrodes. The results of the evaluation suggest that it is important to have electrodes that cover the motor cortex area [46], and that gel-based electrodes are superior to the alternatives, in order to discriminate between movement intentions of the hand and idle activity (median classification accuracy of 77% and 74% on day 1 and 2, respectively). The MRCP morphology was most cleary seen for the headset that covered the motor cortex and used conductive gel. It must be stated that some of the headsets were not developed for recording movement-related activity, which was also reflected in the results. However, it was important to include these headsets, since stroke patients in a comparative study preferred them [15].

The classification between movement-related and idle activity was not significantly higher than chance level (when calculated with 95% confidence limits [50]) for the MyndBand and cEEGrid. This is likely due to the electrode positions, which were too far away from the motor cortex to register the MRCP. The water-based headset covered the motor cortex, and the classification accuracies were significantly higher than chance level (threshold for signficance = 60% [50]). However, many epochs were removed from the analysis due to large amplitudes in the EEG, which potentially arose from electrodes losing contact with the skin. Further processing can be done to remove bad channels and perform the classification based on a single or reduced number of channels. The most reliable classification of movement-related activity was obtained using the Quick-Cap with gel-based electrodes. The median classification accuracy was 77% and 74% for day 1 and day 2, respectively. These accuracies are similar to what has been reported previously [47,48,53]. From a BCI training point of view, the classification accuracy/BCI system perfomance needed for inducing plasticity is not known [7], but the accuracies obtained for the water-based electrodes and Quick-Cap are higher or similar to the BCI system performance that has been reported to induce neural plasticity in previous studies (true positive range: 67%–85%) [11–13]. It has previously been indicated that the BCI system performance and induction of plasticity are positively correlated [11]; therefore, there is an incentive for further improving the classification accuracies and hence the BCI system performance. One way to do this is by applying a spatial filter [20,54]; however, this can only be done when multiple channels are recorded. Besides neurorehabilitation, these results are also relevant for communication or internet browsing for late-stage ALS patients that will be able to produce similar slow cortical potentials. In a simulation study [55], it was shown that an accuracy of 75% can be used to browse the internet, using slowly developing control signals such as the MRCP or ERD. However, to increase the speed of browsing or communication (i.e. increase the information transfer rate) evoked potentials such as P300 should be used if the user is able to operate them [56].

Limitations and Future Perspectives

A limiting factor that could affect the number of rejected epochs in the study was the choice of amplitude threshold. The threshold of 150 μV could be too high to exceed for the cEEGrid electrode, which would affect the classification analysis, since the accepted epochs would still contain noise. This could also account for the high number of rejected epochs for the water-based electrodes and the MyndBand. Another approach could be to use a data-driven threshold using X times the standard deviation; X could be in the range of, e.g., 3–5, depending on how conservative the noise rejection should be. Only the headset with electrodes covering the motor cortex region and with conductive gel applied had clear MRCP waveforms, but both headsets that covered the motor cortex region had significant decoding accuracies; this suggests that it is necessary to use headsets with motor cortex region electrodes in order to decode MRCPs. If gel-based electrodes are used, patients may need to wash their hair after using the BCI, unless a limited number of channels are used. It has previously been shown that a single channel is sufficient to detect the MRCP [53]. In the current study, it was an experimenter with several years of experience within EEG recordings and BCI research who mounted the headsets. If BCI training is going to be performed in rehabilitation clinics and the patient's home then it would be important to test how much time it will take for rehabilitation professionals, caretakers, or relatives to learn to setup the headset correctly and obtain signals of sufficient quality to be classified correctly. In this study, different headsets and electrode types were chosen to test if movement-related activity could be recorded and classified, although they were not designed for that specific purpose. There exist other headsets (and new ones are emerging) that are candidates to record movement-related activity, and, therefore, it would be relevant to perform further comparative studies between headsets to identify the optimal headset in terms of signal quality, comfort, setup difficulty and price, to increase the likelihood of potential end-users adopting the BCI technology.

5. Conclusions

It is concluded that it is necessary to record signals from the motor cortex area to be able to detect movement intentions. The results indicate that the most reliable classification accuracies are obtained with gel-based electrodes. These results may have implications for the choice of headset for BCI applications within neurorehabilitation, or applications that require an estimate of MRCPs.

Author Contributions: Conceptualization, all authors; methodology, M.J. and P.K.; software, M.J.; validation, M.J. and P.K.; formal analysis, M.J.; writing—original draft preparation, M.J.; writing—review and editing, all authors; visualization, M.J.; project administration, M.J.; funding acquisition, M.J. All authors have read and agreed to the published version of the manuscript.

Funding: This research was funded by VELUX FONDEN, grant number 22357.

Conflicts of Interest: The authors declare no conflict of interest.

References

1. Wolpaw, J.R.; Birbaumer, N.; McFarland, D.J.; Pfurtscheller, G.; Vaughan, T.M. Brain-Computer Interfaces for Communication and Control. *Clin. Neurophysiol.* **2002**, *113*, 767–791. [CrossRef]
2. Birbaumer, N.; Cohen, L. Brain–computer Interfaces: Communication and Restoration of Movement in Paralysis. *J. Physiol.* **2007**, *579*, 621. [CrossRef] [PubMed]
3. Huggins, J.E.; Moinuddin, A.A.; Chiodo, A.E.; Wren, P.A. What would Brain-Computer Interface Users Want: Opinions and Priorities of Potential Users with Spinal Cord Injury. *Arch. Phys. Med. Rehabil.* **2015**, *96*, S38–S45. [CrossRef] [PubMed]
4. Biasiucci, A.; Leeb, R.; Iturrate, I.; Perdikis, S.; Al-Khodairy, A.; Corbet, T.; Schnider, A.; Schmidlin, T.; Zhang, H.; Bassolino, M. Brain-Actuated Functional Electrical Stimulation Elicits Lasting Arm Motor Recovery After Stroke. *Nat. Commun.* **2018**, *9*, 2421. [CrossRef]
5. Ramos-Murguialday, A.; Broetz, D.; Rea, M.; Läer, L.; Yilmaz, Ö.; Brasil, F.; Liberati, G.; Curado, M.; Garcia-Cossio, E.; Vyziotis, A. Brain–machine Interface in Chronic Stroke Rehabilitation: A Controlled Study. *Ann. Neurol.* **2013**, *74*, 100–108. [CrossRef]
6. Pichiorri, F.; Morone, G.; Petti, M.; Toppi, J.; Pisotta, I.; Molinari, M.; Paolucci, S.; Inghilleri, M.; Astolfi, L.; Cincotti, F. Brain–computer Interface Boosts Motor Imagery Practice during Stroke Recovery. *Ann. Neurol.* **2015**, *77*, 851–865. [CrossRef]
7. Grosse-Wentrup, M.; Mattia, D.; Oweiss, K. Using Brain-Computer Interfaces to Induce Neural Plasticity and Restore Function. *J. Neural Eng.* **2011**, *8*, 025004. [CrossRef]
8. Frolov, A.A.; Mokienko, O.; Lyukmanov, R.; Biryukova, E.; Kotov, S.; Turbina, L.; Nadareyshvily, G.; Bushkova, Y. Post-Stroke Rehabilitation Training with a Motor-Imagery-Based Brain-Computer Interface (BCI)-Controlled Hand Exoskeleton: A Randomized Controlled Multicenter Trial. *Front. Neurosci.* **2017**, *11*, 400. [CrossRef]
9. Pascual-Leone, A.; Dang, N.; Cohen, L.G.; Brasil-Neto, J.P.; Cammarota, A.; Hallett, M. Modulation of Muscle Responses Evoked by Transcranial Magnetic Stimulation during the Acquisition of New Fine Motor Skills. *J. Neurophysiol.* **1995**, *74*, 1037–1045. [CrossRef]
10. Krakauer, J.W. Motor Learning: Its Relevance to Stroke Recovery and Neurorehabilitation. *Curr. Opin. Neurol.* **2006**, *19*, 84–90. [CrossRef]
11. Niazi, I.K.; Kersting, N.M.; Jiang, N.; Dremstrup, K.; Farina, D. Peripheral Electrical Stimulation Triggered by Self-Paced Detection of Motor Intention Enhances Motor Evoked Potentials. *IEEE Trans. Neural Syst. Rehabil. Eng.* **2012**, *20*, 595–604. [CrossRef] [PubMed]

12. Jochumsen, M.; Navid, M.S.; Rashid, U.; Haavik, H.; Niazi, I.K. EMG-Versus EEG-Triggered Electrical Stimulation for Inducing Corticospinal Plasticity. *IEEE Trans. Neural Syst. Rehabil. Eng.* **2019**, *27*, 1901–1908. [CrossRef] [PubMed]
13. Xu, R.; Jiang, N.; Mrachacz-Kersting, N.; Lin, C.; Asin, G.; Moreno, J.; Pons, J.; Dremstrup, K.; Farina, D. A Closed-Loop Brain-Computer Interface Triggering an Active Ankle-Foot Orthosis for Inducing Cortical Neural Plasticity. *IEEE Trans. Biomed. Eng.* **2014**, *61*, 2092–2101. [PubMed]
14. Cervera, M.A.; Soekadar, S.R.; Ushiba, J.; Millán, J.d.R.; Liu, M.; Birbaumer, N.; Garipelli, G. Brain-computer Interfaces for Post-stroke Motor Rehabilitation: A Meta-analysis. *Ann. Clin. Transl. Neurol.* **2018**, *5*, 651–663. [CrossRef] [PubMed]
15. Jochumsen, M.; Knoche, H.; Kidmose, P.; Kjær, T.W.; Dinesen, B.I. Evaluation of EEG Headset Mounting for Brain-Computer Interface-Based Stroke Rehabilitation by Patients, Therapists, and Relatives. *Front. Hum. Neurosci.* **2020**, *14*, 13. [CrossRef]
16. Ries, A.J.; Touryan, J.; Vettel, J.; McDowell, K.; Hairston, W.D. A Comparison of Electroencephalography Signals Acquired from Conventional and Mobile Systems. *J. Neurosci. Neuroeng.* **2014**, *3*, 10–20. [CrossRef]
17. Radüntz, T.; Meffert, B. User Experience of 7 Mobile Electroencephalography Devices: Comparative Study. *JMIR mHealth uHealth* **2019**, *7*, e14474. [CrossRef]
18. Nijboer, F.; Van De Laar, B.; Gerritsen, S.; Nijholt, A.; Poel, M. Usability of Three Electroencephalogram Headsets for Brain–computer Interfaces: A within Subject Comparison. *Interact. Comput.* **2015**, *27*, 500–511. [CrossRef]
19. International Organization for Standardization. *ISO 9241-11. Ergonomic Requirements for Office Work with Visual Display Terminals (VDTs)—Part 9: Requirements for Non-Keyboard Input Devices*; ISO: Geneva, Switzerland, 1998.
20. Jochumsen, M.; Niazi, I.K.; Mrachacz-Kersting, N.; Jiang, N.; Farina, D.; Dremstrup, K. Comparison of Spatial Filters and Features for the Detection and Classification of Movement-Related Cortical Potentials in Healthy Individuals and Stroke Patients. *J. Neural Eng.* **2015**, *12*, 056003. [CrossRef]
21. Lotte, F.; Congedo, M.; Lecuyer, A.; Lamarche, F.; Arnaldi, B. A Review of Classification Algorithms for EEG-Based Brain–computer Interfaces. *J. Neural Eng.* **2007**, *4*, R1–R13. [CrossRef]
22. Bai, O.; Lin, P.; Vorbach, S.; Li, J.; Furlani, S.; Hallett, M. Exploration of Computational Methods for Classification of Movement Intention during Human Voluntary Movement from Single Trial EEG. *Clin. Neurophysiol.* **2007**, *118*, 2637–2655. [CrossRef] [PubMed]
23. Chi, Y.M.; Wang, Y.; Wang, Y.; Maier, C.; Jung, T.; Cauwenberghs, G. Dry and Noncontact EEG Sensors for Mobile Brain–computer Interfaces. *IEEE Trans. Neural Syst. Rehabil. Eng.* **2012**, *20*, 228–235. [CrossRef] [PubMed]
24. Choi, S.; Han, C.; Choi, G.; Shin, J.; Song, K.S.; Im, C.; Hwang, H. On the Feasibility of using an Ear-EEG to Develop an Endogenous Brain-Computer Interface. *Sensors* **2018**, *18*, 2856. [CrossRef] [PubMed]
25. Duvinage, M.; Castermans, T.; Petieau, M.; Hoellinger, T.; Cheron, G.; Dutoit, T. Performance of the Emotiv Epoc Headset for P300-Based Applications. *Biomed. Eng. Online* **2013**, *12*, 56. [CrossRef]
26. Guger, C.; Krausz, G.; Allison, B.Z.; Edlinger, G. Comparison of Dry and Gel Based Electrodes for P300 Brain–computer Interfaces. *Front. Neurosci.* **2012**, *6*, 60. [CrossRef]
27. Grozea, C.; Voinescu, C.D.; Fazli, S. Bristle-Sensors—low-Cost Flexible Passive Dry EEG Electrodes for Neurofeedback and BCI Applications. *J. Neural Eng.* **2011**, *8*, 025008. [CrossRef]
28. Grummett, T.; Leibbrandt, R.; Lewis, T.; DeLosAngeles, D.; Powers, D.; Willoughby, J.; Pope, K.; Fitzgibbon, S. Measurement of Neural Signals from Inexpensive, Wireless and Dry EEG Systems. *Physiol. Meas.* **2015**, *36*, 1469. [CrossRef]
29. Kappel, S.L.; Rank, M.L.; Toft, H.O.; Andersen, M.; Kidmose, P. Dry-Contact Electrode Ear-EEG. *IEEE Trans. Biomed. Eng.* **2019**, *66*, 150–158. [CrossRef]
30. Käthner, I.; Halder, S.; Hintermüller, C.; Espinosa, A.; Guger, C.; Miralles, F.; Vargiu, E.; Dauwalder, S.; Rafael-Palou, X.; Solà, M. A Multifunctional Brain-Computer Interface Intended for Home use: An Evaluation with Healthy Participants and Potential End Users with Dry and Gel-Based Electrodes. *Front. Neurosci.* **2017**, *11*, 286. [CrossRef]
31. Liu, Y.; Jiang, X.; Cao, T.; Wan, F.; Mak, P.U.; Mak, P.; Vai, M.I. Implementation of SSVEP Based BCI with Emotiv EPOC. In Proceedings of the 2012 IEEE International Conference on Virtual Environments Human-Computer Interfaces and Measurement Systems (VECIMS) Proceedings, Tianjin, China, 2–4 July 2012; pp. 34–37.

32. Mathewson, K.E.; Harrison, T.J.; Kizuk, S.A. High and Dry? Comparing Active Dry EEG Electrodes to Active and Passive Wet Electrodes. *Psychophysiology* **2017**, *54*, 74–82. [CrossRef]
33. Mayaud, L.; Congedo, M.; Van Laghenhove, A.; Orlikowski, D.; Figère, M.; Azabou, E.; Cheliout-Heraut, F. A Comparison of Recording Modalities of P300 Event-Related Potentials (ERP) for Brain-Computer Interface (BCI) Paradigm. *Neurophysiol. Clin. Clin. Neurophysiol.* **2013**, *43*, 217–227. [CrossRef] [PubMed]
34. Pinegger, A.; Wriessnegger, S.C.; Faller, J.; Müller-Putz, G.R. Evaluation of Different EEG Acquisition Systems Concerning their Suitability for Building a Brain–computer Interface: Case Studies. *Front. Neurosci.* **2016**, *10*, 441. [CrossRef]
35. De Vos, M.; Kroesen, M.; Emkes, R.; Debener, S. P300 Speller BCI with a Mobile EEG System: Comparison to a Traditional Amplifier. *J. Neural Eng.* **2014**, *11*, 036008. [CrossRef] [PubMed]
36. Wang, F.; Li, G.; Chen, J.; Duan, Y.; Zhang, D. Novel Semi-Dry Electrodes for Brain–computer Interface Applications. *J. Neural Eng.* **2016**, *13*, 046021. [CrossRef] [PubMed]
37. Zander, T.O.; Lehne, M.; Ihme, K.; Jatzev, S.; Correia, J.; Kothe, C.; Picht, B.; Nijboer, F. A Dry EEG-System for Scientific Research and Brain–computer Interfaces. *Front. Neurosci.* **2011**, *5*, 53. [CrossRef]
38. Zander, T.O.; Andreessen, L.M.; Berg, A.; Bleuel, M.; Pawlitzki, J.; Zawallich, L.; Krol, L.R.; Gramann, K. Evaluation of a Dry EEG System for Application of Passive Brain-Computer Interfaces in Autonomous Driving. *Front. Hum. Neurosci.* **2017**, *11*, 78. [CrossRef] [PubMed]
39. Oliveira, A.S.; Schlink, B.R.; Hairston, W.D.; König, P.; Ferris, D.P. Proposing Metrics for Benchmarking Novel EEG Technologies Towards Real-World Measurements. *Front. Hum. Neurosci.* **2016**, *10*, 188. [CrossRef]
40. Radüntz, T. Signal Quality Evaluation of Emerging EEG Devices. *Front. Physiol.* **2018**, *9*, 98. [CrossRef]
41. Debener, S.; Minow, F.; Emkes, R.; Gandras, K.; De Vos, M. How about Taking a Low-cost, Small, and Wireless EEG for a Walk? *Psychophysiology* **2012**, *49*, 1617–1621. [CrossRef]
42. Rashid, U.; Niazi, I.; Signal, N.; Taylor, D. An EEG Experimental Study Evaluating the Performance of Texas Instruments ADS1299. *Sensors* **2018**, *18*, 3721. [CrossRef]
43. Cruz-Garza, J.G.; Brantley, J.A.; Nakagome, S.; Kontson, K.; Megjhani, M.; Robleto, D.; Contreras-Vidal, J.L. Deployment of Mobile EEG Technology in an Art Museum Setting: Evaluation of Signal Quality and Usability. *Front. Hum. Neurosci.* **2017**, *11*, 527. [CrossRef] [PubMed]
44. Laszlo, S.; Ruiz-Blondet, M.; Khalifian, N.; Chu, F.; Jin, Z. A Direct Comparison of Active and Passive Amplification Electrodes in the Same Amplifier System. *J. Neurosci. Methods* **2014**, *235*, 298–307. [CrossRef] [PubMed]
45. Debener, S.; Emkes, R.; De Vos, M.; Bleichner, M. Unobtrusive Ambulatory EEG using a Smartphone and Flexible Printed Electrodes around the Ear. *Sci. Rep.* **2015**, *5*, 16743. [CrossRef] [PubMed]
46. Shibasaki, H.; Hallett, M. What is the Bereitschaftspotential? *Clin. Neurophysiol.* **2006**, *117*, 2341–2356. [CrossRef] [PubMed]
47. Jochumsen, M.; Niazi, I.K.; Dremstrup, K.; Kamavuako, E.N. Detecting and Classifying Three Different Hand Movement Types through Electroencephalography Recordings for Neurorehabilitation. *Med. Biol. Eng. Comput.* **2015**, *54*, 1491–1501. [CrossRef] [PubMed]
48. Ibáñez, J.; Serrano, J.; Del Castillo, M.; Monge-Pereira, E.; Molina-Rueda, F.; Alguacil-Diego, I.; Pons, J. Detection of the Onset of Upper-Limb Movements Based on the Combined Analysis of Changes in the Sensorimotor Rhythms and Slow Cortical Potentials. *J. Neural Eng.* **2014**, *11*, 056009. [CrossRef] [PubMed]
49. Niazi, I.K.; Jiang, N.; Tiberghien, O.; Nielsen, J.F.; Dremstrup, K.; Farina, D. Detection of Movement Intention from Single-Trial Movement-Related Cortical Potentials. *J. Neural Eng.* **2011**, *8*, 066009. [CrossRef]
50. Müller-Putz, G.R.; Scherer, R.; Brunner, C.; Leeb, R.; Pfurtscheller, G. Better than Random? A Closer Look on BCI Results. *Int. J. Bioelectromagn.* **2008**, *10*, 52–55.
51. Koo, T.K.; Li, M.Y. A Guideline of Selecting and Reporting Intraclass Correlation Coefficients for Reliability Research. *J. Chiropr. Med.* **2016**, *15*, 155–163. [CrossRef]
52. Liljequist, D.; Elfving, B.; Skavberg Roaldsen, K. Intraclass Correlation - A Discussion and Demonstration of Basic Features. *PLoS ONE* **2019**, *14*, e0219854. [CrossRef]
53. Jochumsen, M.; Niazi, I.K.; Taylor, D.; Farina, D.; Dremstrup, K. Detecting and Classifying Movement-Related Cortical Potentials Associated with Hand Movements in Healthy Subjects and Stroke Patients from Single-Electrode, Single-Trial EEG. *J. Neural Eng.* **2015**, *12*, 056013. [CrossRef] [PubMed]
54. Blankertz, B.; Lemm, S.; Kawanabe, M.; Tomioka, R.; Müller, K.-R. Optimizing Spatial filters for Robust EEG Single-Trial Analysis. *IEEE Signal Process. Mag.* **2007**, *25*, 41–56. [CrossRef]

55. Bensch, M.; Karim, A.A.; Mellinger, J.; Hinterberger, T.; Tangermann, M.; Bogdan, M.; Rosenstiel, W.; Birbaumer, N. Nessi: An EEG-Controlled Web Browser for Severely Paralyzed Patients. *Comput. Intell. Neurosci.* **2007**, *2007*, 71863. [CrossRef] [PubMed]
56. Mugler, E.M.; Ruf, C.A.; Halder, S.; Bensch, M.; Kubler, A. Design and Implementation of a P300-Based Brain-Computer Interface for Controlling an Internet Browser. *IEEE Trans. Neural Syst. Rehabil. Eng.* **2010**, *18*, 599–609. [CrossRef] [PubMed]

MDPI
St. Alban-Anlage 66
4052 Basel
Switzerland
Tel. +41 61 683 77 34
Fax +41 61 302 89 18
www.mdpi.com

Sensors Editorial Office
E-mail: sensors@mdpi.com
www.mdpi.com/journal/sensors

www.ingramcontent.com/pod-product-compliance
Lightning Source LLC
LaVergne TN
LVHW071246170726
843515LV00006BA/1347

* 9 7 8 3 0 3 6 5 3 7 2 0 7 *